The workshop was

sponsored by
the German Marshall Fund of the United States, Washington D.C.
and
the Johnson Foundation, Racine, Wisconsin
and
the Commission of the European Communities,
Directorate General XII, Brussels
and
organized by
Pace University School of Law,
Center for Environmental Legal Studies,
Professor Richard L. Ottinger
White Plains, N.Y., USA
and
Fraunhofer-Institute for
Systems and Innovation Research (ISI),
Dr. Olav Hohmeyer
Dr. Hermann Herz
Karlsruhe, Germany

Olav Hohmeyer
Richard L. Ottinger (Eds.)

Social Costs of Energy

Present Status and Future Trends

Proceedings of an International Conference,
Held at Racine, Wisconsin, September 8-11, 1992

With 57 Figures

Springer-Verlag

Berlin Heidelberg New York
London Paris Tokyo
Hong Kong Barcelona
Budapest

Dr. Olav Hohmeyer
ZEW, Zentrum für Europäische
Wirtschaftsforschung GmbH
Postfach 10 34 43
D-68034 Mannheim, FRG

Prof. Dr. Richard L. Ottinger
Pace University School of Law
Center for Environmental Legal Studies
78 Broadway
White Plains, N. Y., USA

ISBN-13: 978-3-642-85122-3   e-ISBN-13: 978-3-642-85120-9
DOI: 10.1007/978-3-642-85120-9

Preface

Although present day politics seems to be preoccupied with questions of economic growth and full employment, the basic environmental problems stemming from the interactions of the economic sphere with global, regional and local environments persist and will have an even greater impact in the future. If economy and ecology are not reconciled in the years to come, mankind will not have a sustainable future on Earth.

The typical negation of environmental problems in times of economic crisis is partially due to the fact that environmental and health damages of economic activities are neither priced nor included in our market price system. This allows politicians to focus their attention on insufficient economic indicators which do not reflect the actual development of the welfare of society. If economic lead indicators like GDP or balance of trade figures were better integrated with information on the environmental and health costs caused by the seemingly beneficial economic development, politicians might have better guidance as to what policy choices would benefit society most.

Recent attempts to identify, quantify, monetize and internalise such cost elements, referred to as external or social costs, can be seen as the first step towards translating the environmental and health impacts of economic activities into monetary units to make them more compatible with the prevailing, flawed economic indicators. Even though such monetized figures will never be absolutely correct, the knowledge of the right orders of magnitude may allow us to integrate economic, environmental and health effects in a somewhat more consistent way. This could help politicians not to forget basic environmental issues while trying to solve the problems of recurring economic crises. Thus, information on external or social costs of economic activities may help to steer the economic system towards a sustainable future.

Since 1988, empirical attempts to derive external or social costs have mostly concentrated on the energy sector as one of the major contributors to environmental problems. At the same time, some work has also been done on the transport sector, as another major source of pollution and negative health impacts. In a number of countries, the results of this research have led to first policy measures which try to internalise some of the identified external costs.

In 1990, the first international workshop on the subject brought together researchers, politicians and administrators from the two leading countries in this field, the United States of America and the Federal Republic of Germany, to exchange ideas and results in Ladenburg, Germany. As this workshop functioned as a catalyst for the acceleration and intensification of the international discussion, all the experts involved saw the need to make it a regular event.

Due to the substantial increase in research activities in this field in other countries, the second workshop became an international conference, still of manageable size, held at Racine,

Wisconsin, in September 1992. This book presents all the papers delivered at the conference reporting on the latest research results as well as recent attempts by governments, administrations and utility commissions to internalise social or external costs in energy prices or energy planning. Again, the prime focus was on the electricity sector due to the fact that most work has been done in this area so far. It is quite obvious that this can only be a starting point, as many other sectors contribute just as much to environmental and health problems.

The second conference was again able to bring together the leading scientific, economic, utility and governmental experts in the field of quantifying the external or social costs of energy and accounting for those costs in energy policy and energy planning. As the papers discussed at the conference give a comprehensive picture of the international discussion in the field, this book, together with its predecessor, the proceedings of the Ladenburg workshop, is a unique source of reference for the relevant contributions to this field of research and political implementation.

It is planned to hold a third international conference on the subject in 1994 or 1995, which will try to broaden the scope of the discussion towards other sources of energy and a better perception and integration of the aspects of sustainability into the debate on social or external costs.

On behalf of all the participants, the editors want to acknowledge their appreciation of the generous financial and organisational support of the German Marshall Fund of the United States, the Commission of the European Communities (DG XII) and the Johnson Foundation (Racine, Wisconsin). The availability of the marvellous conference facilities of the Johnson Foundation provided a splendid setting and superb atmosphere for the intensive discussions.

As the manuscripts of the papers presented at the conference varied significantly in layout and typing, partial retyping and substantial layout editing were necessary and the editors wish to thank Gillian Bowman, Bärbel Katz and Irmgard Sieb for their efforts. We want to thank the staff of the Johnson Foundation for their invaluable support in organising the conference and making it an effective and enjoyable event for all participants. We also want to thank our colleague Herman Herz, formerly with the Fraunhofer-Institute, who was heavily involved with the organisation of the workshop and the collection of all the papers. If he had stayed with us, we would have been a team of three editors. Finally, we want to thank our colleague, Bernd Hartmann, of the Fraunhofer-Institute, who helped in editing and proof reading the final text of this book.

Pace University
School of Law

Richard L. Ottinger

Fraunhofer-Institute for
Systems and Innovation Research

Olav Hohmeyer

CONTENTS

SUBJECT AREA 1:
GENERAL TREATMENT OF THE ASSESSMENT OF SOCIAL COSTS AND THE PERSPECTIVE FOR THEIR INCORPORATION

SUBJECT AREA 2:
EMPIRICAL ESTIMATION OF SOCIAL COSTS OF ENERGY

1.   Introduction and some Conclusions from the Conference

The international conference on the social costs of energy, which is reflected in this volume, concentrated on the social costs of electricity production, although many aspects apply to other areas just as well. The relatively narrow focus of the conference reflects the mainstream of the empirical research conducted so far; it should not be misinterpreted as a value judgement on the relative importance of this specific instance of social costs in market economies.

Throughout this introduction, the term "social costs" will be used for all cost elements of production or consumption handed on to third parties not involved in the specific market transaction. As the different papers show, this is not a generally accepted terminology; two different terms are common. Neo-classical economics would refer to most of these costs as "external costs". However, as the definition of external costs does not allow for important aspects of sustainability, the given broader definition of social costs is used here.

There are many other important instances of substantial social costs apart from those social costs of heat production or transportation arising from electricity generation. There may well be substantial social costs of chemicals' production or of genetic engineering which are not covered in this book. Compared to the first workshop at Ladenburg, the subject has already been substantially broadened, as effects beyond external environmental effects have been addressed such as the sustainable inter temporal allocation of energy resources.

Although the papers have been edited to a limited extent and partially retyped for printing, the texts still vary in a number of specific elements of layout, citation procedures and language proficiency. To arrive at a more uniform text would have taken considerably more time and financial resources than we had available.

The order in which papers are presented in this book is somewhat different from the order in which the presentations were given at the conference. During final editing of the volume, we felt that sorting them into four chapters would make easier reading than leaving them in their original order. Thus the four main subjects referred to in these chapters are:
-    A general treatment of the assessment of social costs and the perspectives for their incorporation.
-    Reports on the actual empirical estimation of the social costs of different parts of the energy system ranging from energy supply to demand side technologies.
-    Discussions of the different instruments and approaches for the internalisation of social costs.
-    Discussions of social cost issues in the light of sustainability considerations.

In this way, general views as well as detailed discussions of estimation and incorporation issues are covered, resulting in the necessary broadening of the present rather narrow view.

We certainly have to make the transition from a simple focus on improved allocation to a more general view of the crucial questions of scale and equity if the social cost debate is to contribute to ensuring a sustainable future of mankind, which is the central task ahead of us. "Scale" refers to the largest possible relative size of the economic system as part of the global environmental system. It is limited by the availability of resources as well as the assimilative capacity of the environment for the waste produced and discarded by the economic system. "Equity" or the inter temporal and international distribution of rights to use the world's resource base and its assimilative capacity for waste will need to be achieved if long term economic development is to be sustainable. Both conditions set the framework under which market allocation mechanisms which are based on better price signals incorporating social costs may function to steer the decisions of individual actors towards a sustainable state of our economic system, even achieving a better allocation of scarce resources.

Most of the reported research draws on four major research activities conducted in Europe and the USA. The first is an ongoing research program on the social costs of different fuel cycles including some demand side technologies as well. This is sponsored by the US/DOE and the Commission of the European Communities in two co-ordinated, parallel research programs. Most of the papers in this volume are based on research work financed by these two programs. Secondly, an extensive German study on the identification and internalisation of social costs had just been completed at the time of the workshop and a number of papers from this research were presented at Racine. Last but not least, the research activities of New York State have proceeded further, producing some new insights presented in a number of papers which are also included in this book. Beyond these papers, there are additional individual contributions reflecting experiences and insights from the specific situation in California, the perspective of a rather progressive utility company on social costs as well as some experiences with emission trading policies at the national level in the US.

What is obvious from the contributions is that a number of different views on the subject remain. There are the 'purists', who concentrate on those effects where the estimation of marginal damage functions as well as the analysis of full cause-effect chains is possible and the 'pragmatists', who prefer the estimation of relatively crude figures for a large range of effects in order to derive better overall estimates of the order of magnitude of the problem. There are the 'damage analysts', who like to arrive at social costs based on estimated damages and the 'control cost supporters', who dismiss the damage cost approach as it does not allow the estimation of marginal cost figures due to lack of data and knowledge and who prefer control costs because the marginal costs of reducing emissions by one unit can be readily derived. Nevertheless, using this approach, the problem of the right level of pollution control remains unsolved. There are the 'neo-classic economists', who insist on a narrow neo-classical definition of external costs as the basis for all analyses and there are the 'ecological economists' maintaining that such a narrow focus will not produce any sustainable development and stressing issues of scale and equity at the same time. Reading these lines it is easy to figure out to which 'gang' at least one of the editors belongs! Furthermore, there are 'supply side advocates' and 'demand side advocates' battling over the appropriate focus of the analysis in the energy system.

As you see, there are a large number of permutations possible taking into account the many different viewpoints. Thus, the editors will restrain themselves from drawing too many conclusions as certainly some participants of the conference may object because of their differing individual points-of-view. So we sum up this introduction with a few hints as to where the development of research in the field of social costs may go:

1. If we want to delay political action until we have produced empirical results on damages and social costs for most of the relevant effects which are agreed upon by all the researchers in the field, action will never be taken.
2. Although we are in a phase of empirical research on many detailed effects, the future of social cost research seems to lie with more general issues.
3. We are progressing fast as to the depth of the energy systems analysed. After concentrating mainly on the operation of energy supply technologies in the early phases of research, we are analysing entire fuel cycles today even including first analyses on demand side technologies for the rational use of energy.
4. The analysis of indirect environmental effects is an upcoming field of research, widening our possible scope of analysis and increasing our coverage of the relevant effects involved.
5. It seems to be likely that the internalisation of social costs will be achieved by a mixture of policy tools ranging all the way from emission trading rights, taxes, environmental adders in planning decisions and emission reduction targets up to control and command policies.
6. The future function of the research and debate on social costs should be seen much more under the aspect of its contribution to a long term sustainable development than it has been so far.
7. However, political action is called for today instead of waiting until the remaining uncertainties and the disputes of the different fractions of researchers have been resolved. What the former Public Service Commissioner of Wisconsin, Mary Lou Munts, said at the first workshop in Ladenburg in 1990 is still valid: **'It is better to be roughly right than precisely wrong'**. When facing the challenge of the endangered long term survival of mankind, this may be even more important today than it was then.

SUBJECT AREA 1:

**GENERAL TREATMENT OF THE ASSESSMENT
OF SOCIAL COSTS AND THE PERSPECTIVE FOR
THEIR INCORPORATION**

2.   The Social Costing Debate: Issues and Resolutions

Alan J. Krupnick, Dallas Burtraw, A. Myrick Freeman III
Bowdoin College

Winston Harrington, Karen Palmer, and Hadi Dowlatabadi
Carnegie Mellon University

0   Introduction

This report is meant to provide guidance to PUCs and other parties interested in the social costing debate, although it will also yield useful information to those concerned with improving environmental policy in general.

Indeed, we begin with the basic theory of environmental policy before turning to issues specific to the social costing of electricity. We view these two types of policies very differently. The first is fundamentally about seeking optimal, first-best policies for controlling pollution and other third party consequences of economic activity (in this are electric utilities) through equating costs of control and their benefits at the margin. The second is about cost-effective, second-best policies--to choose new investment or plant operating characteristics optimally (i.e. that minimize social costs) given policies and regulations faced by the utility.

Within the issues specific to social costing, we identify and examine four: (i) Is social costing a good idea, (ii) How should damages be measured, (iii) When is damage an externality (and therefore an effect in need of an "adder", (iv) How should such externality estimates be implemented by PUCs? Issue (ii) is primarily taken up directly in the DOE-sponsored Fuel Cycle Study (Lee et al., 1992) and so will be pursued very briefly here. Considerable attention is devoted to issue (iii), presenting first a "rule of thumb" analysis that distinguishes between damage and externality in the case of a competitive industry unable to influence electricity price, while in the second part, results of a more sophisticated mode of utility behavior are reviewed to examine how the rules of thumb will hold up. Issue (iv) presents simulation model results for the effects of given "adders" on utility investments, output, and prices. Damages and net welfare effects associated with four alternative "regimes" for implementing social costing -social cost investment, no fees; investment with fees; social cost dispatch; and social cost pricing- are examined.

1 The Basic Theory of Environmental Policy

In this section we explain the terminology to be used in this report and review the basic economics of market failure and pollution control. The context is the electric utility industry which "manufactures" electricity for sale to consumers and other firms.

An emission is a by-product of generating energy services which has an impact, usually adverse, on third parties. By third parties we mean any economic agents who are not engaged in the market exchange. "Emissions" is a generic term referring not just to air pollutants but to any third party impacts.

The impact of an emission on third parties can have the effect of either decreasing the welfare of individuals or increasing the costs of production of other firms.[1] We use the term damage to refer to a monetary measure of this impact. The damage is the sum of money which exactly offsets or compensates for the harm caused by the emission. Since emissions reach third parties primarily through one or more environmental media, the damages are sometimes referred to as environmental damages.[2]

An externality exists when one agent's activity (in this case, the utility) has an effect (either positive or negative) on the welfare or cost of another economic agent and the utility insufficiently takes account of that impact in its own private decision making.[3] Since emissions cause impacts on third parties, they can be the source of negative externalities or external costs.

Economic efficiency (Pareto optimality) requires that the levels of production and consumption of all things that matter directly (or indirectly through their effects on firms and profits) be set so that the marginal social value of the thing equal its marginal social cost, where social value (cost) is defined as the sum of private and external value (cost). For the remainder of this discussion we will focus on external cost. For goods consumed by individuals, economic efficiency requires that each individual's marginal willingness to pay for the good be equal to the marginal social cost of producing it. An equivalent statement for "bads" such as emissions is that the marginal benefit of producing more of the bad equal the

---

[1] Of course, this distinction is artificial. Increases in costs to other firms ultimately decrease somebody's welfare through price-induced changes in producers and consumers surpluses.

[2] Sometimes these environmental damages are called environmental costs. We resist this terminology to emphasize the distinction between these damages and the measure of external costs or externalities to be included in capacity planning decisions.

[3] That is, insufficient for the purpose of achieving economic efficiency. In the economics literature these are referred to as "Pareto relevant externalities."

marginal social cost of the bad. Since the benefit of producing more of a bad is the avoidance of the cost of controlling or reducing the production of the bad, another equivalent statement is that the marginal cost of control for each emission equal the marginal damage that the emission causes.

The efficient level of emissions is shown in Figure 1 as E*. In this figure, MCC is the aggregate marginal control cost curve. It shows the marginal cost of reducing emissions assuming that the responsibility for controlling emissions is allocated among firms so as to minimize the total cost of control. MD is the marginal damage of emissions.

For firms generating electricity, an externality exists if firms insufficiently take into account in their business decisions the damage imposed on others by their emissions. The result of is market failure and an inefficient allocation of resources in the economy. This inefficiency shows up in two places. First, the condition for the optimum level of production of the emission will not be satisfied. The marginal damage of the emission will exceed the marginal cost of controlling it. This is shown in Figure 1 where EMAX is the aggregate level of emissions chosen by firms when they do not take into account the effect of their emissions on third parties. The other place the inefficiency shows up is in the market for electricity. Part of the social cost of generating electricity is the damage caused by the emissions on third parties. But because of the externality, these costs are not recognized by the firm as part of its private costs. This is shown in Figure 2. The market for electricity clears at e1 where the demand or marginal willingness to pay for electricity (MWTPe) is equal to the marginal private cost of electricity (MPCe), but MWTPe is less than the marginal social cost of electricity (MSCe).

There are several ways in which the inefficiency can be corrected and economic efficiency can be achieved. First, it might be possible to create enforceable property rights such that firms cannot generate emissions unless they have obtained the permission of those who feel the impact of the emissions. This would create a market where one was missing before. The MD curve of Figure 1 would become the market supply curve of rights to generate emissions.

As long as the market was competitive, it would clear at a price of t*, and the efficient level of emissions E* would be achieved.

The second way of correcting the market failure is for the government to create a pseudo-price or a charge on emissions by firms. Again, the appropriate charge would be t* in Figure 1. The tax has the effect of forcing firms to take account of the level of their emissions in private decision making. If the tax is set equal to the marginal damage at the optimum point, t*, then firms will limit emissions to the optimum level, E*, and pay total tax revenues of t*E*. What was an externality has now been internalized in the firm's decision making. Third parties would still experience harm or damage equal to the shaded area under the marginal damage curve to the left of E*. These are sometimes known as residual damages, since they remain even after the implementation of a policy to restrict emissions. Efficiency does not require that third parties be compensated out of the tax revenues for these residual damages (Baumol and Oates, 1988).

Figure 1:   The Market for Emissions

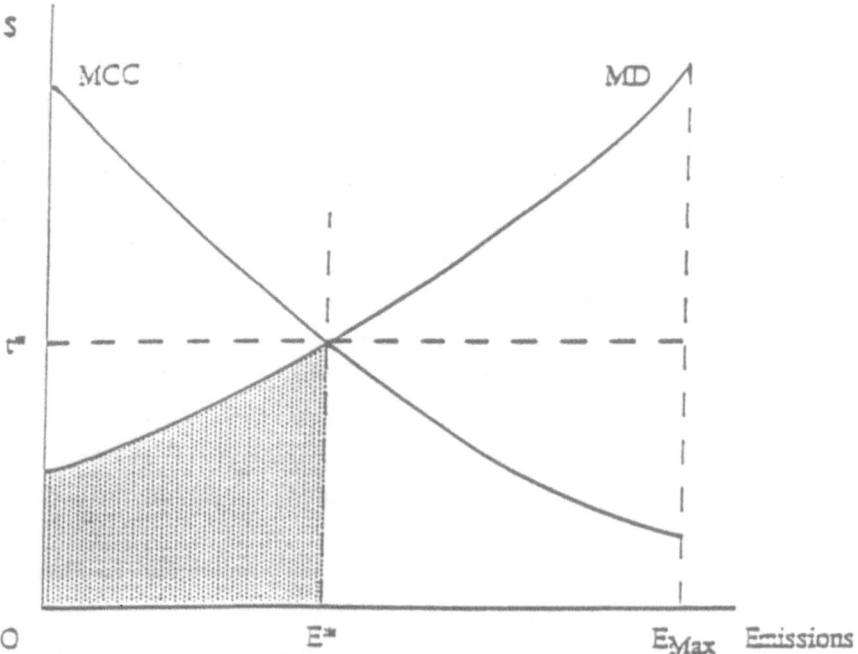

The third alternative is some form of tradable emission permit system (TEP). Permits equal in number to E* are either sold to firms or given away on some arbitrary basis. As long as those whose willingness to pay for an additional permit[4] exceeds the offering price of other permit holders, permits will be exchanged until the market equilibrium is achieved at a price per ticket equal to t*. The opportunity cost of an emission of one unit for a firm is the price of a permit. Firms must take account of this cost in their private decision making, so again, the externality would be internalized.

---

4   Willingness to pay is equal to MCC because it is this cost that is avoided through the purchase of an additional permit.

Figure 2:   The Market for Electricity

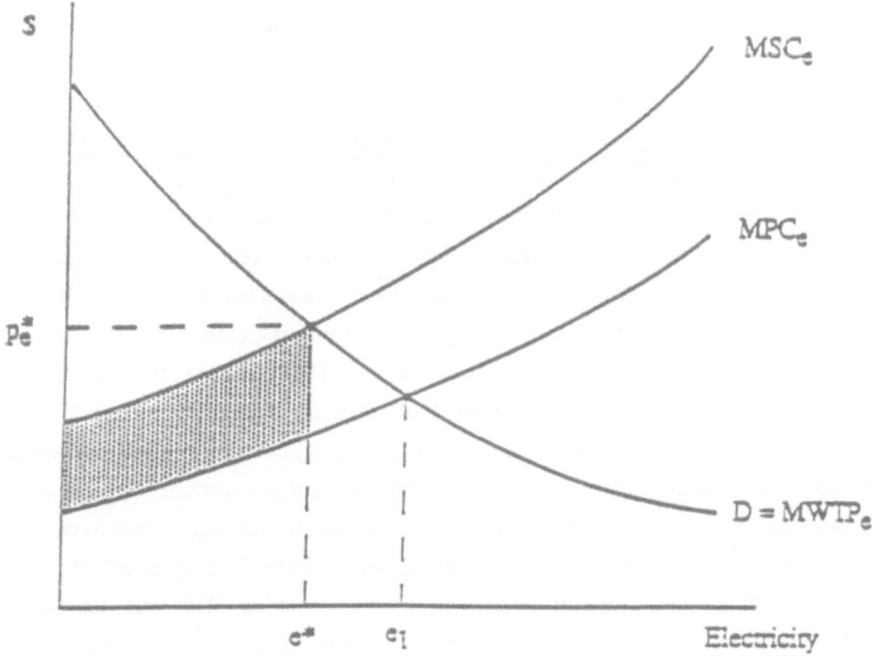

All three of these methods for dealing with externalities have an effect on the market for the good. The required compensation, the tax on emissions, or the price of emissions permits becomes part of the private cost of production. If the costs associated with emissions are correctly internalized to firms, the MPC_e shifts up to coincide with MSC_e. And the market clears at the optimal level of electricity, e^* in Figure 2. The residual damage can also be measured in Figure 2. It is the shaded area below the MSC_e curve. The residual damages are irrelevant and there is no need to compensate those who experience the damages in order to achieve economic efficiency. The presence of residual environmental damages does not mean that there is an externality.

2   Optimal Environmental Policy vs Social Costing as Cost-Effective Policy

A diagram such as Figure 1 in the preceding section is sometimes used to justify statements that externality cost adders are unnecessary when pollution is being optimally controlled at level E*, or even potentially over-controlled. But as we will now show, these diagrams do not capture the essence of the problem of choosing the most economical way of increasing the supply of electricity under alternative forms of environmental policy.

For the most part, U. S. environmental policy does not make use of any of the three approaches to internalizing externalities discussed previously. Rather, the most common policy instrument is specific quantitative restrictions on the level of emissions from each firm through direct regulation, also known as command and control (CAC) regulation. In principle, it would be possible to establish a set of quantitative restrictions at emissions level E*. But even if this level of total emissions was achieved in a cost effective way, it would not be Pareto optimal. This is because the CAC approach to limiting emissions does not deal with the second dimension of market failure, the divergence between MWTPe and MSCe in the market for electricity. The effect of "optimal" CAC regulation of emissions would be to shift the MPCe of Figure 2 up part way toward MSCe. However, the new equilibrium would have a price below Pe* and a quantity above e*. So, the CAC policy does not fully internalize the externality into the price of electricity. The presence of residual damages does have economic significance in this case. Since firms pay nothing for the emissions up to E*, they have no reason to take them into account in private decision making. This can be particularly important in utility capacity planning decisions, wherein major new investments could be made without properly considering the full social costs of each alternative.

All of this has been a prelude to the issue at hand. Social costing has been proposed as a means to deal with decision making when we anticipate an increase in the demand for electricity and must plan for an increase in capacity so that the market will clear and price will equal marginal social cost. Any increase in capacity will be represented by a rightward shift of the MPCe and MSCe in Figure 2 and an upward shift of the marginal control cost curve in Figure 1. The magnitude of these shifts will depend on the means chosen to increase capacity, for example, the fuel type, combustion technology, and location of the facilities. For nonmarginal increases in capacity, these factors could also affect the shape of the marginal damage curve.

In a first best world, the PUC and the environmental agency would work together to achieve a Pareto optimum. The new equilibrium would require adjustments to the pollution charge or the number of TEPs to be consistent with the intersection of the new MCC and MD curves of Figure 1. However, making optimal adjustments to the environmental policy instruments is beyond the power of the PUC. Rather, the PUC should evaluate the alternative supply options taking existing pollution control policy as given. This is a problem of the second best, to which we now turn.

3 Is Social Costing a Good Idea?

It is a good idea but not a great idea. A great idea is to internalize externalities throughout society by replacing our patchwork of command and control policies with economic incentive policies that inherently force internalization. Given that this is unlikely to happen, a second-best option is to seek cost-effective choices of electricity investment and generation, given the policies and regulations in place, where cost-effectiveness here means choosing technologies with the lowest social cost to meet new electricity demand.

Once one makes a distinction between optimal regulation and enters the second-best world of social costing, much of the confusion in the social costing debate dissolves. For instance, economists who feel that social costing is a bad idea justify this position in a way that mischaracterizes the objectives of social costing as a first-best policy. For instance, Joskow says, "their (PUC's) reliance on numerical "adders" to reflect environmental impacts will not achieve the goal of improving the environment at the lowest reasonable cost to society (Joskow, 1992, p.1) [emphasis added]." And "States such as Massachusetts, Nevada, New York, and California have decided they can 'fix' what ails existing environmental regulations by appending various externality "adders" to the private costs of new utility resource options (Joskow, 1992, p12) [emphasis added]."

We believe that in many instances PUCs will not have the authority or the expertise required to set environmental policies and standards. And for those externalities regulated by environmental agencies, PUCs have no business trying to second guess or correct the perceived errors of these other agencies. However, internalizing externalities in electricity system planning is consistent with a much more limited objective, an objective which is entirely appropriate to PUCs. This objective is to minimize the increment to social cost associated with meeting a given increase in the demand for electricity.

We wish to emphasize three points about this objective:

(i) In pursuing this objective, PUCs must take existing environmental policy as given. Therefore, pursuit to this objective does not involve second guessing environmental agencies or attempts to "fix" existing environmental policy. But it does require a concern for the environmental impacts of the electricity supply and demand side management (DSM) choices made by the PUC.

(ii) This objective can be interpreted as a straightforward extension of the traditional rationale for PUC regulation of electricity prices, namely concern for the impact of prices on consumer welfare. A low price for electricity is not a fundamental goal of policy; rather it is one of several instruments to be used to promote the fundamental objective of improving consumers' economic welfare. Since environmental quality also affects consumer welfare, PUCs should be aware of the effects of their decisions on environmental quality. Choosing the electricity supply option with the lowest private cost helps to keep the price of electricity low. But a PUC does consumers no favor by choosing the alternative with the lowest private cost if this choice imposes a hidden "tax" on consumers through its environmental damages.

(iii)   The PUC that wants to promote consumer welfare must take account of both the private costs and the external costs of alternatives in its decision making. This will often require some form of "adder".

However, one feels about the objectives of social costing, there is universal agreement about the existence of a "piecemeal problem" - i.e. the possible losses of efficiency associated with singling out the electricity sector for social costing. (The likely application of this approach in some, but not all, states may also be a kind of "piecemeal problem", to be dealt with in the next section).

These efficiency losses could arise because of distortions introduced in relative prices. Limiting use of social cost adders to the utility industry puts the utility's electricity at a competitive disadvantage relative to self-generation or purchasing power directly from an IPP. Some IPPs may find it advantageous to deal directly with customers rather than go through the PUC review process of bids to connect to the grid. This is the so-called "by-pass" problem. If these potential sources have uninternalized environmental costs associated with them, social welfare could be reduced by requiring externality adders only in the utility system. Electricity, in general, may be inappropriately disadvantaged next to other forms of energy.

The two problems of incorrect electricity price signals and by-pass arise not because social costing is wrong, but rather because it is incomplete in its coverage. PUCs do not presently have jurisdiction over electricity sources that do not connect to the grid. They may not have the authority to set the price of electricity above its private cost in an effort to "pass through" uninternalized environmental costs to electricity consumers. The solution to these problems is to develop a comprehensive system for internalizing environmental costs not only on utilities but on other sources of these emissions as well.

Nonetheless, concern about the piecemeal approach begs the question: why are utilities being picked on? One opinion that is often stated as a justification for singling out utilities is the view that "electric utilities are a major source of our environmental ills and are dragging their feet on demand-side management." The reality is that electric utilities are already more scrutinized and controlled than most industries.

On the other hand is the viewpoint represented by the following: "The utility industry should earn credits for the superior value provided by electricity." The problem with this viewpoint is that the value of electricity is already accounted for in the price. Consumers make consumption decisions based on the marginal value and the marginal price of the product. Since price is fixed from the consumer's perspective, one can vary the quantity one purchases until at the margin the value just equals its price. A fundamental tenet of the economic approach is that we are concerned about the effects of marginal decisions, rather than inframarginal decisions. No one is suggesting that one possibility is to de-electrify America or any such thing. Instead, we are concerned with marginal decisions that may involve new investment, the dispatch of existing capacity, or the proper pricing of energy services. We are concerned with the choice among the various fuel cycles for the delivery of energy services.

The reality as to why utilities have been singled out probably begins with the observation that utilities are an easy target and still an important source of pollution. The belief among certain regulators and their constituencies that social costing will speed penetration of renewables and demand-side management options into utility operations is also an important factor. In any event, electric utilities may not be alone for long. Increased attention already is being directed toward other sectors of the economy, the transportation sector in particular.

Note also the historic role of economic regulation of electric utilities. These organizations have been granted an exclusive franchise, and have accepted along with that a commensurate obligation to serve their communities. The aim of this form of economic regulation has been to improve consumer welfare, with the justification that historically the industry enjoys a declining average cost technology.

There are special responsibilities for a regulated industry and the consideration of full social costs appears to be one. Indeed, the consideration of social costs in utility regulation is nothing new. The historic partnership between regulators and utilities has been interpreted broadly to include social issues such as regional economic development, and this approach has been sanctioned in the courts.

Furthermore, environmental concerns are nothing new to utilities. They are residents of a local area and citizens in their community, just like everyone else. They have always had to comply with relevant environmental laws. And most recognize that regional economic development depends on environmental quality in the first place.

Seen in this light, the concern with full social costs is just an extension of the historic mandate to consider consumer welfare in this regulated industry. Today, analysts recognize that consumer welfare depends not just on the price of electricity, or even on regional economic development, but on the full array of social costs and benefits that attend the delivery of energy services. Consumer welfare achieves a maximum if energy services are delivered not at their least private cost, but at their least social cost.

4   How should Damages be Measured?

First, one should consider all damages and benefits from the entire fuel cycle. For example, in the evaluation of a proposed hydroelectric facility one should consider environmental damages, such as lost trout habitat and associated lost recreation opportunities. In addition, there may be environmental benefits such as the creation of bass habitat behind the reservoir and new recreation opportunities. There may be nonenvironmental damages such as infrastructure costs associated with the damage to roadways done by heavy trucks in the construction of the dam; and, there may be nonenvironmental benefits such as net new job creation, to the extent that previously there existed under-employed labor. All of these effects should be considered on an equal footing.

Second, damages and benefits are generally location specific. Things that differ between locations include ambient concentrations, relevant populations, population density, specific dose-response functions, geology and stream flows, transport conditions, along with nonenvironmental variables such as the local labor market. Furthermore, the conditions that apply throughout the plant's life must be considered. These wide-ranging considerations lead to the determination of the "baseline" from which the marginal facility will have its impact.

The location specificity of damage implies a point which many people have failed to appreciate: there never will be a "big book of values" in which one can look up numbers for the external damage or benefit associated with a kilowatt hour of electricity produced from a fuel cycle. What is important is that analysts use common methods, not common values, in the estimation of these location-specific social costs.

Finally, we must touch briefly on the current debate about methodology. There are two primary approaches to the estimation of social costs. One is termed the abatement cost approach. This approach is founded on the idea that one can use the marginal cost of abatement (compliance with current laws) as a proxy for the residual marginal damage that remains. The basic idea stems from the textbook treatment of economically efficient regulation, which suggests that optimal regulation will equate marginal cost and marginal damage. It is suggested that marginal damage, which the utility does not know, can be approximated to a good first guess by marginal cost, which the utility (or regulators) may know. The advantage of this approach is that it is relatively easy to estimate.

Unfortunately, the disadvantages of this approach are fatal. First, the idea that environmental regulation is intended to be economically efficient is simply not valid. For example, the Clean Air Act actually precludes consideration of costs in setting ambient air quality standards, seeking standards that protect the most sensitive individuals regardless of cost. Second, marginal impacts can be the same across different areas where, as highlighted previously, the damages or benefits (the monetary value of those impacts) can be quite different. It is sometimes offered in defense of abatement costs that they represent at least a lower bound on the damages that may occur from pollution. This idea implies that regulations are always too weak with regard to social efficiency, which recent research suggests is not necessarily true (Krupnick and Portney, 1991). In summary, the abatement cost approach is analogous to looking under the lamppost for the lost car keys. If you find the right answer by looking there, it will be largely a matter of luck rather than science.

The alternative approach to estimating damages is termed the willingness-to-pay (WTP) or damage function approach. This approach begins by identifying residual burdens such as emissions from a power plant. Next, one measures the impacts (injuries) associated with these burdens. From impacts, one estimates damages based on an economic measure of the decrement in consumer welfare that results. There are a variety of valuation techniques that are important in achieving a monetary measure of impacts, depending on the nature of the impact. These include the use of market prices where possible, implicit prices and revealed willingness-to-pay in other circumstances, and techniques such as contingent valuation for impacts that have no analog in market activities. A discussion of these various techniques is beyond the scope of this paper, so it must suffice to report that economists have made tremendous progress in the development of these methodologies. These techniques are generating increasing optimism about the ability to value monetarily in a consistent and valid fashion many of the illusive nonmarket impacts that are associated with various fuel cycles, although there remain large uncertainties and some big holes.

In summary, the WTP/damage function approach is the appropriate measure even though it may require original research. So far, most states have taken a different route. In some cases, states have used the abatement cost approach, though recently they have come under increased criticism for doing so. In other cases, states have used an arbitrary adder to give weight to the consideration of some environmental problems in bidding and resource evaluation. The use of an arbitrary adder has the virtue, from our perspective, that it avoids the false sense of sophistication that is associated with the use of abatement costs. An arbitrary adder may be relatively benign if it is viewed as a rule-of-thumb that serves as a place holder until additional information becomes available.

5   When is Damage or Benefit an Externality?

Given that damages and benefits can be measured, we move to their evaluation with relevance to the policy process in mind. In this section we introduce several criteria that should be useful in circumscribing the set of issues that should be of genuine concern. These criteria are suggested by economic theory, but their precise formulation is sometimes a fundamentally political issue. If regulators first resolve the issues embodied in these criteria, they can serve as rigorous filters to guide analysis.[5]

5.1   Criteria

The first criterion is a fundamentally political question: How should the analysis treat transboundary effects? Transboundary effects pertain to effects of pollution and other impacts that cross over borders to other jurisdictions. One has to decide whether one is altruistic, or hard-nosed and self-interested, in one's approach. That is, should damages to the residents of one's service territory be counted exclusively, or should damages to the state, to the entire U.S., or even to the world be counted as well?

Sometimes it is said that an individual utility can have no impact on a transboundary problem. This is incorrect. One utility has the opportunity to reduce a transboundary problem, such as the global warming problem, to the precise extent of its contribution to the problem. The reason that one might argue that consideration of global warming is inappropriate from an economic perspective is that most of the benefits from individual action to reduce global warming accrue to others. This observation applies to many other issues as well. Whether or not the costs that spill-over to other jurisdictions are to be counted will affect significantly the estimate of social cost.

---

5   Procedurally, one would want to apply criteria such as these to all candidate externalities before completing an estimate of the monetary damages or benefits that may be implied. If the candidate externality is found to be irrelevant for the decision at hand, then the process of valuation can be avoided.

Treating all transboundary effects as costs may have unfortunate consequences. Not only may ratepayers pay more than the benefits they personally receive from the consideration of all costs, but this approach may exacerbate the problems of bypass or industrial flight. If a noncaptive customer decides to bypass the utility's system it may have the opportunity to capture the environmental benefits of the utility's benevolent policies and avoid the cost, even achieving a cost advantage compared to its competitors. Consequently, from an economic perspective one can complain legitimately that state public utility commissions are an inappropriate forum for addressing issues such as global warming. Political analysis can vary from that, to the extent that legislatures want to consider strategic and political issues or to try to prompt federal leadership. But it seems to us that as important as transboundary problems may be, they lie outside the purview of utility regulators absent direction from state or federal governments.

A second criterion asks: What is marginal in the decision at hand? If an increment in capacity is being considered (i.e., the external costs of one additional plant), many potential external costs will be irrelevant. For example, it is often pointed out that historic subsidies to research and development have given certain fuel cycles a price advantage. But, clearly, these are sunk costs and not marginal to the decision at hand. Energy security externalities identified with the military dimensions of foreign policy and the costs of the Strategic Petroleum Reserve are not relevant because these national policies are invariant with the decision of an isolated utility. Whatever the cause of the 1991 Persian Gulf War, it is unlikely that the U.S. response would have been different if even 20% of imported oil demands were reduced (let alone the addition or deletion of a single facility). As with the case of transboundary issues, many social costs are fixed from the perspective of individual utilities and properly should be the subject of policy debate in a different forum.

A third criterion is whether the candidate externality would differentiate between fuel cycles. If an externality affects all fuel cycles in a comparable manner, then it is not worthy of extended consideration in many policy contexts because it is not helpful in differentiating the options according to social cost. One example of such a possible externality is the macroeconomic costs that may result from the short-run inflexibility of prices of electricity, which results from price regulation. This inflexibility may lead to dislocations in the economy when resource prices change but electricity prices do not reflect these changing costs. However, since all fuel cycles are afflicted in a similar manner, what is sometimes called the regulatory externality of fixed prices is not a useful criteria in a siting decision.

A fourth criterion asks: what is the relevant regulatory environment? In the case of environmental pollution, two general regulatory environments are relevant in the U.S. One is command-and-control, which describes the approach of setting minimum technology performance standards for polluters. Command-and-control is by far the most common type of pollution control in the U.S.[6] A second approach is marketable permits, which describes the imposition of a quantity constraint on the total amount of pollution that can be released and allows polluters to buy and sell pollution permits among themselves.

In a command-and-control regime, the rule of thumb we recommend is that the relevant externality is any residual marginal damage that is observed after the firm has complied with all applicable environmental laws.

The second approach to environmental regulation that we consider is tradable emissions permits (TEPs). This is the approach that is embodied in the 1990 Clean Air Act Amendments pertaining to the regulation of sulfur dioxide and the control of acid rain. In this case the quantity of sulfur dioxide emissions is constrained on a national basis.

The most important issue concerning tradable permits is whether the pollutant in question is "uniformly mixing." This means that within an air shed or a water basin the concentration of the pollutant and its impact does not vary with the location of the emission.

If the pollutant in question is uniformly mixing then no marginal damage can occur as the result of operating a new facility or changing the dispatch of existing facilities. The reason is that there is a quantity constraint on the total amount of pollution that can be emitted. In order to emit pollution at the new facility, the utility must obtain offsetting permits at another facility. Consequently, the marginal impact is zero, marginal damage is zero, and there is no externality. On the other hand, if the pollutant is not uniformly mixing (or at least approximately so) then there may be a change in local environmental quality that should be addressed.

To appreciate these results more fully, the next section provides a detailed presentation of the underlying logic and the following section a more rigorous and realistic model yielding a somewhat more qualified conclusion regarding externalities vs damages in a command and control regime.

---

[6] We do not discuss the role of liability and tort awards as incentive devices.

5.2 Rules of Thumb

We make the following assumptions about the PUC. First, the PUC has no control over environmental policy, per se. It must take such things as emission standards, pollution tax rates, or numbers of TEPs as given. Despite the absence of power over environmental policy, we assume the PUC has the objective of minimizing the increment to social cost (inclusive of environmental concerns) associated with meeting the increase in the demand for electricity.[7] It achieves this objective through its choice of the mix of new capacity.

We assume that the PUC has before it a set of proposals from a utility and (perhaps) IPPs. Each proposal is characterized by a quantity of electricity, a price or bid which we take to represent private cost, and a description of its emissions. In the presence of CAC regulation, private cost will include only the costs of controlling emissions to meet standards. In the presence of other approaches to regulation the private cost will also include the pollution taxes to be paid, the cost of permits, etc. The PUC wishes to rank the proposals by social costs, so any external cost should be added to each proposal's private cost to determine its social cost. This external cost is the "adder." In the rest of this section we identify the correct adder under each of three alternative environmental regulatory systems: command and control (CAC), a tax on emissions, and tradable emissions permits. In each case, we also consider the possibility that environmental regulation may not be optimal. We will show that in some circumstances, the magnitude of the adder will depend on whether existing environmental policy is optimal, over-controlling, or under-controlling.

Command and Control Regulation

The CAC case is the simplest of the three that we consider. With CAC, there is no mechanism for internalizing the marginal damages in the private cost or bid of each option. Thus, the adder should be equal to the marginal damages for each source. Whether the CAC policy is over-controlling or under-controlling is irrelevant because the PUC cannot do anything to make matters better. In fact, any effort on the part of the PUC to adjust its ranking of supply options in an effort to "correct" for inappropriate environmental policy will necessarily make things even worse.

---

[7] For a different view of responsibilities and objectives of public utility regulation, see Agathan (1992).

Suppose that the PUC has two alternatives, a clean source (no environmental damages) and a dirty source in the sense that it has large environmental damages even with the presently required controls on emissions. For the clean source, MSC(clean) = MPC(clean). And for the dirty source, MSC(dirty) = MPC(dirty) + MD(dirty). Finally, suppose that their marginal private and social costs are related in such a way that the marginal social cost of the clean source is greater than the marginal private cost of the dirty source, but less than the marginal social cost of the dirty source:

$$MSC(clean) > MPC(dirty)$$

$$MSC(clean) < MSC(dirty)$$

The utility or PUC looking only at private costs would choose the dirty source. But since the PUC wishes to minimize the social cost of the new capacity, it should choose the clean source. Including an adder equal to the marginal damage of the dirty source reverses their ranking on the basis of cost. Alternatively, if MSC(dirty) < MSC(clean), the PUC should choose the dirty source.

Could the PUC "correct" for the effects bad environmental policy by using a different adder? The answer is "no." The PUC can do no better than to choose the source with the lowest social cost. Suppose that environmental policy involves setting emissions standards that are too strict from an economic efficiency perspective. If the PUC ignores or discounts the adder in an effort to tilt the scale in favor of the dirty source, it winds up choosing a source that imposes additional external damages and excessive abatement costs on society.

Now consider the case where existing emissions standards are not strict enough. Can the PUC "correct" for this under-control by choosing the clean source even though its marginal social cost exceeds that of the dirty source? Again, the answer is "no." Given that a dollar of private cost has the same significance for economic efficiency and welfare as a dollar of external cost, the PUC should choose the source with the lowest social cost.

The bottom line of this discussion is that if the PUC is trying to minimize social cost the PUC should treat the costs imposed on society by inappropriate environmental standards for existing sources as sunk costs. There is nothing the PUC can do about them. They should be ignored. However, the environmental consequences (given existing environmental regulation) are not sunk but variable with an incremental investment in capacity. If excessive control cost exists it is already built into the bid or private cost, so the PUC should find the external cost and add it to the private cost. The appropriate adder is marginal damage.

A Tax on Emissions[8]

With an emissions tax system, a potential new source's bid will be based on the sum of production cost and the tax on the remaining emissions. The production cost will include any cost associated with controlling emissions in response to the incentives created by the tax. The PUC will wish to rank the supply options by marginal social cost, which will be the sum of the production costs (including costs of control) and the marginal damages imposed by the remaining emissions. Marginal social costs will differ from bids to the extent that marginal damages differ from the tax on emissions. The PUC should adjust the bid by including an adder equal to MD - t.[9]

If existing environmental policy is too strict, the emissions tax is greater than marginal damages. The PUC should correct for this by incorporating a negative adder equal to MD - t. This negative adder corrects for the fact that the part of the tax included in the private cost or bid is not reflective of a social cost, but rather it is just a transfer payment. Similarly, if the environmental policy is not strict enough, the PUC should adjust the bid upward by an amount equal to the excess of marginal damage over the emission tax.

Some adjustment may be necessary even if the emissions tax is just equal to marginal damage at the existing equilibrium. Suppose the marginal damage function is upward sloping and the proposed increment to supply involves a nonmarginal change in emissions. In a first best world, the emissions tax would be adjusted upward to reflect this fact. And the bid would be based on the anticipated higher emissions tax rate. But in the second best world, bidders and the PUC take the emissions tax rate as given. Therefore the PUC should include an adder equal to the excess of marginal damages in the new equilibrium over the unchanged emissions tax rate.

Tradable Emissions Permits

This case is of special interest because of the offset requirements for new sources in air pollution non-attainment areas and the SO2 allowance trading system being established under

---

[8] Currently there are no examples of these in the U.S., although they are prevalent in Europe (OECD, 1989).

[9] The statement that the adder is equal to MD - t is an over simplification. MD and t are usually expressed in units of $/E, while the adder is expressed in units $/kwh (or $/mw). So the adder is calculated as (MD - t)E/kwh.

the Clean Air Act Amendments of 1990.[10] The calculation of the correct adder in the case of tradable emissions permits depends on whether the marginal damage per unit of emission varies across sources, and if so, the extent to which the rules which determine how permits are traded reflect any variation in marginal damages. In general, the marginal damage per unit of emission depends both on the effect of emissions on ambient environmental quality and on the marginal damage associated with a decrease in ambient quality. Both of these relationships can vary significantly across sources of emissions at different locations. The origins of this variation include differences in the ability of the environment to absorb emissions, differences in the characteristics and numbers of agents experiencing the impacts of the emissions (e.g.., urban vs. rural areas), and nonlinearities in the damage function.

We first consider a simple case where the marginal damage per unit of emission is same for all sources. An example would be a globally or regionally well-mixed pollutant with a uniform ambient concentration. The effect of a one unit increase in emissions on ambient concentration and damages is therefore independent of the location of the source of the emission. The key feature of this case is that any new source must purchase permits for every unit of emission, and since the total number of permits available is fixed, emissions from other sources are reduced by an equal amount as they sell permits to the new source.

Given these assumptions, the marginal social cost of a new source is the sum represented in the following Table 1. If the pollutant is uniformly mixed the two marginal damage terms are equal and cancel out.[11] In equilibrium, the price of a permit will equal the marginal control cost at the source selling permits. The new source must include the price of its permits in its bid. So the bid will equal the sum of marginal production cost and the cost of permits, and this will be equal to the marginal social cost of the new source. In other words, the price of permits internalizes the cost of reducing emissions by an equal amount at other sources. The appropriate adder is zero. This is true regardless of whether the number of permits available in the market is optimal, too large, or too small.

We now turn to the case where the marginal damage per unit of emission varies across sources. We will first assume that the environmental agency has adopted optimal trading rules to reflect this variation. These rules will require that any source i which desires to increase its emissions by purchasing permits from source j must purchase d_{ij} permits per unit of its

---

[10] Proposed new sources of air emission in regions presently not attaining primary national air quality standards must certify offsetting reductions in emissions from other sources as a condition for obtaining an emissions license. They may obtain offsets by purchasing reductions in emissions from other sources.

[11] There is a change in the incidence of damages. Some gain as their damages are reduced while other lose. But the net effect is zero.

emission where dij is the ratio of its marginal damages to marginal damages at source j (MDi/MDj).[12] We will then examine how our results change if the environmental agency has not adopted the optimal trading rules.

Table 1: Marginal Social Costs of Production at a New Source in the Case of Tradable Emissions Permits when the Pollutant is Uniformly-Mixing

1. Marginal production cost at the new source including its control costs.(+)
2. Marginal control cost at the source selling permits to the new source. (+)
3. Marginal damage of emissions at the new source. (+)
4. Marginal damages avoided at the source selling permits and reducing emissions. (-)

Suppose that it is known that the proposed new source i would purchase its required permits from existing source j and that the ratio of the marginal damages from the two sources is three. In other words, dij = MDi/MDj = 3. Let the present price of permits sold by source j be P*; and suppose that the incremental demand of the new source is small enough that P* can be assumed to be constant. Given this ratio of marginal damages between the new and existing source, the optimal trading rules will require that the new source purchase three permits from source j for every one additional unit of emission. Thus the price of one unit of emissions at the new source, Pi, is 3P*. The new source will optimize by purchasing enough permits so that its marginal control cost is equal to the price it must pay (Pi = 3P*).

As in the case of the well-mixed pollutant, the marginal social cost of the new source is the sum represented in the following Table 2.

Since the optimal trading rules require that the new source purchase 3 permits for each unit of its emissions, the last two components are equal to each other and cancel out. And the marginal control cost at the source selling the permits is internalized in the bid of the new source through the necessity to purchase permits at Pi = 3P*. So, the adder is zero in the case of a permit system with optimally designed trading rules.

---

[12] For details concerning optimal trading rules, see, for example, Tietenberg (1985), or Bohm and Russell (1985).

Table 2:   Marginal Social Costs of Production at a New Source in the Case of Tradable
            Emissions Permits when the Pollutant is not Uniformly-Mixing

1. Marginal production cost at the new source including its control costs. (+)
2. Marginal control cost at the source selling permits to the new source. (+)
3. Marginal damage of emissions at the new source. (+MDi)
4. Marginal damages of emissions at the source selling permits and reducing emissions. (-3 MDj)

Note that this result was obtained without making any assumption about the optimality of the pattern of environmental quality or the number of permits issued for trading. As long as the permit trading ratio is equal to the ratio of marginal damages, the last two components of the above summation cancel out. And the private bid of the new source internalizes the marginal control cost at the source selling permits.

Now let us see what happens when the trading rules are not optimal. Suppose that in the above example, permits traded one for one, even though MDi = 3 MDj. Then, the last two components of the above summation would not cancel out. Although the private bid would internalize the marginal control costs imposed on the selling source through the permit price, the difference between the marginal damages at the two sources would have to be added to the bid to reach the correct marginal social cost. In other words, the environmental cost adder would be the difference between the marginal damages at the new source and the old source. If the marginal damages at the selling source were greater than the marginal damage at the new source, the adder would be negative.

Alternatively, suppose that the marginal damages per unit of emissions are the same at both sources but that permits do not trade on a one for one basis. For example, suppose that the new source must purchase more than one permit for each unit of its emissions so that there will be a reduction in total emissions.[13] Then, the correct adder is based on the product of marginal damages and the net change in emissions. If total emissions are reduced, the adder is negative.

---

[13] This requirement is a feature of the present federal offset program.

Conclusions

We can summarize the results of our analysis in Table 3.

Table 3:   Level of Environmental Control

LEVEL OF ENVIRONMENTAL CONTROL		
Environmental Policy Instrument	Under-Control	Optimal Control	Over-Control
Emission tax	Adder = MD - t > 0	Adder = MD - t = 0	Adder = MD - t < 0
TEPs: Optimal Trading Rules	----------------------- Adder = 0 -----------------------	
TEPs: NonOptimal Trading Rules	------------ Adder = MD(new) - MD(old) ------------	
Command and Control	---------------------- Adder = MD ----------------------	

Table 3 shows that the correct adder depends on the nature of the environmental policy instrument and the details of the implementation of that instrument. The adder will be zero only when the true costs of the new source, broadly construed, are fully internalized in the private cost calculus of the bidder. For example, with a TEP system and optimal trading rules, the adder is always zero because the cost that the new source imposes on society is the additional control cost at the selling source and this cost is already internalized through the cost of permits. But with a TEP system and trading rules that are not optimal, the differences in marginal damages across sources are not properly internalized. Nonetheless the correct costs to be internalized are not necessarily the external damages.

5.3   A More Sophisticated Approach

The previous section offered an informal analysis of the issue of social costing for planning purposes, and the relationship between an optimal adder and exogenous, preexisting policies. In the case of command-and-control regulation, we suggested rules of thumb for approximating these relationships. In other cases we suggested how an optimal adder would relate to marginal damages.

In this section ,we offer results based on a simple but fairly general formal model of social welfare maximization (see Burtraw et al, 1992), under the assumption that the goal of the PÙC is to maximize social welfare, within its ability to do so. In this paper, we investigate the design of an optimal adder that could be used for either planning or dispatch purposes (depending on the definition of the cost functions). We consider two technology options for the regulated utility, and one option for utility customers to bypass the grid and generate their own energy services (including fuel substitution) which would have its own externalities. Within the formal model, however, we do not consider the possibility of substitution away from energy services to other factors of production such as capital, labor, or resources.

We find that the optimal (second-best) adder depends on the market structure of the industry, that is, whether the firms are subject to cost-recovery rate setting and whether prices deviate from marginal costs. Since we assume average cost pricing is already second-best, the optimal adder is zero when there are no external damages. However, in the presence of external damages such as environmental pollution, the calculation of the optimal adder must take into account the ramifications of moving toward or further away from marginal cost pricing.

Second, we formally model an opportunity for bypass of the electricity grid by the utility's customers. This opportunity may have deleterious environmental effects which must also be taken into account. In some special cases the optimal adder is shown to approximately equal marginal damage. One such case is when the opportunity for bypass is nonexistent, or when emissions from independently generated energy services are zero. In this case, whenever price and marginal social cost are within 40% of each other, and elasticity is greater than or equal to -0.7, the optimal adder is within about 25% of marginal damages. The difference between the optimal adder and marginal damage is smaller the smaller is the elasticity of demand and the smaller the deviation between price and marginal social cost.

In a competitive market, such as may characterize the independent power sector supplying power from various technologies to a regulated utility for transmission and distribution, and against a backdrop of command-and-control environmental regulation at the federal or state level, the optimal adder differs from marginal damage only due to potential emissions and associated damage resulting from bypass of the electricity grid. The smaller are either these emissions or the potential for bypass, the closer is the optimal adder to marginal damage. In the absence of an opportunity for bypass, the optimal adder precisely equals the marginal damage from the residual pollution that occurs after the firms have complied with all relevant environmental laws.

The command-and-control setting is the most important for the consideration of environmental issues because it is the form taken by most existing federal and state environmental regulations in this country. The rules we identify for these adders are invariant with regard to the rigor of preexisting regulation, that is whether emissions are over-controlled or under-controlled from the standpoint of economic theory. Against the backdrop of command-and-control regulation, the simple rule of thumb that an adder be set equal to marginal damages has the potential to be quite robust. In further research we intend to determine whether this rule can be a useful guideline for the use of adders for planning or dispatch decisions.

Setting the fee equal to the marginal damages appears to be a standard result in the economics literature, as found in Baumol and Oates (1988) and elsewhere. Note this important difference, however. In the standard model, where there is no preexisting environmental standard that must be met, the optimum effluent fee must be set equal to what the marginal damages would be at the optimum. If one starts at a nonoptimal point hoping to use the fee instrument to reach optimality, the proper fee is not obvious. This analysis establishes that the optimal adder does not depend on whether existing regulation is inefficient or efficient, and offers a theoretical framework for comparison of the optimal adder with observed marginal damage given existing regulation. While environmental damage estimation is not easy under any circumstances, it is at any rate easier to estimate marginal damages than to estimate what damages would be at some as-yet-unattained optimum.

6   Implementation Issues

Proposed and existing social costing regulations--both of which focus primarily on the selection of new sources of generating capacity--require utilities to invest in the project(s) with the lowest social costs of electricity supply, but do not require them to actually pay the environmental damage costs of electricity generation. Ranking potential sources of new supply by social cost per unit of electricity produced is referred to as "ranking with grandfathering." This is because new generation facilities, but not existing ones, are covered by this form of regulation.

Another possible approach to social costing would be to impose the environmental costs of power production at new generating units on a utility by means of a tax on emissions. The tax rate would be set equal to the marginal damage cost associated with an additional unit of pollution. This form of social costing, referred to as "taxation with grandfathering," would pass to consumers the full social cost of generating electricity at new generating units only.

Under a third form of social costing, a utility would be required to internalize the external costs of emissions from both new and existing generating units. This regulatory regime would govern not only investment behavior, but also the schedule of use or dispatch of all generating units. When a utility has to consider the complete social cost of each kwh of electricity generated, it will dispatch all units in order of unit social cost, a practice referred to as "social cost dispatch."

Modeling Utility Decision Making

To explore the effects of imposing each of these three social costing regimes on a utility's investment and dispatch decisions, researchers at Resources for the Future (RFF) developed and simulated a model of utility planning and dispatch for a representative mid-Atlantic utility. The objective of this hypothetical utility which uses nuclear, coal steam, oil steam, gas steam, gas turbine and hydro technologies to generate electricity was to build and operate generating units in a way that minimizes the cost of satisfying a given expected level of demand for electricity plus a 20 percent reserve margin. The model allowed for demand to vary over time and specified a realistic load duration curve which depicts the time duration of electricity demand at or above a particular level. In the simulation, the utility had to increase its generating capacity by deferring retirements and by investing in new generating units to meet increased demand. The range of technologies available for investment included four fossil-fueled technologies and three renewable technologies: hydro electric, solar and wind. Ranked from least expensive to most expensive in terms of private cost per kwh generated, the technologies are hydro-electric, integrated gasification combined cycle (IGCC), gas turbine combined cycle (GTCC), pulverized goal with wet flue gas desulfurization (PC/FGD), atmospheric fluidized bed coal (AFBC), wind and solar.

To simulate how the hypothetical utility would respond to social costing in its investment and dispatch decisions, the RFF study inserted estimates of the environmental costs associated with three pollutants that are emitted when fossil fuel is burned to generate electricity: sulfur dioxide (SO_2), nitrogen oxides (NO_x), and carbon dioxide (CO_2).

The RFF study considered four sets of illustrative estimates of the unit environmental costs of NO_x and SO_2 emissions. In the first set of estimates based on illustrative estimates of environmental damage costs, NOx emissions are valued at $0.25 per pound and SO2 emissions at $0.50 per pound. A second set of estimates multiplies these values by a factor of 10. In a third set of estimates, based on abatement costs, NOx emissions are valued at $3.25 per pound and SO2 emissions at $0.75 per pound, while the fourth set multiplies these values by 10. The social costs of CO2 emissions were not based on estimates of environmental damage or abatement costs, but on two tax rates proposed for limiting these emissions: $25 per metric ton (MT) and $100 per MT of carbon content of the fuel which translates to $6.82 per MT and $27.27 per MT respectively of CO2 emitted. The RFF study considered the effects of social costing both with and without valuing CO2 emissions.

Results of the Model Simulations

The RFF study compared the utility's investment decisions, its dispatch of generating units, and the related social costs under the various social costing regimes to a base case scenario with no environmental costing. With no social costing, the utility builds new IGCC and GTCC generating capacity and uses these new generating units to generate 27 percent of its total electricity output. The utility builds a small amount of new hydro capacity and runs its existing hydro unit at capacity throughout much of the year. It does make some life-extending investment in its aging coal-fired units, but allows its expiring oil-fired unit to shut down.

Under this base scenario, the social costs of the utility's investment and operating plan depend on the environmental cost assumptions adopted. Using relatively high external costs for SO2, the study revealed that the ratio of social costs to private costs is 1.2 in the absence of a carbon tax. With a CO2 tax of $6.82/MT, the ratio rises to 1.3. With a tax of $27.27/MT, the ratio is 1.8. At ten times the relatively high SO2 cost level, the ratio is 2.6. Using relatively high NOx costs, the ratio of social costs to private costs is 1.7 in the absence of a CO2 tax, 1.9 with a CO2 tax of $6.82/MT, and 2.4 with a CO2 tax of $27.27/MT. At ten times the relatively high NOx external cost levels, the ratio rises to 8.4.

Under a social cost program in which new electricity generating units are ranked by the social cost per unit of electricity produced (ranking with grandfathering), the utility's dispatch and investment behavior is unaffected unless CO2 is assigned an external cost of $27.27/MT. Under this set of scenarios, new units using renewable energy are built, but account for less than 3 percent of the electricity generated. Under any environmental cost assumptions, this regulatory regime does not decrease the social costs of electricity generation.

As compared with ranking, a social cost program in which emissions from new generating units are taxed (taxation with grandfathering) has a more substantial effect on the utility's investment behavior and on how generating units are dispatched. The impact on dispatch is especially pronounced when carbon emissions are taxed, leading the utility to increase significantly its reliance on existing fossil- fueled units. This tendency to avoid new generating units may lead to an increase in social costs, because the asymmetric tax creates a stronger incentive to substitute existing sources of electricity for new sources. The social costs of electricity generation never fall under taxation with grandfathering.

Under a social cost dispatch program, the utility's investment and dispatch decisions vary depending on the environmental costs adopted. With the high SO_2 set of external costs, the utility's investment and operating behavior is not changed substantially from the base case unless CO_2 emissions are assigned a cost of $27.27/MT, in which case existing gas stream units are employed more intensely, or external costs for SO_2 and NO_x emissions are set at $7.50 and $32.50 per pound respectively, in which case the utility shuts down all of its existing coal units and relies heavily on new coal-burning units. (See Figure 3.)

With relatively high NO_x, the utility shuts down its existing coal- fired boilers at the lowest external cost levels and increases its reliance on new coal-fired units. When CO_2 emissions are included, the utility increases its reliance on renewables and reduces its use of existing oil steam units. Social cost dispatch never leads to an increase in social costs. Under all of the illustrative environmental cost assumptions, the utility chooses generating resources that result in lower social costs than in the base case. As the assumed marginal environmental cost estimates increase, the social cost savings from the imposition of social cost dispatch rise.

Implications of Environmental Costing for Electricity Prices

The effect of social costing on electricity prices varies with the environmental cost assumptions and the scope of the regulatory regime. The average cost of generating a kilowatt-hour of electricity with no social costing is $0.052. This unit cost estimate includes the embedded capital cost of using existing generating units but does not include the overhead costs of operating the utility.

33

Figure 3:   A Hypothetical Utility's Sources of Electricity Generation under an Environmental Costing Program of Social Cost Dispatch

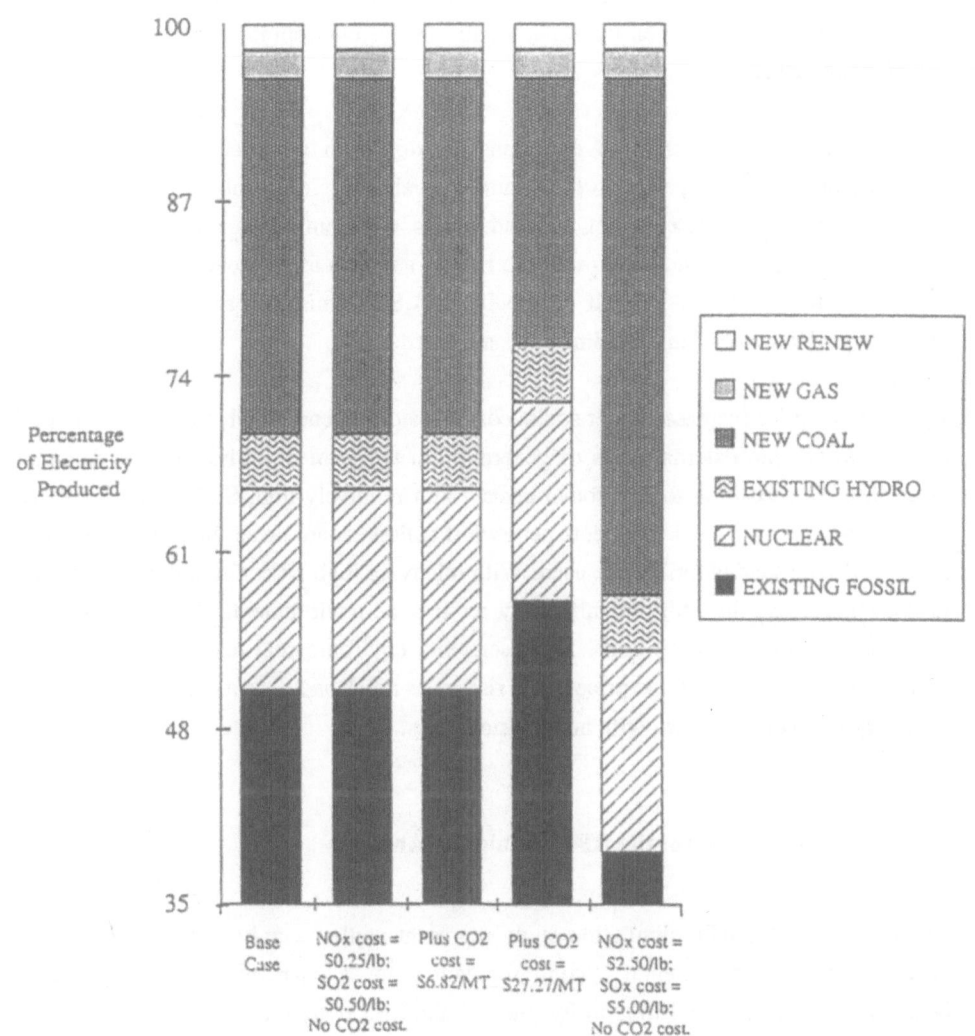

Assuming that demand for electricity does not change, the findings of the RFF study suggest that the unit generation cost and the consumer price for electricity would rise between 0 and 0.05 cents under a social costing program of ranking with grandfathering. This virtually non-existent price effect reflects how little a requirement to invest in new sources of generating capacity with the lowest social costs of electricity supply would affect a utility's employment of generating resources, given the range of illustrative environmental cost estimates for the included pollutants.

Under a social cost program of emissions taxation with grandfathering, the size of the electricity price effect depends on the assumed level of environmental costs. With relatively high SO_2 costs, the price is not affected unless CO_2 emissions are taxed. At a tax of \$27.27/MT, the price could be expected to rise by as much as 0.5 cents. With relatively high NO_x costs, the price could rise if only NO_x, and SO_2 emissions are taxed; however, it is likely to rise by more when CO_2 emissions are taxed.

The potential price increases under social cost dispatch depend on whether or not the utility is required to pay the external costs of generation in the form of emission taxes or to simply dispatch units according to full social costs. With relatively high SO_2 costs, the consumer price would rise by from 1 to 3 cents per kwh if emissions are taxed, but only up to .75 cents per kwh if only shadow prices are used. With relatively high NO_x costs, the increase in price is smaller because the utility significantly reduces its environmental tax bill when it shuts down its existing coal-fired units. However, with environmental taxes set at ten times the basic abatement cost level, the price could rise by as much as 3.5 cents per kwh with emission taxes and 1.25 cents per kwh with no emission taxes.

Limitations and Results of the RFF Simulation Analysis

The findings of the RFF simulation study represent neither a universal prediction of utility behavior under environmental costing regulation nor a characterization of the expected behavior of a particular existing utility. Instead, the RFF study provides an empirical example of the response of a representative mid-western utility to illustrative environmental costs. The applicability of the findings of this study to either the general class of electric utilities or to a specific utility is restricted by the limited range of environmental externalities that are included in the analysis and by the specification of environmental costs.

With respect to externalities, the study focused only on emissions of a small number of air pollutants and, therefore, failed to include other environmental externalities, such as emissions of water and soil pollutants or environmental damages associated with the construction of various types of power plants. Moreover, a constant emission factor for each pollutant at each type of generating unit was used to estimate the levels of those emissions that were considered. While this assumption is accurate for some pollutants, NOx emission rates vary with the level of capacity utilization at each generating unit.

With respect to environmental costs, the illustrative external cost estimates adopted in this study were intended to indicate a range of possible environmental costs that might be used for implementing EC regulation. However, these costs do not reflect state-of- the-art estimates for environmental damages for the selected pollutants. If the studies of environmental damages of electricity generation currently underway in several states estimate damages that lie outside of the ranges included in this study, then a particular utility's response to a social costing regime based on these estimates may be at variance to the responses suggested in the simulation study.

Keeping these limitations in mind, the RFF study indicates that the outcome of a social cost program depends on the breadth of the program and on the environmental cost estimates on which it is based. The most effective environmental costing regime is social cost dispatch. Such a regime causes utilities to internalize the environmental damage costs of all the electricity they generate and, thus, leads them to reduce the social costs of electricity generation. A less comprehensive regime, such as ranking with grandfathering or taxation with grandfathering, could lead to higher emissions levels than those that would occur in the absence of environmental costing. Under partial environmental costing, utilities may not have an incentive to invest in cleaner technologies as long as the lives of existing generating units that use more polluting technologies may be extended with no regulatory penalties. Indeed, in RFF's empirical analysis, social costs never decrease under partial environmental costing, whereas they never increase under social cost dispatch. The incentive to substitute existing dirty units for new clean units can be even stronger when emissions from new supply sources are taxed than when utilities are required to invest in new sources with the lowest social cost per unit of electricity produced.

A striking result of the RFF study is that social cost dispatch, based on the study's illustrative relatively high SO_2 environmental cost estimates, leads to no change in the utility's investment and dispatch decisions unless high CO_2 costs are assigned or unless the external costs of SO_2 and NOx are set at ten times their initial levels. While these results reflect only a limited range of environmental externalities, they suggest that comprehensive social costing based on air pollution damage costs may have only a small effect on utility investment and dispatch behavior.

7 Conclusion

Introducing social costs into utility decision making is not the first best policy for internalizing damages associated with energy use. If this approach is applied to electric utilities only, energy markets could become distorted. It introduces possible anti-new source bias if applied to only new sources. It requires that other policies, such as potentially inefficient environmental laws, be taken as a given. It offers an inappropriate jurisdictional control for many issues, such as global warming or foreign policy, which will be a source of frustration for many advocates. And it could even result in increases in pollution. It would be preferable for federal and state laws to be set and designed efficiently affecting all sectors of the economy.

Nonetheless, application and investment of the concept of social costing of electricity can lead to more efficient electricity generation choices. While the piecemeal problem is potentially significant, so are the benefits of social costing.

References

Agathan, Paul A. 1992. "Dealing with Environmental Externalities," Public Utilities Fortnightly, February 15, pp. 23-24.

Baumol, William J. and Wallace E. Oates. 1988. The Theory of Environmental Policy (Cambridge and New York, Cambridge University Press).

Bohm, Peter, and Clifford S. Russell. 1985. "Comparative Analysis of Alternative Policy Instruments," in Allen V. Kneese and James L. Sweeney, eds., Handbook of Environment and Resource Economics vol.1 (New York, North Holland).

Burtraw, Dallas, Winston Harrington, A. Myrick Freeman III, and Alan J. Krupnick. 1992. "The Analytics of Social Costing in a Regulated Industry," unpublished manuscript (Washington, D.C., Resources for the Future).

Burtraw, Dallas, and Alan J. Krupnick. 1992. "The Social Costs of Electricity: How Much of the Camel to Let into the Tent?" Quality of the Environment Division Discussion Paper QE92-15 (Washington, D.C., Resources for the Future).

Freeman, A. Myrick, Dallas Burtraw, Winston Harrington, and Alan J. Krupnick. 1992. "Accounting for Environmental Costs in Electric Utility Resource Supply Planning," Quality of the Environment Division Discussion Paper QE92-14 (Washington, D.C., Resources for the Future); and Electricity Journal, vol. 5, no. 7 (Aug./Sept.), pp. 18-25.

Freeman, A. Myrick and Alan J. Krupnick. 1992. "Getting Externalities RightA Response to Prof. ," Electricity Journal, vol. 5, no. 7 (Aug./Sept.), pp. 61-63 (a reply to Joskow's "Dealing with Environmental Externalities: Let's Do It Right" paper).

Joskow, Paul L. 1992. "Dealing with Environmental Externalities: Let's Do It Right!" Edison Electric Institute Issues and Trends Briefing Paper No. 61.

Krupnick, Alan J. and Paul R. Portney. 1991. "Controlling Urban Air Pollution: A Benefit-Cost Assessment," Science, vol. 252, 26 April, pp. 522-528.

Lee, Russell, et al. 1992. U.S. - EC Fuel Cycle Study: Background Document to the Approach and Issues, ORNL/M-2500, Oak Ridge National Laboratory, Oak Ridge, Tenn., and Resources for the Future, Washington, DC, for the U.S. Department of Energy under Contract No. DE-AC05-840R21400, November.

Organisation for Economic Co-operation and Development [OECD]. 1989. Economic Instruments for Environmental Protection (Paris, OECD).

Palmer, Karen L. and Alan J. Krupnick. 1991. "Environmental Costing and Electric Utilities' Planning and Investment," Resources, 105, Washington, D.C., Resources for the Future based on Palmer, Karen L. and Hadi Dowlatabadi, "Implementing Environmental Costing in the Electric Utility Industry," Quality of the Environment Discussion Paper QE91-13 (Washington, D.C., Resources for the Future).

Tietenberg, Thomas H. 1985. Emissions Trading: An Exercise in Reforming Pollution Policy (Washington, D.C., Resources for the Future).

3. Perspectives on Incorporation of Environmental Externalities

Jeffrey D. Tranen, Vice President
New England Electric System, Westborough, Massachusetts 01582

Summary

Incorporating environmental considerations including externalities in utility planning and operation is best accomplished by an overall system approach. The approach should balance criteria such as environmental improvement, economics, and reliability of supply. In addition, utilities must also take into account thresholds for health, safety, and other significant factors.

New England Electric System companies have incorporated this balanced approach in its strategic long-range plan, NEESPLAN 3 - Environment, Economy, and Energy for the 1990's. NEESPLAN 3 sets forth specific, balanced goals for improving the environment, controlling the cost of electricity, and maintaining a reliable energy supply.

Processes now used to estimate the cost of "environmental damage" should continue to be evaluated as a component of cost-benefit analyses. Information from such evaluations may provide useful data to influence new and revised environmental laws and regulations. The appropriate role for utility rate regulators is to insure that utilities anticipate the potential for such changes and factor them into their resource plans.

Approaches for Incorporating Environmental Externalities

Incorporating environmental considerations including externalities is best accomplished by overall system approaches. These are more likely to result in least cost approaches to pollution reduction than fragmented approaches or those that do not directly address total cost and/or total pollution reduction.

As noted in the NEESPLAN 3 discussion later in this paper, it is important to balance cost, environmental improvement, and reliable energy supply. Approaches that focus more directly on total cost and total pollution reduction are better able to accomplish this necessary balancing. Ideally actions should be taken in a coordinated fashion within the appropriate region of impact.

By definition an environmental externality has not yet been internalized as a law or regulation. It is not merely residual pollution since current laws or regulations may have considered and rejected the cost effectiveness of further pollution reduction. The current level of internalization varies by pollutant; some in our view being fully internalized while others not at all. Various approaches have been used to incorporate environmental considerations or are currently under consideration to incorporate environmental externalities. These include:

1) "Command and Control" unit-specific limits;
2) Tradeable permits;
3) Offset programs;
4) Taxes and fees; and
5) Adders.

Discussion of Approaches

Following is a brief discussion of these approaches from the perspective of how well we believe they are able to achieve a balanced, cost-effective approach to environmental improvement.

1. Command and Control:

Generating sources, new and existing, are subject to a wide array of unit-specific limits. On the air quality side, these include limits to comply with National Ambient Air Quality Standards, Reasonably Available Control Technology (RACT), Best Available Control Technology (BACT), Lowest Achievable Emission Rate (LAER), New Source Performance Standards (NSPS), and operating permit limits. On the water quality side, there are water quality standards, National Pollutant Discharge Elimination System (NPDES) effluent limits, and water usage or withdrawal limits. Likewise, there are unit-specific limits on solid waste production and disposal.

These types of limits are often fragmented and tend to cost more than necessary to reduce overall pollution, because they don't necessarily affect all units and don't allow for flexibility to implement least cost approaches to overall pollution reduction. "Best technology" requirements are particularly costly because they mandate emission reductions at extraordinarily high marginal costs, while existing facilities that could have reduced pollution at relatively lower cost have no incentive to install new or improved pollution control technology.

2. *Tradeable Permits:*

The most notable program of this type is the sulfur dioxide allowance trading program under the acid rain title in the Clean Air Act Amendments of 1990. In the allowance trading program, overall utility sulfur dioxide emissions are first reduced and then capped. Generating units are able to trade their allocation of allowances (or must purchase allowances if they are allocated none).

In this type of approach, caps can be set after consideration of the costs and benefits of achieving various levels of pollution reduction. Even if total cost is not precisely known when the cap is set, the concept of the approach allows flexibility to achieve least cost overall pollution reduction.

3. *Offset Programs:*

Offsets refer to emission reductions achieved at a location away from a generating source. Offsetting certain types of air emissions (e.g. carbon dioxide) may be far more cost-effective than controlling at a generating source. Also, new facilities must offset all of their emissions of nonattainment pollutants in order to construct. The key point of offsets is that they provide flexibility to the source owner to find cost-effective means to reduce overall pollution, with the result that the desired environmental effect (net level of pollution) occurs at least cost.

4. *Taxes and Fees:*

Assessment of sufficiently high pollution taxes or fees can serve as an incentive to stimulate action to reduce the pollution (e.g. air emissions). In particular, carbon or carbon dioxide taxes have been suggested at the Federal or multi-national level to stabilize or reduce greenhouse gas emissions. Some studies have shown that a large tax would be necessary to stabilize or reduce carbon emissions.

This approach can be implemented on a broad base of polluting resources and therefore overcomes many of the "piecemeal" flaws of other approaches. However, it does not focus directly on total cost or total pollution reduction and, therefore, would likely require continuous refinements over time. Also, the approach could be viewed as merely a revenue enhancer for the taxing authority; if redistribution of the taxes were to occur, the question arises as to the proper way to implement it. However, we believe that a low tax, coupled with caveats and the ability to use offsets, would be an appropriate long-range signal to energy markets.[1]

5. Adders:

The adders approach to resource planning evaluation is based on the theory that the total (or "social") cost of producing electricity is comprised of two components. The first is the direct revenue cost, consisting of carrying charges on capital costs, operation & maintenance costs, and fuel costs. The second is the environmental "damage" cost attributable to the generating facility, usually determined in units such as dollars per ton of a type of air pollutant emitted.

While simple in theory, actual applications and effects of the second, or "adders" component, varies very widely. For example, the variations include:

- adders based on the cost of estimated environmental damages;
- adders based on marginal cost of control for the pollutant;
- adders based on least cost of control;
- adders used only on an advisory basis;
- adders used only if they are a significant part (e.g. over 10%) of the total (social) cost;
- adders that are constrained by local concerns, such as employment effects;
- adders that are constrained by application to only new resources; and
- adders that are constrained by a court decision.

Once set, adders are relatively easy for rate regulators to administer and enforce. Adders also serve as a "placeholder" or reminder of environmental considerations during the planning process. However, adders suffer from many recognized shortcomings. Adders are not developed and applied consistently among states, leading to inconsistent planning criteria for multi-state utilities and power pools. Adders do not lead to least cost ways of reducing overall pollution, as the focus is not on total pollution or its total control cost. Also, because adders are only applied to regulated electric utilities, a "piecemeal" approach results with potentially unintended consequences such as bypass of the electric utility by its customers.

The NEES Approach

As discussed in the following section, New England Electric System companies have developed a strategic long-range plan that incorporates desirable features of the above-described approaches for incorporating environmental externalities in the resource planning process. The plan represents a least cost way of achieving environmental improvement.

The NEESPLAN 3 Approach

About New England Electric System

New England Electric System (NEES) is a registered public utility holding company.[2] Our retail subsidiaries serve almost 1,300,000 customers or 2,700,000 people in most of Rhode Island, a substantial part of Massachusetts, and smaller parts of New Hampshire. Figure 1 shows our service area and the location of many of the generating facilities that supply our customers. NEES subsidiaries own or purchase power from generating sources in all six New England states and in Canada, about two-thirds of which comes from fossil fired sources. Because of the diverse locations of generating sources and customers, NEES companies are subject to a wide variety of federal and state legislation and regulations - dealing with aspects such as the environment, rates, and energy supply.

NEESPLAN 3 Goals

During 1990, as part of our quality improvement process,[3] we determined that we should develop and implement a long-range plan to address and balance three criteria improving the environment, managing costs and improving service. In 1991, we launched NEESPLAN 3 - Environment, Economy, and Energy for the 1990's, which set three major goals:

Goal No. 1 is to continuously reduce the environmental impact of our electric service. More specifically, we have set a target to reduce our weighted net air emissions by an estimated 45% over the decade[4]. Figure 2 depicts the estimated emissions in 1990 and 2000.

Goal No. 2 is to maintain overall rate increases at or below inflation for the balance of the decade, and to keep our rates the lowest among the major utilities in the states we serve. Figure 3 shows the average rates for the NEES retail companies in comparison to other major New England utilities.

Goal No. 3 is to strengthen the diversity of the energy sources that serve our customers, through the competitive marketplace. We will do this by bringing more non-utility generation into service predominately fueled by gas, and by increasing the use of natural gas

43

Figure 1:   New England Electric System

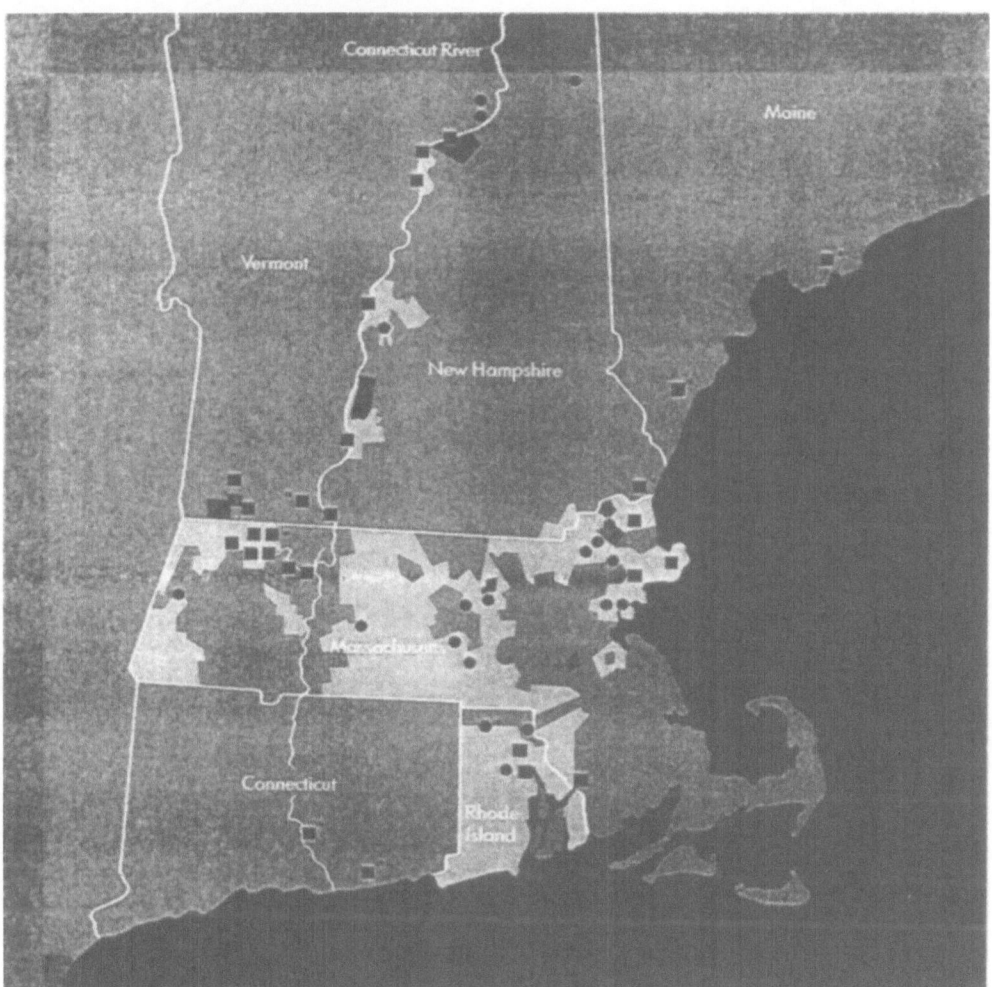

Service Area-Retail   Light Shaded Area

Service Area-Wholesale   Dark Shaded Area

■ Generating facilities owned by NEES
(* indicates partial NEES ownership)

● Non-utility generating facilities
from which NEES purchases power
(* indicates partial NEES ownership)

Figure 2:   NEES Net Air Emissions 1990 vs. 2000

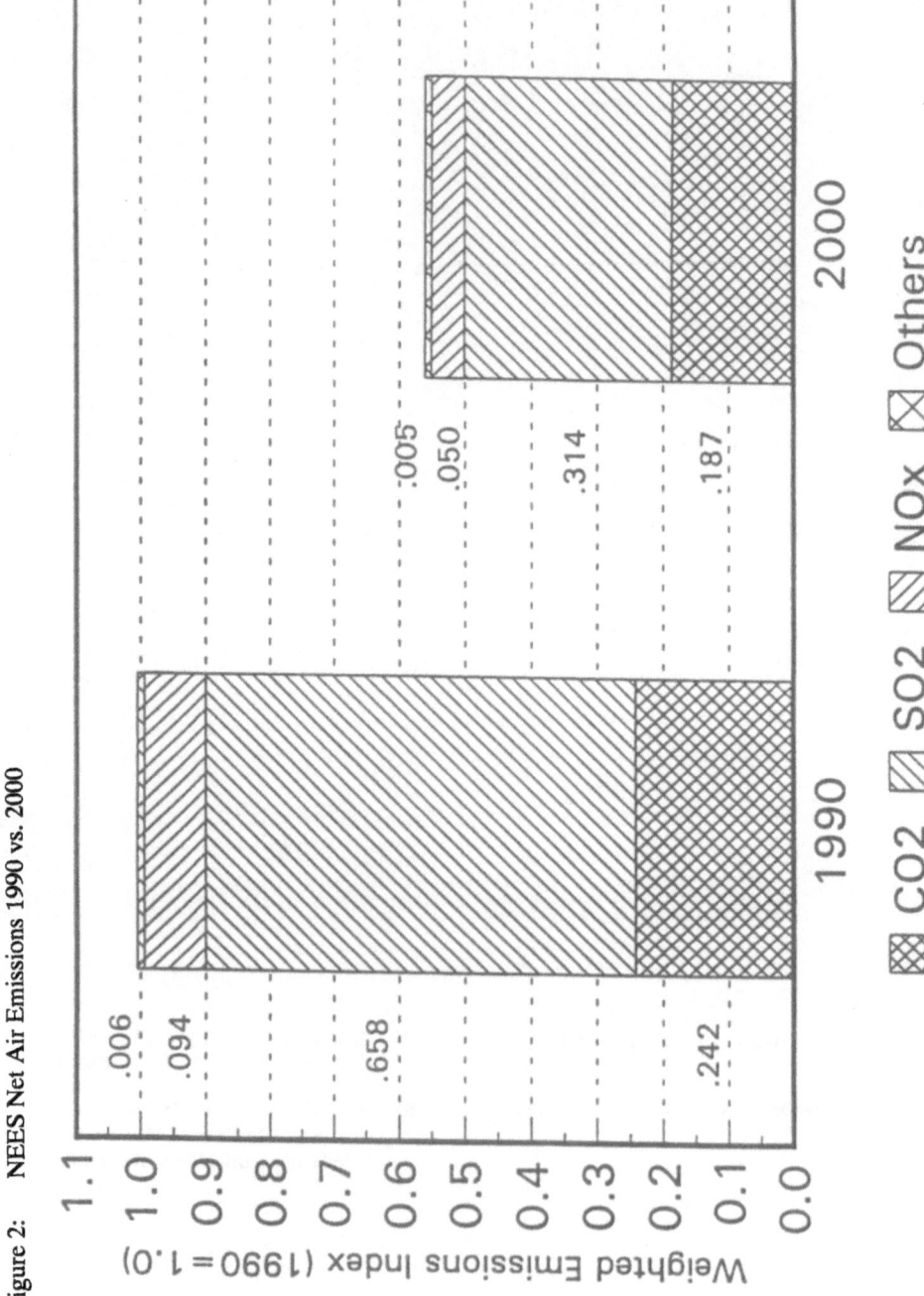

Figure 3:   Cost Per KWH - 1991 for Selected England Utilities

Company	Cents per KWH
Commonwealth Energy	11.25
United Illuminating	11.06
Eastern Utilities	10.64
Northeast Utilities	9.54
Public Service of New Hampshire	9.46
Boston Edison	9.29
Central Vermont Public Service	9.06
NEES	8.99
Central Maine Power	8.43
Municipals	8.39

as a generating fuel at our existing stations. Figure 4 shows how natural gas will increase significantly as an energy source over the decade. Figure 5 shows the increasing diversity of operating sources in our energy mix.

Key aspects of NEESPLAN 3 regarding incorporation of environmental externalities include:

- the plan addresses environmental improvement as a key component and is not limited to compliance;
- the plan balances environmental improvement with the other two major goals; and
- the plan considers environmental improvement from an overall system standpoint taking into account all sources of electric energy, i.e., existing as well as new sources.

The Air Emissions Reduction Goal

The air emissions reduction goal takes into account emissions from all sources of energy that supply our customers, regardless of location or ownership. The net emissions reduction includes the effects of emissions offsets and renewable energy sources. The major contributors to the net weighted reduction are:

- Sulfur dioxide: A 53% tonnage reduction, representing a reduction of about 74,000 tons per year. This reflects the implementation of Federal and state acid rain laws along with the increased use of natural gas, at existing stations and in new units.

- Nitrogen oxides: A 50% tonnage reduction, representing a reduction of about 30,000 tons per year. This reflects new NO_x controls at existing units to reduce ambient ozone and best technological controls on new units.

- Carbon dioxide: A 20% tonnage reduction, representing a reduction of about 3,000,000 tons per year. This includes the effects of emission offsets of greenhouse gases.

The Innovative Environmental Programs

NEESPLAN 3 includes two innovative environmental initiatives. The first deals with adding renewable resources to our energy mix. The second deals with using emissions offsets to reduce net air emissions.

Figure 4:   NEES Energy Mix by Fuel Type

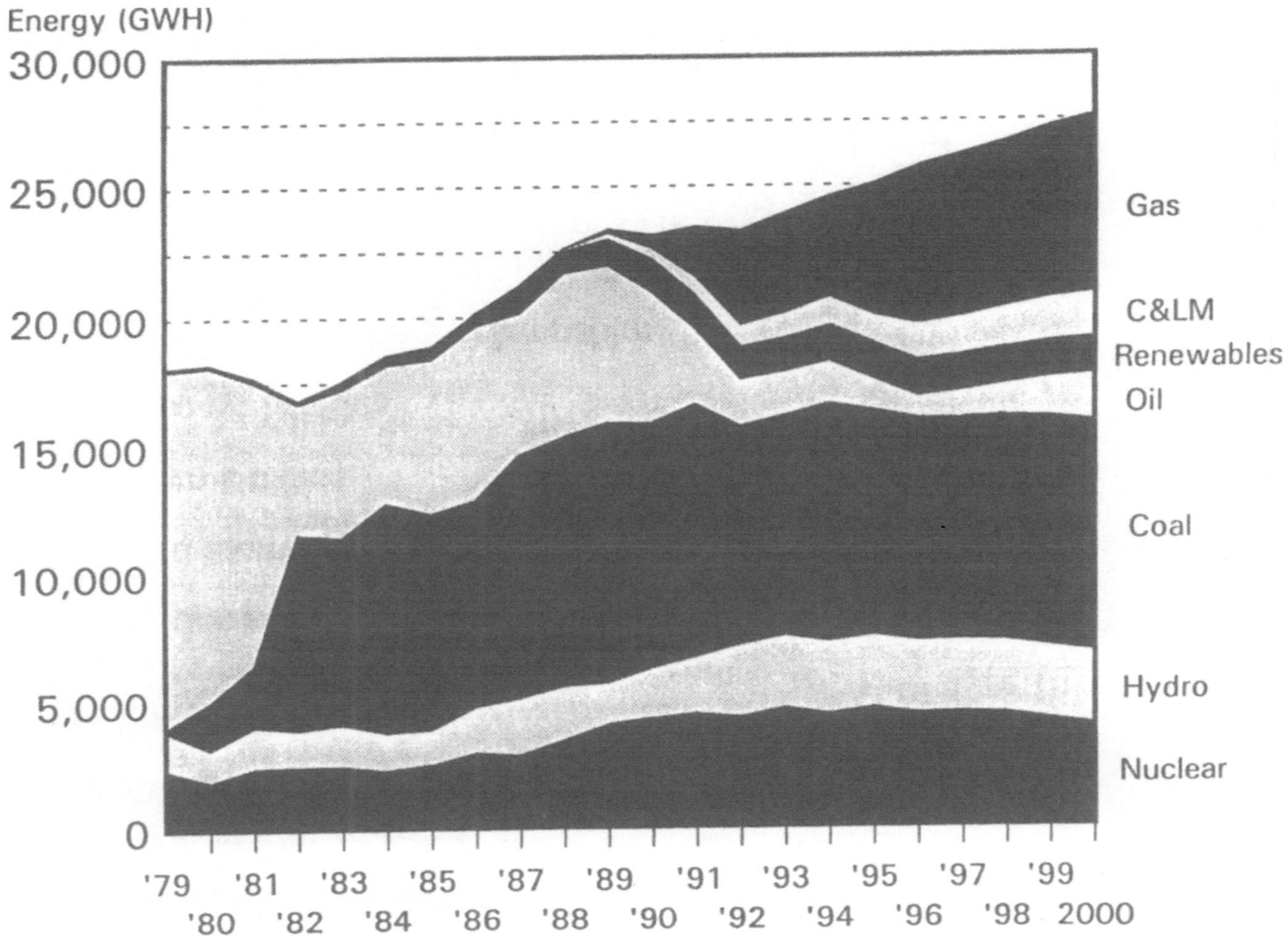

Energy (GWH)

30,000

25,000

20,000

15,000

10,000

5,000

0

'79 '80 '81 '82 '83 '84 '85 '86 '87 '88 '89 '90 '91 '92 '93 '94 '95 '96 '97 '98 '99 2000

Gas

C&LM
Renewables
Oil

Coal

Hydro

Nuclear

48

Figure 5: NEES Energy Mix by Operation (GWH)

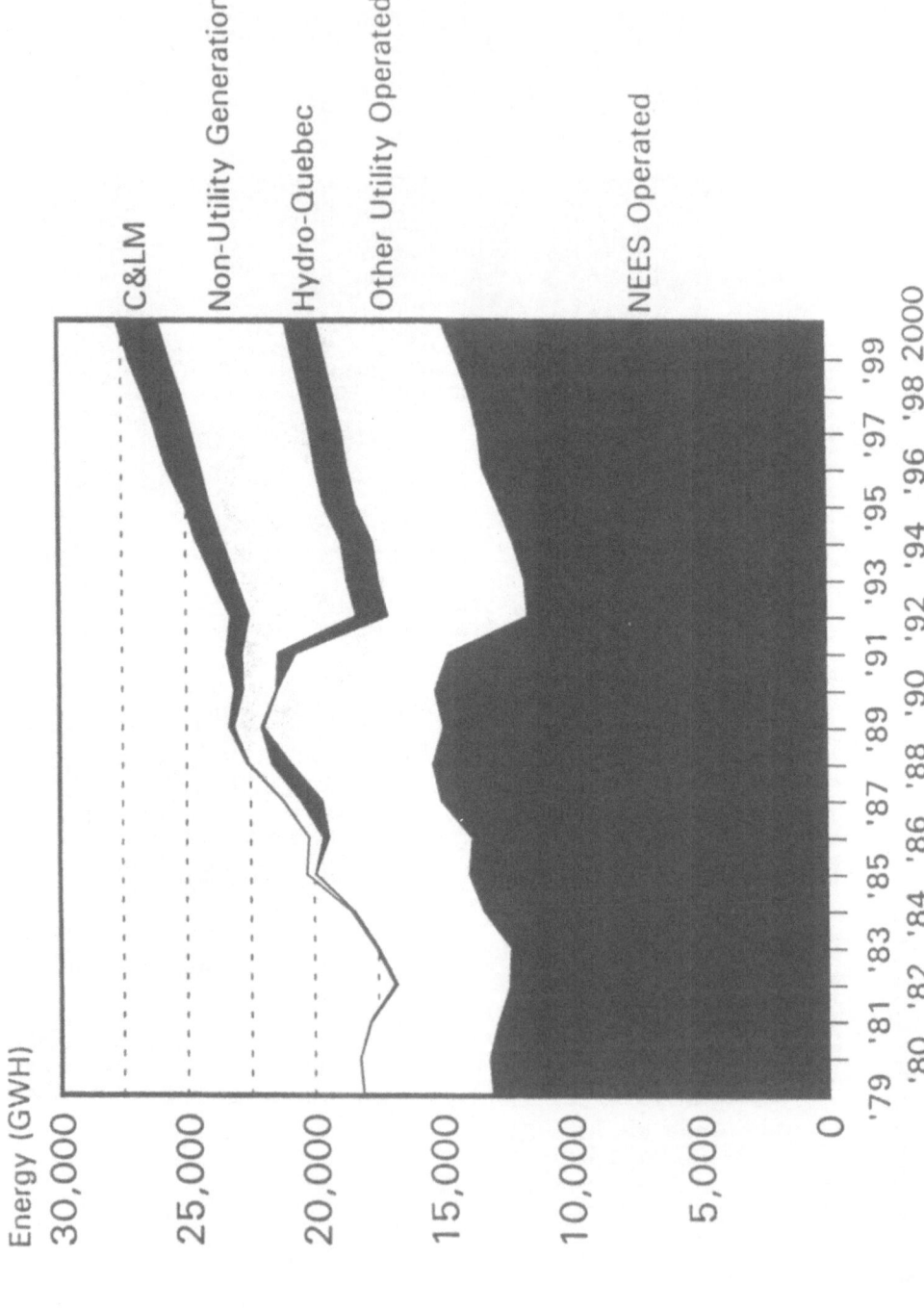

"Renewable resource" technologies are methods of generating electricity without using fossil fuel. To spur the use of renewables in New England and to learn about their commercial viability for future larger development,we are seeking to purchase up to 200 million kWh annually (equivalent to about 40 megawatts of capacity) from new low or non-emitting sources. Our solicitation has come to be known as the "Green RFP."

In response to our solicitation, we received 41 bids, for facilities located in all six New England states, as well as Canada. The total energy represented by the bids is seven times the energy we are seeking. As Figure 6 shows, we received bids from a wide variety of types of renewables, including biomass, wind, solar, small hydro and landfill gas. We are currently evaluating the proposals on the basis of cost-effectiveness and viability. We are also placing special emphasis on each project's environmental advantages and the technology's potential to significantly contribute to the future energy supply in New England. We are targeting to award contracts by the end of 1992.

The "emissions offsets" initiative is also underway and is making substantial progress. Offsets refer to the control or counteraction of air emissions that occurs away from electricity generating sources. We believe that offsetting emissions of greenhouse gases will be less costly than controlling carbon emissions at fossil fuel generating sources, while being equally effective since the range of concern is global.

We are following the recommendation of the National Academy of Sciences and others that utilities and others implement low-cost measures to mitigate the further buildup of greenhouse gases. We are now developing a series of pilot programs to analyze how best to offset greenhouse gases. To determine which pilots to develop, we consider:

- the environmental effectiveness of the offset strategy;

- the cost effectiveness of the strategy, usually in units of dollars per ton of carbon-dioxide equivalent reduced[5];

- the long-term viability;

- the public and regulatory acceptance; and

- synergies with other company environmental improvement programs such as conservation and renewable resources.

Our pilot programs, which are in various stages of development, include:

1. Forest Management - to enhance carbon dioxide "absorption" by improving forest management practices or enhancing vegetation growth.

Figure 6:   Renewable Resources Initiative Annual MWH by Energy Source Type from Proposals

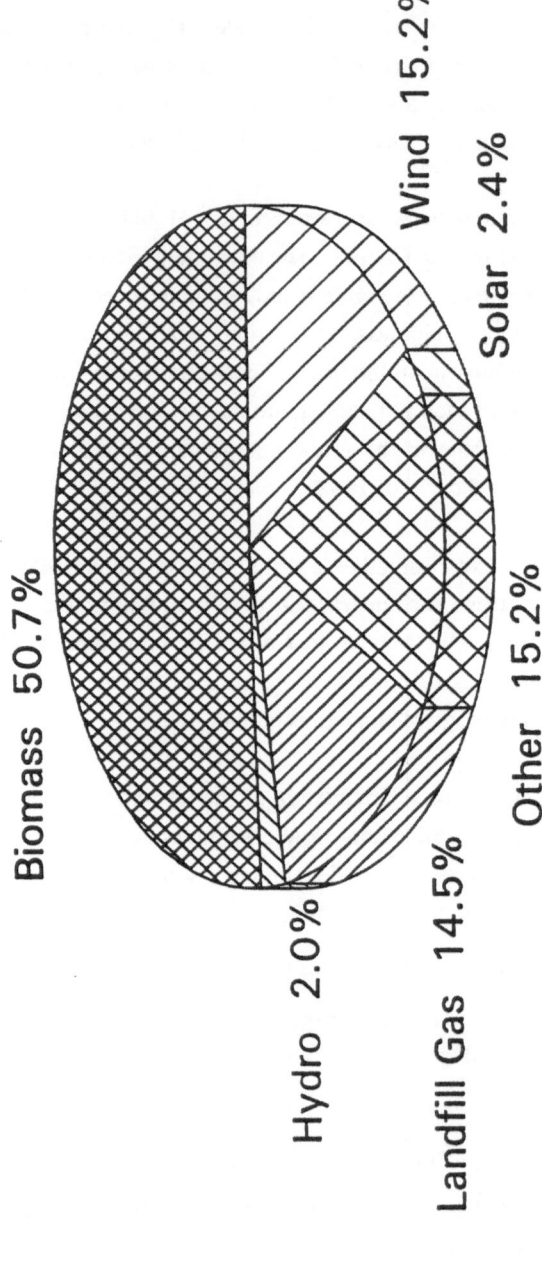

2. Coalbed Methane Recovery - to enhance the capture and use of methane emitted during coal mining; methane is considered to be a much more potent greenhouse gas than carbon dioxide.

3. Coal Ash Utilization for Cement - to recycle coal ash as a replacement raw material for cement, reducing carbon dioxide emissions that would otherwise occur.

4. Chlorofluorocarbon (CFC) Recycling/Destruction - To capture and recycle or destroy a gas that has been linked to global warming, by collecting old refrigerators and freezers from our customers and safely handling CFC's and other waste products.

5. Marine Algae Fertilization - to enhance research on the concept of stimulating carbon dioxide absorption in the oceans, which theoretically could absorb megatons of carbon dioxide.

In particular, the forest management and coalbed methane recovery projects look favorable in terms of costs and expansibility. It appears that these types of offsets could contribute significantly to the 20% carbon dioxide reduction goal in NEESPLAN 3, at a cost of less than $2/ton of carbon dioxide equivalent.

Communicating Progress and Results

As we move forward with the NEESPLAN 3 environmental programs, we are making special efforts to communicate with and receive comments from various parties. For example, we have formed an Environmental Collaborative to help us develop the details of initiatives of using renewable resources and emissions offsets. The Environmental Collaborative includes representatives from the environmental, consumer, business and academic communities. Also, we inform interested regulatory agencies (ratemaking, environmental, energy) of our progress on the programs. We also publicize our programs via the media, bill stuffers, and other means so that the general public is well-informed.

Value of Efforts to Estimate the Cost of Environmental Damage

General Observations

Processes that estimate the cost of environmental damage do provide useful information, in the form of establishing the value of avoiding such damage. Scientists, regulators and businesses should work cooperatively to continue to refine damage estimates and produce information on relative costs and benefits. The information provided by these cooperative efforts can then be used to influence new or revised environmental laws and regulations as well as permitting decisions by environmental regulators.[6] The appropriate role for utility rate regulators is to insure that utilities anticipate the potential for such changes and factor them into their resource plans.

The process should clearly focus on using environmental damage as the basis. Processes that use proxies such as cost of control are scientifically inconsistent and inappropriate. Proxies only exacerbate the flaws associated with the adders approach. Proxies are particularly problematic in environmental areas where scientific knowledge and legislation are evolving, such as for global warming and greenhouse gases.

Status of Incorporation of, Externalities for Key Air Emissions

For three key air emissions that are often the subject of externalities evaluations, following are our perspectives on the status of incorporation of externalities with respect to utility resource planning.

Sulfur Dioxide

Sulfur dioxide emissions from existing and new utility units are already regulated by environmental regulatory agencies under one or more programs such as National Ambient Air Quality Standards (NAAQS) attainment, New Source Performance Standards (NSPS) and Best Available Control Technology (BACT). Some states have additional environmental protection requirements, such as state emission limits and acid rain laws.

The Federal acid rain law in the Clean Air Act Amendments (CAAA) of 1990 requires a massive sulfur dioxide reduction by utilities. The law then caps nationwide utility emissions along with establishing allowance (emissions) trading to meet the cap.

The effect of the cap and trading program is to internalize all costs related to emissions of sulfur dioxide, including the amount of residual "damages" deemed appropriate by the CAAA. New generating units that emit sulfur dioxide (and therefore also have to meet NAAQS, NSPS and BACT requirements) will have to acquire allowances from other generating units. Therefore, any new unit emitting sulfur dioxide will not create any "net damages," and the cost for the allowances will directly appear in its cost to supply electric energy.

Therefore, the appropriate course of action is not to use an "adders" approach at all. Instead the focus should be on making the CAAA work, such as by completing sulfur dioxide trading rules and encouraging cost-efficient allowance trading.

Nitrogen Oxides

Nitrogen oxides emissions from many existing and all new utility units are already regulated by environmental regulatory agencies, Federal and State.

The Federal acid rain law and ozone NAAQS attainment requirements of the CAAA of 1990 require a massive nitrogen oxides reduction by utilities. Trading and banking programs may be part of the implementation mechanisms. In ozone nonattainment areas, all new sources must offset nitrogen oxides emissions at more than a 1:1 basis, meaning a net reduction in emissions will result.

The cost of offsets will be directly reflected in a new unit's cost to supply electric energy. In addition, the massive reductions in emissions at existing units will provide more environmental improvement than the small differences in emissions among new units that "adders" could force.

Therefore, the appropriate course of action is not to use an "adders" approach at all. Instead, efforts should be focused on developing and implementing cost-effective and environmentally-effective rules under the CAAA. This would include encouraging emissions banking/trading and cost-effective controls on existing units to meet the CAAA ozone and acid rain requirements.

Greenhouse Gases

There remains a vast uncertainty associated with global climate change. Likewise, there is substantial uncertainty as to the relative effects that various types of greenhouse gas emissions have on global climate. It is clear, however, that the range of the impact is global and that it makes no difference where greenhouse gases are controlled or offset. As previously discussed, the National Academy of Sciences has, therefore, recommended that parties begin to implement low-cost measures to mitigate the further buildup of greenhouse gases.

Appropriate legislative activity is underway at the federal level. The CAAA of 1990 requires that utilities begin to monitor their carbon dioxide emissions. In H.R. 776 (the "Comprehensive National Energy Policy Act"), the U.S. House of Representatives has addressed greenhouse gases in several ways:

- Section 1104, "Assessment of Alternative Policy Mechanisms for Addressing Greenhouse Gas Emissions," requires a comparative assessment of approaches such as emission caps, efficiency or emissions standards, and emissions trading before taking further action.

- Section 1105, "Voluntary Reductions of Greenhouse Gases," establishes an accounting system for voluntary emissions reductions from methods such as forest management, fuel switching, methane recovery, CFC capture and destruction, and methane recovery.

Given the scientific uncertainties, the NAS recommendation, and the existing and proposed Federal legislative direction, the most appropriate course of action is to encourage programs such as we are doing under NEESPLAN 3, namely:

- Pilot programs for greenhouse gas emissions offsets; and
- Pilot programs to increase the use of renewable energy resources.

Because the effects of carbon dioxide emission are not internalized yet in laws, it is reasonable for utilities and commissions to anticipate possible federal actions in the future in decisions being made now, e.g. approving new generating units that will have a long life. A "least cost of control" adder of no more that $2.00 per ton is supportable for evaluation purposes.[7] This value however, should be used among other judgment factors (revenue cost, fuel diversity, fuel efficiency) in the overall process of resource planning, not as the final decision-making method.

Carbon dioxide and other greenhouse gas emissions are examples of where efforts should be focused on improving the understanding of the value of avoiding damages and the cost of controls or offsets.

Conclusions

In conclusion, incorporating externalities in utility planning and operation is best accomplished by an overall system approach, balancing environmental improvement, economics, and reliability of supply. NEESPLAN 3 incorporates this balanced approach and sets forth specific goals for these criteria.

Processes now used to estimate the cost of "environmental damage" should continue to be evaluated as a component of cost-benefit analyses. Information from such evaluations may provide useful data to influence new and revised environmental laws and regulations. The appropriate role for rate regulators is to insure that utilities anticipate the potential for such changes and factor them into their resource plans.

References and Endnotes

[1] See written testimony of John W. Rowe, President and Chief Executive Officer of New England Electric System, for a March 3, 1992 hearing on domestic and international climate change policy, held by the U.S. House of Representatives Committee on Energy and Commerce, Subcommittee on Energy and Power. In the testimony, Mr. Rowe stated:

"The United States Should Not Take Unilateral Action to Reduce Greenhouse Gas Emissions. The immediate role of the United States should be to support scientific efforts, both national and international, to understand the global climate phenomena, and to seek agreement on implementing a "no regrets" strategy. Now is the time to learn, as we are doing in NEESPLAN 3, how greenhouse gases can be controlled or offset in an economical way. A well-defined action plan, that commits all nations, should be part of any international agreement.

If there is an international agreement which is verifiable, the United States should consider a modest but broad-based tax on carbon dioxide and other greenhouse gas emissions from all sources (e.g., transportation, industrial/commercial, and utilities). This would send a useful long-range signal to energy markets, and should be substituted for counterproductive externalities regulation. A broad-based tax of no more than $2.00 per ton of carbon dioxide-equivalent, rising at a real rate, would begin to implement the "no regrets" strategy. This level of tax represents prudent estimates of both the "cost of damage" and the "cost of control" of greenhouse gases. This would encourage sources to develop and implement relatively low cost emission reduction programs, consistent with scientific evidence on the cost of damage. Credits should be given for emissions offsets. Each affected party could then assess its control and offset opportunities, and determine the best economic course of action. The United States should specifically avoid large carbon emissions taxes that are not scientifically justified and would significantly increase the price of domestic fossil fuels and weaken our economy."

[2] The name "New England Electric System" means the trustee or trustees for the time being (as trustee or trustees, but not personally) under an agreement and declaration of trust dated January 2, 1976, as amended, which is hereby referred to, and a copy which as amended has been filed with the Secretary of the Commonwealth of Massachusetts. The principal subsidiaries of New England Electric System (NEES) are The Narragansett Electric Company, Massachusetts Electric Company, Granite State Electric Company and New England Power Company. As used in this document, the terms "NEES" and the "System" refer to one or more of the subsidiaries of New England Electric System.

[3] NEESPLAN 3 evolved from the quality management process that NEES began in 1990 with a comprehensive service commitment:

The NEES companies pledge to provide our customers the highest possible value by continuously improving electric service, managing costs, and reducing adverse environmental impacts.

[4] In 1990, the Massachusetts Department of Public Utilities (MDPU) established initial monetary values for eight different air emissions to be used in assessing the total social cost of new capacity or energy resources. The values established by the MDPU reflected some of the most expensive control technologies that had been required in the United States. Those values overstate the actual cost and benefit of reducing emissions and, in May 1991, the NEES Massachusetts retail subsidiary, Massachusetts Electric, submitted a filing requesting that the MDPU revise its values downward. In early 1992, Massachusetts Electric proposed alternate values based on the estimated damages actually caused by residual air emissions, which were generally lower than those filed in May 1991.

NEESPLAN 3 estimated 45% weighted net air emission reduction is based on monetized values filed by Massachusetts Electric in May 1991, which generally fall between the MDPU's initial values and the damage-based values proposed by Massachusetts Electric in early 1992. Because the index is a weighted composite of the eight constituent emissions, changes in the relative valuation of air emissions may result in a modified overall goal. The values used in NEESPLAN 3 are: NO_x, \$200/ton; SO_2, \$600/ton; volatile organic compounds, \$250/ton; particulates, \$300/ton; CO, \$14/ton; CO_2, \$2/ton; Methane, \$20/ton; and N_2O, \$360/ton.

5 In its 1991 report, Policy Implications of Greenhouse Warming, the National Academy of Sciences lists the estimated relative per-ton effects of greenhouse gases:

Carbon dioxide	1
Methane	21
CFC-11 and -12	5,400
Nitrous Oxide (N_2O)	290

6 For example, Federal and state regulations require that new major emissions sources install Best Available Control Technology (BACT). By law, BACT is a case-by-case analysis that takes into account energy, environmental, and economic impacts and other costs. Environmental damage costs and costs of avoiding such damage would be information of great value to those who propose and determine BACT.

7 See testimony of Dr. William D. Nordhaus for Massachusetts Electric Company, Integrated Resource Management Draft Initial Filing, Testimonies and Exhibits on Environmental Externalities (May 20, 1991). Also see Investigation into Environmental Externalities, D.P.U. Docket No. 91-131, Initial Brief of Massachusetts Electric Company (February 26, 1992).

4.   The Prospects for the Use of Environmental Benefit Assessment in the EC

P. Valette / L. DeNocker
Commission of the European Communities
General Direction for Science, Research and Development

1   Introduction

The increased attention towards the identification and quantification of external costs of energy use is justified for several reasons. Firstly, economic science provides well-founded arguments to internalise externalities in pricing and decision making. Secondly, a number of countries have already started to develop and/or implement certain policy instruments to do so. Thirdly, there are both in the E.C. and in the U.S. legal requirements to take account of externalities in decision and policy making.

The core of these arguments is that if environmental externalities are not taken account of they are passed on to others which is neither efficient nor fair. Notwithstanding all scientific uncertainties and methodological and data problems, the available figures on the external costs of energy use clearly indicate that the involved potential loss of efficiency and fairness may be important.

In addition to these scientific and legal reasons, there is the consideration of equity : intergenerational equity as well as equity between different countries. The current degradation of the stock of natural capital passes environmental costs on to future generations. To ensure that intergenerational equity is taken into account in decision making not only poses the problem of measurement of current and future environmental impacts, damages and preferences, but also how to discount them properly and how to tackle uncertainties. Equity between countries regards the question of a fair distribution of the access to the limited quantities of natural assets and resources (including the cleaning capacity of the environment), and of the socio-economic burden of environmental protection.

The requests for the use of comprehensive cost benefit analysis in the treaty establishing the European Community (and its changes) as well as in the new environmental policy plan for the EC indicate that environmental benefit assessment will have to play a key role for future policy making. At the same time these expectations forward a great challenge to the scientific community to elaborate uncontroversial, transparent and flexible methodologies.

2 Legal Requirements for Analysis of Externalities in the Treaty

2.1 A Historic Perspective

The environmental policy of the EC should be seen from a historic perspective. In the 35 years between the signature by 6 member states of the treaty of Rome in 1957, establishing the European Economic Community and Euratom, and the signing in 1992 by 12 member states of the Maastricht Treaty, awareness over environmental issues has grown considerably while at the same time the process of European integration has moved forward. As an illustration, it has to be remembered that the treaty of Rome did not explicitly include anything for an EC environmental policy, whereas it is clearly apparent in the Single European Act.

2.2 The Single European Act

Although the official start of EC environmental policies was in 1972 with the first action program, it was only in 1987 with the Single European Act that it received a clear basis in the treaty.

The single European Act of 1987 changed the treaty to create the impetus for the completion of the internal market by 1992. Within this overall revision, an environmental chapter was introduced. This not only reflects the need to accompany the internal market with environmental protection measures but also created a clear mandate for the development of an EC environmental policy by indicating objectives and criteria.

The objectives indicate that it concerns environmental policy in a broad sense, in order to contribute to: (art 130 R, 1)

- preserving, protecting and improving the quality of the environment;
- protecting human health;
- prudent and rational utilisation of natural resources.

The treaty further specifies the principles for EC action, of which integration into other policy areas and the polluter pays principle (Art. 13R,2) are especially relevant for this paper. Furthermore, it lists the criteria the EC should take into account in the elaboration of its environmental policy : (art. 130 R,3).

- available scientific and technical data;

- environmental conditions of the various regions of the Community;

- the potential benefits and costs of action or of lack of action;

- the economic and social development of the Community as a whole and the balanced development of its regions.

A simple reading of this mandate clearly indicates the need for a better understanding and quantification of environmental externalities.

2.3   The Confirmation in the Maastricht Treaty

This role could be further reinforced by the modifications of the treaty following the so-called Maastricht treaty. This new revision aims to create the institutional setting towards a monetary, economic and political union. It was signed by all member states in February 1992 and is aimed to enter into force in 1993, after the agreement of national parliaments.

This revision introduces the principle of *"sustainable and non-inflationary economic growth respecting the environment"*, as a principle objective of the Community (art 2). It further adds that environmental policy *"shall aim at a high level of protection taking into account the diversity of situations in the various regions of the Community"* (art 130 R,2).

The need for a better evaluation of external costs of energy use is further reinforced by two changes in the decision making process.

- Firstly, whereas the Maastricht Treaty introduces qualified majority voting for environmental policies as the general rule, decisions of a fiscal nature and those that may affect the fuel choice of a member state and its energy security, will still require unanimity of all member states. This indicates that these two areas, for which external cost accounting of energy use could have a specific contribution to policy making, are at the same time, from a political perspective, very sensitive issues.

- Secondly, the treaty foresees the possibility of temporary derogations or financial support for measures that create disproportionate costs for the public authorities of a member state (art 130 S,5).

Both the S.E.A. and the Maastricht treaty clearly require the E.C. to develop its understanding and quantification of the impact of environmental and other policies on the environment, public health, and natural resources. It also requires the evaluation of the impact of environmental protection on the economy and socio-economic cohesion. This does not only require the development of cost-benefit analysis and of benefit assessments of environmental externalities, but also that these are built on scientific consensus, are transparent and take regional differences and socio-economic cohesion into account.

2.4 European Environmental Agency

The importance of externalities accounting was also officially recognised when establishing the European Environmental Agency[1]. As its mandate is to set up a European environment information and observation network, one of the derived tasks will be *"to stimulate the development of methods of assessing the cost of damage to the environment and the costs of environmental preventive, protection and restoration policies"* (art.2, viii).

3 The New Impetus for Valuation in the New 5th Action Program "Towards Sustainability"

Until now the use of cost benefit analysis has been restricted to the evaluation of physical impacts on the one side and the identification of pure economic costs at micro or macro level at the other side, or to risk assessment (especially relevant for nuclear power). Contrary to the US, the EC and its member states have hardly monetarized the environmental impacts for decision making. A study made for the EC indicates that both the current practice of decision making and practical considerations are evaluated to be an obstacle for an enlarged used of benefit assessment[2]. Therefore, both the importance attached to monetarization in the new

---

1 Council regulation (EEC) n° 1210/90 of 7.5.90 on the establishment of the European Environmental Agency and the European environment information and observation network.

2 O.J. Kuik, F.H. Oosterhuis, H.M.A. Jansen, K. Holm, H-J. Ewers, Assessment of Benefits to Environmental Measures, Graham & Trottman for the Commission of the European Communities, 1992, 121p.

environmental policy programme and the joint EC-US research programme on the externalities of the fuel cycles, can offer new perspectives for the use of environmental benefit assessment.

In 1992, the Commission presented its 5th environmental policy program[3]. Its title, "Towards Sustainability", suggests its ambition to create the necessary changes of society's patterns of behaviour and decision making in order to ensure a sustainable *"growth management"*. This will require a significant broadening of the range of instruments of environmental policies.

One of the key elements of this program is to *"get the prices right"* and to ensure that environmental externalities are accounted for in market mechanisms (taxes, subsidies) and in key economic indicators like GNP. Consequently, the program requires a broad list of research actions to be undertaken to provide for the necessary methodologies and information *"to get prices right"*:

- evaluation of the natural and environmental resource stocks in economic terms,

- development of renewable resource indicators,

- adaptation of economic indicators (GNP) for environmental externalities,

- development of meaningful cost benefit analysis methodologies and guidelines,

- redefinition of accounting rules, concepts, conventions and methodologies for environmental externalities.

At the level of the firm, a similar integration of information of environmental impact and performance would be required (in annual reports, provisions in the accounts for environmental risks).

This new environmental policy program is however not accompanied by a comprehensive cost-benefit analysis of the action program itself. The EC is well aware of the difficulties involved, especially regarding the evaluation of environmental impacts and its monetarization. To the extent that the latter would not be comprehensive or complete, the cost of "lack of action" would be underestimated which risks that *"decision making will tend to be biased against a sustainable optimal policy response"*. This wording stresses the importance of research programs to stimulate benefit assessment in monetary terms. The 5th action program nevertheless promises to undertake an analysis of costs and benefits when specific proposals are forwarded.

---

[3] CEC, Towards Sustainability, A European Community Program of Policy and Action in relation to the Environment and Sustainable Development, as proposed by the Commission of the European Community, CEC, COM (92)23, 30.3.1992.

Consequently, the 5th environmental policy program proposes a 5 point action programme for the coming years in order *"to devise an appropriate and effective costing mechanism"*.

This program includes:

- improvement of information on the state of the environment, including tolerance capacities.

- valuation of environmental assets, based on international cooperation.

- development of a Community wide cost-benefit methodology.

- execution of comprehensive cost-benefits analysis of all EC environmental policies and other policies with an environmental dimension.

- development of environmentally adjusted national accounts.

As a conclusion, the 5th Environmental action program clearly shows the importance the EC attaches to the environmental benefit assessment: *"we must urgently develop a methodology for cost-benefit analysis at EC-level, and apply it to all projects and policies with an environmental impact..."*. This will however require a large research effort. This needs to pay special attention to intergenerational equity and regional diversification.

4 The Joint EC/US Programme on the External Costs of Fuel Cycles

The joint EC/US program on the external costs of fuel cycles aims at developing such methodological frameworks as the 5th action program calls for.

Its objective is threefold:

- to develop a comparable methodology for the quantification and monetary assessment of external costs and benefits of fuel cycles.

- To apply these methodologies, using the best information from existing sources, for the major fuel cycles, including energy conservation.

- To make recommendations for priority areas for further research.

The project will result in an operational accounting framework for the quantification and monetarization of priority environmental and other externalities. It offers a tool to calculate the marginal external costs and benefits for a specific power plant, at a specific site using specified technologies.

The methodology needs to take account of the fact that externalities are site specific while allowing a maximum exploitation of the transferability of existing studies in all areas, ranging from technology characterizations, identification of physical impacts including modelling of dispersion and dose-response functions, to economic valuation. Therefore, transparency and flexibility are two major characteristics of the accounting framework.

The methodology to measure environmental and health impacts follows the damage function approach which is applied to the priority impact pathways of emissions and impacts for the different stages of the fuel cycles. This way the accounting framework allows for a systematic, coherent and transparent evaluation of physical impacts. The same applies to the economic valuation of these impacts which will rely on the results obtained from the different methods to measure the individual's willingness to pay for health and environmental goods or to avoid health or environmental damages.

The objective is to implement the accounting framework for all fuel cycles, including renewables and energy conservation, in all member states. Therefore, the accounting framework will allow flexibility to adopt these different steps of the methodology to either regional or site specific conditions, to assumptions about base-line emissions or to the user's preference (e.g. for other discounting rates).

Once it is fully developed and demonstrated, it will offer an operational tool and support system to execute complete and comprehensive cost benefit analysis which will allow decision making to take account of external costs.

Quantification of Non-Environmental Externalities

As explained before, the major objective of the accounting framework is to quantify environmental and health externalities. Nevertheless, this valuation process will be extended to non-environmental effects, that are not properly valued by the market.

Security of supply is the first example that needs to be treated; it has different facets. One is related to the cost of major disruptions in the primary supplies of the Community. This involves an assessment of the costs and damages associated to these disruptions. Costs derive from self restrictions on fuel consumptions, they also depend on how these are implemented. They may also result from investments made for security purposes. Damages can be obtained by simulating the short term equilibria that result from coping with the disruption through a mix of market forces and their regulation. Networks play a particular role in this overall issue of security of supply; their development has an obvious bearing on the overall security of supply of the Community. The fact that networks could be the source of externalities can be included in the notion of "european interest".

Other non-environmental externalities will be considered in the future: regulatory and employment externalities are the most significant examples.

The Benefits of International Cooperation

The profits to be reaped from international cooperation are obvious. A comprehensive and complete analysis of the externalities of the fuel cycle is facing in most areas problems of uncertainties or of lack of scientific understanding and quantification. The first objective of the international cooperation is to work together to bridge these gaps, especially in these areas were information is missing and new methods have to be developed. Cooperation will increase cost-efficiency of research while promoting consensus on scientific issues on both sides of the Atlantic. The latter will also contribute to ease acceptance of the methodologies by policy makers.

The Modelling Framework

The internalisation of external costs into energy policies, requires not only a better assessment of these external costs but also tools to identify the optimal mix of instruments for internalising external costs. Therefore, the linkage of the accounting framework with a new generation of an energy-environment-economy (E3) modelling framework is another priority research action.

Such a framework would represent the different driving mechanisms of the E3 systems and would integrate their interfaces as well as the policy instruments. The accounting framework would provide the blueprint and the information basis of the representation of the externalities associated with the energy system. This new generation of general equilibrium models are under development within the Commission's research programme JOULE and is named SOLFEGE (Systems for Externalities and General Equilibrium).

Such models would also be well suited to tackle two other issues which are also relevant for integration of environmental externalities. The first relates to regional differences. The Internal Market will not result in a fully homogenized geographic entity. A lot of different national legislations with possibly significant impact on economic and energy development will still prevail. Modelling the different member states, their regulation and competitive environment and their interrelations is thus a second key point of the modelling activity, whereas the accounting framework allows to take into account ecological differences between regions as well as the differences in preferences for environmental and health protection.

Secondly, it is necessary to take the relations between different policies into account, especially since not all of them will have the same degree of harmonization at Community level. The evaluation of the impact of non-environmental policies on the energy system and the environment is therefore a key challenge of the modelling activity.

The Relevance of the External Costs Programme

In this context, the large effort which is currently devoted to the evaluation of the external costs of energy and to the development of the accounting framework is fully justified. It should provide all the information necessary to embed external costs evaluation in the new generation of models which will have to answer these complex questions.

In addition, the accounting framework is focusing on a difficult and sensitive area. Energy externalities involve a broad range of different impacts, ranging from global impacts on ecosystems, local health impacts, transport accidents with a direct impact as well as long term impacts on health, etc. Each of them is accompanied by its own problems of scientific uncertainty and lack of data and information. At the same time, the outcome may affect important policy choices like fuel choice or taxation and may be subject to sensitive political discussions.

However, these choices are key elements for the development of a sustainable growth and benefit assessment can play an important contribution to our understanding of the trade offs to be made. The use of valuation techniques in USA for CBA analysis in this field is encouraging in this respect.

5 Conclusions

It is clear from both the analysis of the treaty and of the new environmental policy program that the prospects for the use of benefit assessment for environmental policy are encouraging.

They create the obligation for the European Community to undertake research and develop methodologies and tools in this area.

The joint EC-US project on the external costs aims to answer these questions. It places the great number of existing dispersed scientific data and models in a common and coherent framework; it will provide a tool to allow for a complete, transparent and comparable assessment of the different fuel cycles. Furthermore, it will be completed by a modelling framework necessary for a better appreciation of the effectiveness of the policy instruments which would be implemented for the internalization of the external costs.

This obligation creates at the same time a major challenge to the scientific community. As policy making needs a hard core of both comprehensive and uncontroversial analysis, the key question will be to what extent consensus can be raised on operational accounting tools, the interpretation of their results and priorities for further research.

5. Why Utilities Should Incorporate Externalities

Stephen Wiel
Lawrence Berkeley Laboratory
Berkeley, CA 94720

Introduction

Should utility companies incorporate environmental externalities in their planning and operations? Of course they should, so long as our society has failed to fully internalize the impact of power plants and energy distribution facilities.

Why should utilities incorporate externalities? So that they generate electricity and supply natural gas and other fuel in ways which provide energy services at the lowest cost to society. The decision by Sierra Pacific Power Company, for example, of whether to conduct the proposed 180MW Pinon Pine Clean Coal Demonstration Project, to purchase 180 MW of long-term geothermal baseload capacity and energy from QF's or to construct 180 MW of natural gas combustion turbines should account for all factors including environmental externalities. Likewise, Sierra Pacific's daily decision of whether to curtail production of its Valmy coal-fired power plant in favor of its own gas-fired combustion turbines or purchased hydropower economy energy from the Pacific Northwest should also account for environmental externalities. Only this way will consumers select the level of energy services which best suits their needs and values, and only this way will society provide these energy services for the lowest total cost.

The Role of State Regulators

The involvement of state PUCs in the long range planning and resource selection of utility companies began in 1975. Since then, through a process known as integrated resource planning, regulators have attempted to influence the long-range planning decisions of the investor-owned utilities they regulate to ensure that the utility companies will meet the energy needs of their constituents in an economical and safe manner. Although environmental impact has always been one factor that PUCs have considered, it was not until 1989 that New York took the bold step and monetized the values of pollutants to be used in evaluating the resource acquisition choices made by its utilities. A year later, in 1990, Massachusetts followed suit, but chose different numbers for its "adders." And in 1991, followed suit, but chose different numbers for its "adders." And in 1991, Nevada and California joined the movement to internalize externalities. All four states have included adder values for both carbon dioxide and sulfur dioxide. Right now, some two dozen more states are considering adding this market mechanism to their integrated resource planning requirements for utilities.[1]

At the Nevada hearings, when we considered our rule while I was still a Public Service Commissioner, I heard all of the same arguments that I have subsequently heard about why we should not adopt adders. Quite frankly, I have to admit that many of the criticisms are true. States are not the right place to do it; it should be done at the federal level. If Nevada uses adders and other states do not, we may get trans-boundary dislocations where all new facilities are located out-of-state. We also might get inter-fuel dislocations -- inadvertently promoting the use of dirty fuel technologies we were trying to discourage. For example, if we increase the price of electricity too much, we may encourage some customers to bypass the regulated system and generate their own electricity using diesel fuel.

Nevertheless, not all the criticisms are valid. First, regulatory commissions are an appropriate forum for addressing environmental problems. To the degree that environmental externalities in the utility industry can be addressed at the state level, then an appropriate place to do it is before the PUCs. Second, state regulators may not have the expertise we might like them to have for addressing this issue, but they have the expertise they need -- the ability to judge between the various technical experts and make the right political decisions. In fact, in Nevada, they did not establish a methodology for internalizing environmental externalities; rather, we evaluated the alternative put before us and chose the one we thought would best accomplish our policy goals.

It is important to understand environmental costing from the point of view of a state regulatory commission. There are a number of factors which led the Nevada PUC and others to move forward on this issue. First, many of us have recognized, or at least believed, that the wrong resources were being selected by the utility companies. They were under-investing in energy efficiency and renewable fuel sources. This prompted us to evaluate whether there was something wrong with the selection process which could be fixed. Nevada's adders were not intended to justify specific resource selections, but rather they were to ensure that the best resources are selected after environmental factors are duly considered in the process. Second, we have recognized that people are willing to pay more for a cleaner environment. That is a political judgment we have made. Finally, we have decided that the absence of federal leadership on this issue should no longer be a reason for delay.

In fact, state initiatives often provide the impetus for federal action. An example of this occurred recently when the federal government adopted appliance standards that had percolated out of a patchwork quilt of state regulations. Hopefully, federal action will follow the various state efforts to establish environmental costing in the utility industry. And when the federal government has all the numbers correct, state regulators will adjust their adder values to zero to reflect that all the externalities have been internalized. The current efforts of state regulators will simply have seen to it that better investments have been made between now and then.

The Role of the Federal Government

I have been involved in many discussions on whether or not federal support for the new state environmental accounting would be worthwhile. I have mixed feelings about such support. The reasons regulators are taking action at the state level to account for damage from residual power plant pollution is because our energy prices don't already account for such damage. In my mind, this clearly is a failure of the federal government. The best solution would be carbon, uranium and other pollution taxes, perhaps combined with a general energy tax, which would result in prices that correctly reflect the true economic cost to society of each type of energy (coal, natural gas, fuel oil, nuclear, gasoline, solar, wind, geothermal or whatever) without regard to artificial geographic boundaries. An acceptable alternative would be allowance trading globally for all pollutants in the manner recently instituted for sulfur dioxide by the Clean Air Act Amendments. State-by-state environmental accounting only within the regulated utility industry is actually a poor compromise. From my point of view, such state-by-state utility accounting is clearly better than nothing, but it is not what the federal government should be advocating. Furthermore, I worry that federal involvement, rather than being devoted to finding true economic efficiency in our society, would be politicized to continue to support energy prices below their true economic costs.

Responses to Critics

Treatment of environmental externalities is the most controversial issue to arise in the electricity industry in the U.S. during the eight years I served on the Nevada Public Service Commission. Two challenges stand out in my mind and deserve a response.

In his article "Weighing Environmental Externalities: Let's Do It Right,"[2] Paul Joskow provides regulatory utility commissioners some much needed advice on how best to institute environmental adders. Unfortunately, he accompanies it with an unfounded and absurdly illogical plea. Joskow states that the concept of what state regulatory commissioners are doing today in monetizing residual pollution is far from ideal and that regulators are sometimes making mistakes. He goes on to argue that regulators should therefore abandon such efforts. The fact is that what regulators are doing is better than what exists today and therefore should be pursued.

As to the mistakes regulators have been making as they forge their way into new regulatory territory, we should learn from them -- we should do it right:

- Regulators should apply their adders to all sources, both old and new, and should apply them to operation of power plants, not just to expansion of capacity.

- Regulators should base the value of residual pollution on estimates of environmental damage, not on control costs.

- Regulators should only value residual pollution not already accounted for in utility company decisions.

- Regulators should make sure the adders will have the practical impact they expect -- actual environmental benefits that exceed actual costs on a global scale.

- Regulators should create a regulatory environment that provides incentives for utilities to satisfy environmental constraints at the lowest possible cost.

Finally, there are been criticism by Ben Hobbes[3] and others that the values placed by state regulators on sulfur dioxide (SO_2) emissions are inappropriate because of the allowances market created by the recent Clean Air Amendments." The truth is that these values may soon be inappropriate, but for now they are appropriate.

The values that were established for sulfur dioxide by New York, Massachusetts, Nevada and California were chosen before the passage and implementation of the 1990 Clean Air Act Amendments. The value of residual sulfur dioxide from power plants, whatever it may be, remains an externality until a utility company accounts for it in its resource selection and plant operation decisions. When properly priced, the residual pollution becomes internalized, even though it is still emitted. The emission allowance requirements of the Clean Air Act Amendments has not yet resulted in utility companies internalizing the value of residual sulfur dioxide from power plants.

Some argue that even after full implementation of the Clan Air Act Amendments we will still have to place an added value on residual sulfur dioxide because the actual amount of damage will still be greater than the market value of emission allowances. Others argue that the value of residual sulfur dioxide will drop to zero after full implementation of the Clean Air Act Amendments, either because the market value of the emission allowances will be approximately equal to the damage value or because, as Hobbs argues, placing an additional value on residual sulfur dioxide would actually have no effect on net sulfur dioxide emissions. Time and debate will undoubtedly shed more light on this subject.

But the issue raised by such critiques is important and their major thrust is valid. States which require their utility companies to adopt the new environmental accounting to internalize environmental externalities, will have to be first on their feet over the next few years to reassess and reset values for residual sulfur dioxide as the allowance trading market matures.

References

[1] Stephen Wiel, "The New Environmental Accounting: A Status Report," The Electricity Journal, November 1991.

[2] Paul Joskow, "Weighing Environmental Externalities: Let's DO It Right," The Electricity Journal, May 1992.

[3] Benjamin F. Hobbs, "Letter to the Editor," The Electricity Journal, March 1992.

6. Internalization of External Costs During the Crisis of Environmental Policy or as a Crisis for Economic Policy

Eberhard Moths
Bundesministerium für Wirtschaft
Postfach 14 02 60, D-5300 Bonn

Introduction

This paper discusses the general reluctance towards internalizing external costs. It thus focuses on a phenomenon of economic irrationalism. Those who have set out to stabilize on a sustainable basis the very foundations of economic and social life must brace themselves for a good deal of irrationalism.

This reluctance concerns the social and political dimensions of external costs and it has not precisely been the focus of the academic world's attention so far. But this fact does not change anything about the volume of external costs and their further expansion (not only of electricity generation) which are being crucially determined by the global reluctance towards internalizing external costs.

Those who know the economic history of the Federal Republic of Germany will recall the term 'reluctance to investment' of which German companies were found guilty some 20 years ago when showing themselves unwilling to invest. Psychologists interpret such reluctance as a pathological unwillingness to act or lack of interest.

Ecological IQ?

The currently diagnosable reluctance towards internalizing external costs depends on the general frame of mind (intelligence quotient?) as well. However, the economy's state of health is said to be the crucial factor:

Where the economy is in a bad state of health, either regionally or globally, there is no money available for "clearing debris". That is what people say. On the other hand, a brisk rate of economic expansion fills both the public and the private purse. Such money could then well be spent on climate protection and other forms of ecological benefits ("growth is the only way towards more environmental protection!").

The validity of this thesis must be doubted. Reluctance towards internalizing external costs dates back to before the beginning of the ecological age. Where would the many "legacies" of environmental damage come from if not from such reluctance?

In the Federal Republic of Germany, the policy on environmental protection does not yet have teeth. How then can entrepreneurs and consumers be afraid of the cost-effective dimensions of such a policy that do not yet make themselves felt?

Against this background, it is rather more likely that the environmental policy's missing teeth will be responsible for new phase of economic weakness: incessant overexploitation of resources does not only affect one's state of health or frame of mind, but also means permanent erosion of assets, resources and environmental quality. This fact does not leave the economic superpowers unscathed.

Difference Analysis below the Perceptibility Threshold?

There is no empirical evidence for any interrelationship between the extent of voluntary and, as the case may be, decreed internalization of external costs on the one hand and the different growth cycles of the German economy on the other. Such evidence is, if at all, construed. The reason therefor is rather simple: In the past "fat" years of the old Federal Republic of Germany, expenditures on ecological repair work and ecological prophylaxis were factually below the threshold of perceptibility justifying efforts of interpretation.

To substantiate this statement, I wish to refer to the budget of the German Federal Ministry for the Protection of the Environment. It amounts to DM 1339 million (about US$ 950 million) in 1992. This minute budget will be even smaller by DM 47 million in 1993. Measured by the Federal budget, the share is 0.3 %. As a percentage of the German Gross Domestic Product (GDP), this budget is roughly 0.08 percent. Against this background, specific impact analyses are statistical nonsense.

In other areas (experts would be pleased, as I think, to draw my attention to the cost-relevance of German limiting values or safety requirements), evaluation would all too soon also be deadlocked. How informative is an erratic bar diagrams, whose increase in size is to impress the reader when its basis is small or wrong? The political art of subsumption (in other words:

the impertinence with which success stories are put together in the field of environmental protection) often shows close similarities with balance-sheet falsification.

Illusioning oneself and others would be much more difficult if private and public spendings on measures aggravating ecological problems had to be weighed against measures to redress them. Better two messages of success than one (negative) horizontal checksum!

Single reference to such systematic and political blemishes of public reports such as private balance sheets must be deemed sufficient for this purpose.

Paranoia as a Principle of Prophylaxis?

Where the diagnosis is correct that actual financial burdens arising from requirements for direct internalization of external costs are still small, other factors must be deemed responsible for the reluctance towards internalizing external costs.

Industry's nervous to aggressive response to internalizing external costs is to be explained by primarily historical and psychological reasons (e.g. reallocation debate). And what matters most is that there is method to this. One is: beware of new financial burdens and restrictions on entrepreneurial freedom from the outset!

We all know that the normal entrepreneur "prosecutes" costs as a source of potential losses or reduced earnings day and night. Incessant awareness and attentiveness is the very basis of the survival of this species. This philosophy explains why entrepreneurs aggressively take prophylactic action to nib potential cost burdens in the bud (their behaviour is not basically different from that of motorists in Germany). It is to be assumed that they have instrumentalized as a strategy this innate "paranoia" as regards the internalization of external costs ("principle of prophylaxis").

A craving for predictability is another characteristic of private entrepreneurs in regard of both the outcome of their calculations and the profitability of investments. Any consistent internalization of external costs is thus bound to cause them bad dreams.

A consistent internalization of external costs would according to the present state of knowledge - exceed the framework that is otherwise set by product innovation and low-wage countries. Price increments reflecting internalized external costs to take account of the "polluter-pays principle" would present themselves to entrepreneurs as a "change of system" not only in the world of dream.

It would represent a new type of world economic system indeed. But they have not cared either to take precautions or to do some advance thinking in order to take care of such a case. After all, the way in which the "greenish" entrepreneur Stephan Schmidheiny (Business Council for Sustainable Development) apparently thinks a change of policy can be brought about is regrettably not that simple. Not even the profit-and-loss concept has fully won the upper hand so far against the concept of victory and defeat. This holds true not only for countries like Iraq or Serbia.

Entrepreneurs know their figures. They will find out without much effort what an increase in price per litre of standard gasoline to DM 5,-- would mean (Such proposals have really been made in Europe. And they are even being discussed in all seriousness. However, this is so not so much for ecological, but for fiscal reasons). In the Federal Republic of Germany, this would allow DM 100 billion in terms of purchasing power to be skimmed off all of a sudden in order to be "mis-allocated" for the purpose of repairing ecological and human damage. In any event, the money would come under the control of government and its bureaucracy, and it could no longer change hands over the counter. Production capacities would be less well utilized.

Closing One's Eyes to Reality

But this is not yet the end of the bad dreams: The situation would turn out to be even more dramatic if Professor Ewers, who is well versed in balance-sheet techniques, made his calculations in the way in which the external costs arising in the operation of nuclear power stations are calculated in Germany. As much as DM 500 billion would have to be paid in terms of annual risk insurance premiums to take care of a nuclear meltdown; this amount would have to be added to the electricity bill (DM 3,60 per kWh multiplied by 147 billion kWh per annum). This would be more than the annual budget of the Federal Republic of Germany, which stands at DM 425 billion at present.

Such enormous amounts of money explain why entrepreneurs and consumers reject any liability for external ("disguised") costs. They pretend to be rich rather than admit that they are bankrupt. Out of sight, out of mind! This is, in principle, not any different from a man who has fathered a child born out of wedlock and who refuses to pay child support.

But how can the mother and the child cope with the situation? It is not that easy for them to get a fair share of the money paid up by the honest taxpayer. Apart from this, it is the general wish to bring down the public sector share in GNP. This is what not only the American electorate is being told time and again.

Against the background of this pathological practice of pushing inherited and new burdens as well as mixtures thereof aside in Germany as in other industrial countries, Michael Gorbachev's saying of "those who come (who understand/who respond/who internalize external costs) too late will be punished by life" may again forebode evil. Just think of the dying forests, poor ground water quality and damaged genes. But our ability to close our eyes to the existence of these dangers is as good as ever.

The "loss of reality" is apparently so great that we do not even notice that we are falling victim to an international process of expropriation, hedonistic, not communist in character this time. The persistent neglect or muddling under of external costs severely damages the market-based system's claim of efficiency at the same time. When prices are meaningless and do not tell anything that comes close to the "truth", the allocation effect of the prices and costs systems is rendered inoperative. To this danger innate in the system we close our eyes as well instead of discussing how to get it under control.

MS Titanic as a Vertical Communication Problem

The reluctance towards internalizing external costs depends on where one stands. Whether one perceives external costs and their consequences depends on the view one has adopted. Such views may be interest-guided: either the sunny side of the road or the bed next to the landfill.

No doubt, from the upper deck of a luxury liner the world looks different from what it does from the lower deck. Also, the fact that the communication between those above and those below has not been functioning any more for a long time makes matters worse; there is a world of difference between them in many cases. You can have Champaign and enjoy yourself for quite some time to come while on the upper deck, whereas this is not so when you are in one of the lower decks and the ship has begun to sink after it has hit the iceberg. Even though people may be sitting in the same boat, they perceive the dangers and irritations as quite different in quality.

However, these traditionalized patterns are rather limited in analytical and political value nowadays. Being rich cannot be equated with ecologically bad behaviour nor can being poor be interpreted as ecologically noble or ecologically more considerate behaviour.

I am commenting on these interrelationships because this may help pour some water into the wine of the omniscient economists who claim that they and the models they have developed for really internalizing external costs can provide the solution. Being a civil servant of an economics ministry, I know that the trade in economistic models, rather brisk by now, has not brought the reduction of external costs even one inch further. And a quick reduction of such costs - precisely in economic terms - is the crucial question. It would be false modesty and the wrong policy if the EC is only willing to "stabilize" the output of CO_2 by the year 2000. Will the ozone layer be less depleted by then? Or is the entire discussion on CO_2 output just a tactical move in continuing to sell the "survival" of nuclear energy in socially acceptable terms.

Intellectually, it is preposterous in my view that the EC is only prepared to levy an energy tax if the USA and Japan are prepared to do the same. Europe's regional climate, including the local climate of Bonn, initially depends crucially on what the private individual can bring about in terms of containing pollution. If - in spite of our highly developed awareness of environmental needs (corresponding polls have shown that 70 percent of the German population have been that strongly aware for over ten years) - we do not drive our private motor cars perceptibly less frequently and at a slower pace, we are moving in the wrong direction. But the wrong direction is no function of cyclical development.

SUBJECT AREA 2:

**EMPIRICAL ESTIMATION OF SOCIAL COSTS
OF ENERGY**

7. Estimating the Impacts, Damages and Benefits of Fuel Cycles: Insights from an Ongoing Study[1]

Russell Lee
Oak Ridge National Laboratory
Oak Ridge, TN 37831-6205

1 Introduction

1.1 Purpose and Background

The purpose of this paper is to share some insights from an ongoing study of fuel cycle externalities. The fuel cycles being studied involve the use of coal, biomass, oil, hydro, natural gas, uranium, wind, and photovoltaic sources to generate electric power. Conservation options are also to be addressed.

The primary objectives of the study are:

(1) to implement the Damage Function Approach as described in the Background Document for this study as a means of estimating external costs and benefits of fuel cycles, and by so doing, to demonstrate their application to the above fuel cycles;[2]

(2) to use existing information to develop, given the time and resources, estimates of marginal damages and benefits associated with selected impacts of new electric power plants using benchmark technologies, at two reference sites in the United States; and

(3) to assess the information available to estimate externalities, and by so doing, to assist in identifying gaps in knowledge and in setting future research agendas.

The paper focuses on insights that have conceptual implications for estimating the externalities of generating electric power and for incorporating social costs in decisions about the use of alternative energy sources. The insights are *not* about specific numerical results, for reasons that will be evident later in this paper.

1.2   The Damage Function Approach

Our study uses the Damage Function Approach (DFA) to estimate fuel cycle externalities. Well known studies by Hohmeyer, Pace University (Ottinger), Bonneville Power Administration, Tellus Institute, Hagler, Bailly Inc., and others have also estimated social costs and externalities of fuel cycles.[3] Our approach is similar in spirit to the Pace study and to the Bonneville Power Administration studies, though our study is in some respects more detailed in its modeling and analytical content.

The Damage Function Approach consists of the following steps:

(1)   describe the major stages and activities of a fuel cycle and estimate their residual emissions, or other altered physical or chemical conditions (such as soil erosion in the biomass fuel cycle);

(2)   analyze the chemical transformations, transport, and deposition of these emissions and other residuals; and estimate the resulting changes in the concentrations of pollutants and other materials;

(3)   estimate the effect on ecological, human, and social resources that are exposed to these changes in concentrations;

(4)   translate these environmental and other impacts into the economic value that is placed on these impacts by the individuals affected; and

(5)   distinguish between the externalities, and the social costs and benefits which are "internalized" in the market; social costs are internalized in the sense that they are reflected in the price of electricity.

This paper discusses insights gained in the process of implementing these steps to estimate the externalities associated with a single, new power plant.

2   Establishing Boundaries on the Fuel Cycles

Before implementing the Damage Function Approach, it is necessary to identify the components of the fuel cycle. In general, it consists of the activities necessary to generate electric power, beginning with the extraction or provision of the fuel or resource, and ending with the disposal of the activities' residuals. It is necessary to establish boundaries on each fuel cycle -- that is, to decide what is to be included and what is not. There are two significant issues.

2.1 Upstream, Secondary, and Indirect Emissions

The first issue is about the upstream, secondary, and indirect emissions. Total fuel cycle externalities include those associated with the electric power plant itself, the "upstream" activities that must take place in order to supply fuel to the plant (such as mining coal or growing biomass feedstock), the secondary activities that must take place for the plant (and possibly other structures) to be built, and the indirect activities which are the upstream and secondary activities of the secondary activities (and so on).

The problem of assessing upstream, secondary, and indirect emissions is to identify the activities and their process cycles, and then to estimate the residuals from all these activities. There are too many activities and residuals to completely identify and estimate their magnitude. In our study, upstream activities were included in the analysis to the extent possible, secondary emissions were assessed in terms of their order of magnitude, and indirect activities were not addressed. Our analysis of a number of the fuel cycles indicates that externalities from upstream activities are the same order of magnitude as those from the power plant itself. Our analysis of the importance of secondary emissions is ongoing. For the coal fuel cycle, secondary emissions are two or three orders of magnitude less than those from the coal-fired plant.

2.2 Incremental Effects

A second issue in establishing boundaries for the fuel cycle concerns the activities and infrastructure to include as part of the fuel cycle. The context of the study -- that of considering a single power plant, rather than some national energy plan, or even rather than a utility's full set of integrated resource planning options -- provides a basis for establishing consistently defined boundaries for the fuel cycles. If a structure is likely to be in place prior to the building of a power plant, then the existing structure is not part of the fuel cycle. For example, it is unlikely that a new coal mine would be opened just because a new power plant is built (though there are some so-called "captive" mines). Therefore, the emissions, impacts, and externalities of developing a new mine are generally not part of the damages and benefits from the new power plant.

The rule of thumb is to consider only the net incremental impacts. For example, damages and benefits associated with tree plantations used to grow feedstock for a biomass-fired power plant would be included as part of that fuel cycle's damages and benefits. However, the calculations are in the sense of *net* impacts, such as the *change* in soil erosion between the biomass plantation and the land uses which it replaced.

Controversy arises, however, in considering the use of coal, for example, in a long run context. At some point, a new mine will surely be developed. An argument can be made to pro rate the impacts of mine construction to the coal fuel cycle, but following the rule of thumb of considering only the incremental effects, we do not do not consider those impacts in the context of our study. In general, each energy resource has a unique set of questions about establishing the boundaries for its fuel cycle.

3   Inability to Address all Impacts

In developing the methodological approach for the study, we used an accounting framework to describe the sequence of steps of the Damage Function Approach (as summarized in Sect. 1.2). Figure 1 is an example of one of the matrices developed for the accounting framework. The matrix lists the stages and activities of the fuel cycle along the rows of the matrix, and the emissions and other residuals of the fuel cycle in the columns of the matrix. Similar matrices were developed to link these emissions to possible impacts, and to translate these impacts into social costs and benefits.

As work progressed on the study, we realized that while the accounting framework is a useful concept, it is misleading to present it as a central part of our analysis. People viewing the matrices developed the impression that we were going to fill in all (or at least many) of the cells of the matrices. While our study would *start* filling in some of the cells, the matrices would still be extremely sparse. Most of the cells would be left blank. Any one activity does not cause all of the residuals listed in the columns of the matrix, and each residual does not give rise to every possible ecological or health impact. There are two other major reasons for our inability to address all impacts of a fuel cycle.

3.1   Level of Effort

One reason for the inability to address all impacts is the great number of fuel-cycle activities, emissions, impacts, and economic valuations. There are many thousands of possible impact-pathways, each defining a particular fuel-cycle activity, emission, impact, and economic value. Many of these impact-pathways require extensive literature review and modeling to estimate the changes in concentrations of pollutants, the resulting impacts, and their economic values.

The methodological approach is not simply one of "grabbing numbers" from the literature. A considerable amount of modeling and calculations is required. For example, for the oil fuel cycle, modeling is required for the dispersion of the primary pollutants from the plant, the formation and concentration of ozone caused by the plant, emissions such as leaks from refinery operations, oil spills, and emissions at the wells. Each modeling task is itself a significant level of effort, which must be addressed for each technology for each time frame, for each site, and for each fuel cycle. Each modeling effort generally requires not only the model runs and their analysis, but the collection of the all-important input data that are specific to different power plant sites, and to different sites for each of the upstream activities.

3.2   Lack of Knowledge

The second reason why most of the cells of the accounting framework matrices are empty is that there are important gaps in knowledge. In particular, there are severe limitations in scientific knowledge about exposure-response relationships with regard to ecological impacts. There are two important reasons for this fact. First, power plant-specific and site-specific impacts frequently can not be generalized to other projects or sites. Second, impacts that are distributed over large regions are inherently difficult to quantify. Quantitative regional impact assessments have been performed for only a few ecological resources and types of stress. Despite the large effort devoted to research on acid deposition over the last decade, quantitative regional assessments of the impacts of acid deposition on fish populations and on forests are possible for only a few especially well characterized regions.

With regard to health impacts, those associated with the air inhalation pathway are better characterized than others. Environmental transport of pollutants to and through the food chain and aquatic impact-pathways are significant areas for further scientific research.

Figure 1: Example of accounting matrix used to conceptualize the Damage Function Approach

A mapping of stages of the coal fuel cycle into emissions and other residuals																
Stages of the coal fuel cycle and activities within stages	Air emissions												Water emissions			
	Primary										Secondary					
	HC	Particulates	Metals	SO_2	NO_x	CO	CO_2	Methane	Radon	Heat	Acid rain	Ozone	Suspended Solids	Organics	Heat	Metals
Coal Mining																
Mine construction																
Mining of coal																
Waste water management																
Solid waste management																
Post - mining																
Coal Cleaning and Beneficiation																
Plant construction																
Treatment																
Waste water management																
Solid waste management																
Decommissioning																
Coal Transportation																
Facility construction																
Loading																
Transit																
Unloading																
Decommissioning																
Generation																
Plant construction																
Generation																
Waste water management																
Solid waste transportation																
Solid waste management																
Decommissioning																
Power Transmission																
Construction																
Transmission																

4 Major Areas of Uncertainty

What is the source of the greatest uncertainty in estimates of externalities? Some people answering that question think that it is in the ecological impacts, since so few exposure-response functions have been estimated. Other people feel that the uncertainties concerning economic valuation dominate the uncertainties in the technical data. Yet other people state that many important impacts are intrinsically non-economic and simply cannot be monetized.

There is no final answer to this question at this point -- though our study is providing information on the uncertainty and the quality of data used, and we will be in a better position to answer that question at the end of the study. For now, it appears that the answer may be that the uncertainty is very great everywhere, at every stage of the analysis. The rest of this section discusses the major sources of uncertainty.

4.1 Data on the Reference Environments

Before any calculations can be done, data are needed on where the fuel cycle activities take place. Some data, such as population data, are relatively accurate. Yet, those data become less accurate when expressed in terms of a spatial distribution relative to a power plant. Other data, while conceptually easy to compile, are too detailed and costly to collect. An example is a complete inventory of buildings and structures, and their materials (which may be affected by acidic deposition and particulate matter).

Ecological data are the most difficult to compile and generally do not exist. For example, the spatial distribution of different types of plants, by species; fish populations, by species and by river reach; and the number of spotted owls, etc. are data that generally do not exist. Yet, these are the types of things that are valued in economic terms. Without such data, it is impossible to accurately estimate impacts, and of course damages and externalities. General statements such as "altered stream flow as a result of retrofitting a dam with a powerhouse affects riparian environments" are too vague and qualitative to value in any quantitative sense.

In addition to the sites of the power plants themselves, the sites of the upstream (and to a lesser extent the secondary and indirect activities) are also important to determining the type and magnitude of impacts and externalities. For example, underground coal mines have greater risks in terms of exposure to coal dust, radon, and occupational hazards than surface coal mines. Conventional uranium mines and mills involve processes very different from in situ leaching. The exact locations of the biomass plantations, and the land uses they replace, will greatly affect the next change in environmental effects associated with the different land uses. Data are difficult to compile, not only for the power plant sites, but for the upstream activities as well.

4.2 Technology Characterization

One thing evident from our study thus far is that impacts, and thus damages, benefits, and externalities, are very project-specific. For example, emissions expected from an integrated gasification combined cycle coal plant are considerably lower than from a pulverized fuel plant (Table 1 gives an illustrative comparison). In the case of hydroelectric fuel cycles, impacts vary considerably depending on whether the project entails building a new dam, retrofitting an existing dam, or diverting water from its natural course into penstocks.

Table 1: Comparison between total fuel cycle emissions for pulverized fuel and integrated gasification combined cycle technologies (lb/MBtu)

	Pulverized fuel	Integrated gasification combined cycle	Ratio
Carbon dioxide	214	184	0.86
Nitrogen oxide	0.589	0.044	0.075
Sulphur oxide	0.589	0.065	0.11
Particulate matter	0.03	0.0	0.0

Source: ORNL/RFF (1992, Table 9.5-1)

Estimates of emissions can vary, even for conventional technologies such as pulverized fuel coal plants. The specifications of the plant will affect the magnitude of the impacts and externalities. These specifications include not only the installed pollution abatement equipment and their efficiencies, but also stack height and other source parameters which are used in atmospheric transport modeling. For future technologies, which are important in the context of longer-run energy planning, obtaining data on their design specifications and emissions is even more problematic.

The characterization of the upstream activities and technologies also affects estimates of impacts and externalities. For example, the amount of sulphur and ash in coal will affect the level of emissions from the power plant. Different types of crude oil also have varying amounts of sulphur and other constituents. Several different data sources may be required to obtain estimates of emissions of different pollutants; these data sources may not be consistent in their assumptions.

4.3 Atmospheric Modeling

A number of fuel cycles involve the combustion of fuels -- coal, biomass, oil, and natural gas. Pollutants are emitted from the stacks and elsewhere. Estimating the dispersion of pollutants in the atmosphere is of considerable importance in assessing their ecological and health impacts. Also, the uranium/nuclear fuel cycle includes the possibility of reactor accidents which, if they occurred, would involve the dispersion of radionuclides. The expected concentration of pollutants (or radionuclides) at different locations relative to the point-source, and the spatial distribution of the population and environmental receptors, will have a great bearing on the severity of the impacts from these pollutants. While considerable attention has rightly been given to the level of emissions of pollutants, knowing the resulting local and regional concentrations is also of great importance. Thus, modeling the dispersion of pollutants is crucial to estimating externalities.

The Ozone Isopleth Plotting with Optional Mechanisms (OZIPM-4) model is one of the more important models being used in the study. The model is being used to estimate the increased concentration of ozone due to the emission of ozone precursors from power plants, principally NO_x.[4] Regional-scale photochemical models could be used instead of OZIPM-4. They include the Urban Airshed Model, the Regional Oxidant Model, and the Reactive Plume Model.[5] For these models, a very large commitment of resources is required to compile accurate inputs. Since these models require such extensive data inputs and since they still simulate only short-term concentrations, rather than seasonal concentrations, the OZIPM-4 model was chosen for its simple application and current chemical kinetic mechanism.

Yet, OZIPM-4 has many significant limitations -- for example, the spatial distribution of emissions is treated as being uniform. All emissions are assumed to mix uniformly within the mixed layer. The assumption obviates the need to specify wind speeds or the dimensions of the air parcel (other than height). The model's estimates of ozone formation are very sensitive to the area over which the emissions are assumed to be averaged. After estimating the single-valued increase in ozone concentration, another calculation (apart from OZIPM-4) is required to account for the relative frequency of the prevailing wind direction and for the spatial distribution of the local population. Thus, considerable uncertainty exists about the increased exposure of the population to ozone formed from precursor emissions from the power plant.

4.4   Estimating Impacts

Many studies stop at estimates of emissions, without atmospheric modeling or estimates of impacts. Many of the reasons for such limited analysis were discussed in previous sections. Data on the baseline environmental conditions are usually unavailable. Exposure-response functions are few in number, especially any that are widely applicable (refer to Sect. 3.2). Impacts are generally site-specific. For example, Figure 2 provides a qualitative comparison of the size of some of the impacts of a coal fuel cycle. The figure compares some of the estimated social costs for two rural sites in the United States.[6] There is a large difference between the sites. Although this finding is intuitive, the implication is that there is simply no *single* number for the externality associated with any fuel cycle.

Another source of uncertainty is the fact that the quality of information about health and ecological dose-response functions stands to be improved. Considerable recent research suggests the likelihood of some health impacts, even though emissions are within regulatory standards.[7] This issue of impacts below standards is controversial, however, and is a source of uncertainty about fuel-cycle health effects. Estimates of ecological impacts are even more imprecise, and in many cases are impossible to quantify.

91

Figure 2:   Illustrative comparison of selected residual fuel-cycle damages estimated for different sites (different shaded segments represent different stages of the fuel cycle)

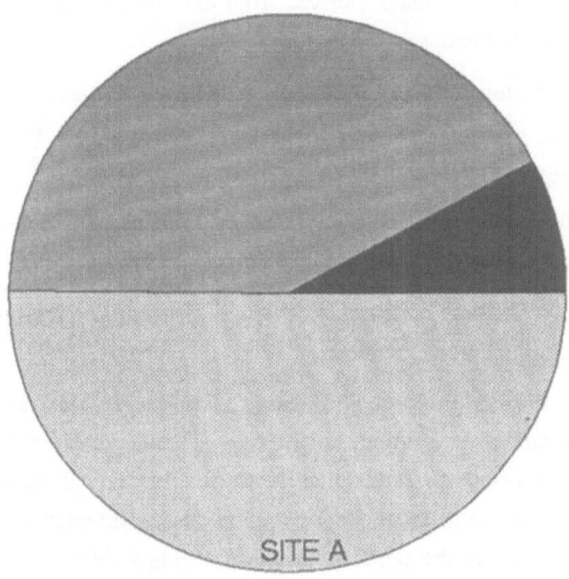

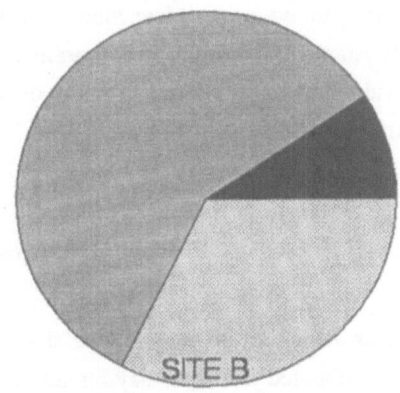

4.5   Economic Valuation

There is considerable uncertainty in the economic valuation of impacts, more in some cases than in others. There are many fundamental issues in economic valuation:[8]

(1)   the transferability of estimates derived from one location and set of circumstances to a similar, yet not identical, set of conditions;

(2)   the valuation of non-use and existence values in which individual welfare is, according to the premise, affected by an environmental condition (e.g. the existence of a pristine wilderness area), even though they do not use or experience it;[9]

(3)   asymmetry in the willingness to pay to avoid impacts, versus the willingness to accept payment as estimates of economic value;

(4)   the influence of income constraints on individuals' willingness to pay and their impact on estimates of the value of something;

(5)   the differences in valuing potential catastrophic events, such as major nuclear reactor accidents, compared to common accidents -- should valuation be based on perceived probabilities of accidents rather than scientifically derived estimates, are catastrophic deaths valued more than those from common causes, do catastrophic deaths have "non-use value" in terms of the population at large being willing to pay to reduce the risk to others?

(6)   the changes in social values over time and across regions; and

(7)   the aggregation of damages and benefits estimated for different types of impacts, the potential redundancy in adding these damages and benefits, and the limitations of studies in which bundles of impacts are not valued simultaneously.

Also, for some impacts, there are no, or very few, studies of their economic value. Estimates of values need to be substantiated through corroborative studies. Finally, there is the opposition to economic valuation itself, at least of many of the impacts, which according to some people, by their intrinsic nature, can not be monetized.

4.6   Identifying the Externalities Portion of the Overall Social Costs and Benefits

Estimating the social costs and benefits themselves does not provide estimates of the externalities. Some of the estimated damages may already be internalized. For example, in the U. S., with the Clean Air Act Amendments, there are provisions for trading of SO_2

pollution permits and overall limits on total SO_2 emissions. Thus, emissions from a newly-constructed plant would be offset by reductions at another plant owned by that utility or by some other utility. A utility could purchase pollution permits to allow additional emissions, and the utility selling the permits would reduce its emissions. The costs of complying with the regulations would be reflected in the prices of the electricity from those utilities. The impacts of the SO_2 emissions would be internalized, at least *in part*. However, since impacts are site-specific, changes in the geographical locations of emissions would affect the impacts, so that there would generally be a net change in impacts and thus externalities in any system that allowed the trading of pollution permits.

As another example, an argument may be made that the health effects of working in coal mines are internalized. Miners receive a fairly high wage, in part for the added occupational health risks. This added labor cost is reflected in the price of the coal and ultimately in the price of electricity. However, for a number of reasons, part of the social costs associated with those health impacts may not be internalized -- possibly because of the lack of information by miners about the actual magnitude of the risks, or possibly because of the lack of labor mobility in mining regions.[10] The final step in our DFA is to discern the portion of the estimated social costs and benefits that are truly externalities. Because of the complexities, it would be mistaken to label social costs as externalities without careful study.[11]

5   Implications

5.1   Knowing the Limitations

In this paper, we have pointed out that it is impossible to quantify every impact of a fuel cycle. There are insurmountable problems in terms of both the immense level of effort required to undertake such an analysis, and the lack of scientific knowledge for potentially important impacts. With those constraints on our study, the emphasis is on local rather than on regional or global impacts. Adding only quantitative estimates of externalities associated with individual impact-pathways could be dangerously misleading. All of the potentially important impacts are not quantified. Some of those not quantified, such as potential global warming from CO_2 and other emissions, may overwhelm the others in importance. Thus, it seems pointless to suggest some specific number as *the* overall externality of a fuel cycle.

Even if most of the potentially important impact-pathways are addressed in a study, estimates of most impacts and externalities are specific to the power-plant project and proposed site. There is no *generic* value for the externality of any fuel cycle. Thus, in our study, we have emphasized that its primary objective is to compile and assess analytical tools and exposure-response functions, and to demonstrate their use. We do not attempt to compile the "big book of numbers," where one could look up the universal externality value for the natural gas fuel cycle, for example, or *the* externality due to impingement of adult rainbow trout fish in the turbines of a hydro powerhouse. Emphasizing methodology rather than specific numbers follows the idea that rather than impose national command and control externality standards, the federal government (in the U.S.) could better assist States' Public Utility Commissions in their regulatory planning by developing a methodology and analytical tools that they can use for their own studies.

The DFA is best-suited to addressing incremental impacts from a single power plant and its associated fuel cycle. In the context of long term energy planning at some regional or national scale, the great difficulty in using the DFA is in treating the cumulative effects of many power plants. Site-specific effects and any nonlinear interactions in the emissions of power plants and other activities could mean that a simple additive, linear extrapolation is overly simplistic.

With the international interest in fuel cycle externalities, what general conclusions can be drawn from our and from other studies? The general conclusions would have to be more about concepts, principles, and analytical methods than about numerical truisms. Not only do baseline environments and populations vary across countries, so do individual social values. Most of the research on externalities has been done in Europe and the U. S. In the absence of many studies in other countries, perhaps the only recourse is primarily to use information from Europe and the U. S. Any estimates of externalities would have to be expressed relative to some norm such as the total cost of generating electricity, to account for international exchange rates and differences in income levels.

While the tone of the paper may sound negative in terms of discussing the many limitations in estimating externalities, the intent of that tone is to caution those anticipating or expecting some "silver bullet" for the problem of estimating the externalities of fuel cycles. On the other hand, uncertainty and analytical limitations are *not* necessarily a rationale for policy inaction. Policy, planning, and decision-making should proceed appropriately on the basis of the best available information.

5.2 Preliminary Ideas on Using Results of Other Studies

The preliminary results of our study and the discussion in this paper point to certain ideas worth considering in any plan to account for fuel cycle externalities:

(1) results of prior studies should not be used without assessing their underlying assumptions -- specifically the power plant technology and design, and the site of the plant and of the upstream activities; externalities are generally site-specific, except of course those related to CO_2 and other greenhouse gases which have global impacts;

(2) it is preferable to have site-specific analysis for most impacts, to the extent possible; if results of previous studies are used directly, it is important to use results from reference sites that are similar in terms of their baseline environmental conditions (such as ambient concentrations of hydrocarbons, NO_x, and ozone);

(3) scaling previous results, perhaps linearly, can be considered for certain health effects; the justification for this scaling is that the air transport and dose-response models, which were used to estimate pollutant concentrations and their effects on health, are linear or near-linear;[12] as a result of these functional relationships, health impacts are generally proportional to the size of the population, holding constant the spatial distribution of the local population, topography near the plant, meteorological conditions, baseline ambient concentrations, and of course plant design and emissions; all of these parameters certainly vary from site to site, but as a first-order approximation, linear scaling of selected impact-pathways may be reasonable (though the point of using sites with comparable baseline ambient conditions is particularly important for ozone impact-pathways); notwithstanding the possibility of scaling, it is preferable to have site-specific analysis;

(4) the baseline ecological conditions must also be taken into account to ascertain which impact-pathways are relevant for a particular site;

(5) regional and global impacts (viz., acidic deposition and CO_2 impacts) are more difficult to quantify; potential externalities for which no quantitative estimates exist should be taken into account, at least in a subjective qualitative way.

(6) an externalities tax that is directly based on estimates of external costs is one way to internalize externalities; this tax would be imposed on both existing and new plants; it would be based not just on emissions, but also on site-specific attributes such as the size of the local population that would be affected by fuel-cycle activities;

(7) one-for-one trading of emissions allowances would generally not be an economically-efficient way of internalizing externalities; these allowances do not take into account

site-specific differences; the magnitude of externalities generally depends on the attributes of the site(s).

Since our study is ongoing, these ideas are preliminary. They should be regarded as an initial basis for discussion, and *not* as a definitive prescription for policy.

References and Endnotes

[1] This paper is based on an ongoing study by Oak Ridge National Laboratory (ORNL) and Resources for the Future, for the U. S. Department of Energy (DOE). The study is a collaborative effort with a European research team funded by the Commission of the European Communities (EC). ORNL is managed by Martin Marietta Energy Systems, Inc. for DOE under contract DE-AC05-84OR21400. The author acknowledges the ideas and the many contributions of the whole research team, which are reflected in this paper. However, the views expressed in the paper are solely those of the author. Filenames: OTTING_3.PAP, APPB.BD, SITE.DRW.

[2] Oak Ridge National Laboratory and Resources for the Future (1992) *U.S.-EC Fuel Cycle Study: Background Document to the Approach and Issues*. Report No. 1 on the External Costs and Benefits of Fuel Cycles: A Study by the U.S. Department of Energy and the Commission of the European Communities.

[3] Hohmeyer, O. (1988) *Social Costs of Energy Consumption: External Effects of Electricity Generation in the Federal Republic of Germany*. Berlin, Germany: Springer-Verlag; Pace University Center for Environmental Legal Studies (Ottinger, R. L. et al.) (1990) *Environmental Costs of Electricity*. New York, NY: Oceana; Bonneville Power Administration-RPPC (1991) *Environmental Costs and Benefits: Documentation and Supplementary Information*, Portland, OR. February 22; Tellus Institute (S. S. Bernow and D. B. Marron) (1990) *Valuation of Environmental Externalities for Energy Planning and Operations*, Boston, MA; and Hagler, Bailly Inc. (Rae, D. et al.) (1991) *Valuation of Other Externalities: Air Toxics, Water Consumption, Wastewater and Land Use*, prepared for the New England Power Service Company, Westborough, MA.

[4] U. S. Environmental Protection Agency (1989) *User's Manual for OZIPM-4 (Ozone Isopleth Plotting With Optional Mechanisms), Vol. 1*. Office of Air Quality Planning and Standards, RTP, NC, EPA-450/4-89-009a.

5 Discussions of these models are in Appendix C.1 of Oak Ridge National Laboratory and Resources for the Future (1992) *Damages and Benefits of the Coal Fuel Cycle: Estimation Methods, Impacts, and Values*. Prepared for the U. S. Department of Energy and the European Commission. Oak Ridge, TN: unpublished preliminary draft. The report is to be published after review and revision.

6 Data for the figure were based on preliminary calculations that have since been revised, though the point of the figure still holds: fuel cycle damages are generally site-specific.

7 For example: Krupnick, A. J. and R. J. Kopp (1988) "The Health and Agricultural Benefits of Reductions in Ambient Ozone in the United States," Appendix to Office of Technology Assessment, U. S. Congress, *Catching Our Breath: Next Step for Reducing Urban Ozone*, Washington, D.C.; McDonnell, W. F., et al. (1991) "Respiratory Response of Humans Exposed to Low Levels of Ozone for 6.6 Hours," *Archives Environ. Health*, 46(3), 145-150; and Larsen, R. I., et al. (1991) "An Air Quality Data Analysis System for Interrelating Effects, Standards, and Needed Source Reductions: Part 11, A Lognormal Model Relating Human Lung Function Decrease to O_3 Exposure," *J. Air Waste Manage. Assoc.*, 41(4), 455-459.

8 Most of these issues are discussed concisely in Section 1.4 of the draft coal report (ORNL/RFF 1992).

9 For a critical assessment of contingent valuation of non-use values, see Cambridge Economics, Inc. (1992) *Contingent Valuation: A Critical Assessment*, proceedings from a conference held in Washington, D. C., April 2 and 3, 1992.

10 Further discussion is provided by Alan Krupnick in Section 5.3 in ORNL/RFF (1992).

11 Freeman, A. M. III, et al. (1992) "Accounting for Environmental Costs in Electric Utility Resource Supply Planning," Discussion Paper QE92-14, Resources for the Future, Quality of the Environment Division, Washington, D. C.

12 Discussion of the air transport modeling and dose-response functions are given throughout the ORNL/RFF (1992) draft coal report.

8.   Identification and Incorporation of External Costs Associated with Energy Use

Klaus P. Masuhr

Prognos AG

Missionsstr. 62, 4012 Basel, Switzerland

1   Introduction

(1) The energy sector is a key area of any modern economy. Our material prosperity depends to a decisive extent on the energy-supply situation. The availability of energy is one of the most important prerequisites for nearly all industrial production processes.

Nonetheless, the energy-supply system and, more than that, our overall approach to energy in general has become increasingly questionable. As a result of the extremely wasteful energy consumption in all areas of the economy, in households, in the industrial and the services sectors as well as in the transport sector our air has been polluted, our health has been damaged, our fauna and flora are coming under increasing threat. The risks connected with the use of nuclear energy are leading to impassioned political and social controversies. Energy-related emissions of gases such as carbon dioxide, nitrous oxides, methane etc. constitute a threat to the world climate.

(2) The existence of environmental damage of this kind has been known to us for some time now. Still, totally insufficient consideration is given to it in strategic as well as everyday decision-making. Our notorious blindness to environmental damage poses a threat to a key economic steering mechanism, i.e. the allocation effect of the price and cost system. Since the energy prices formed as a result of the play of market forces disguise the "true costs" of energy production and use, the result can be massive economic misallocations. This has a key effect on the efficiency of a market economy. Those affected by environmental damage are also the victims of a disguised collective expropriation process currently not being offset by any regulatory policy compensation or avoidance strategy. Economic distributory justice cannot be guaranteed in this manner.

Since processes of this kind are taking place constantly without being incorporated in any way in the allocational and regulatory processes of the economy and of society, it goes without saying: these processes are external effects of the energy-use.

(3) Scientific analysis of the various dimensions of external effects is still a very young field. Similarly, the question as to whether and with what instruments the energy policy sector should attempt to incorporate external effects, particularly external costs, has been a subject of discussion for only a few years now.

The government sector has shown increased interest in environmental issues and has formulated a number of specific environmental objectives. However, a considerable information-related and scientific deficit exists at the empirical/analytical as well as at the instrumental/administrative levels with regard to the implementation of these objectives. The "reconciliation of ecology and economy" is, in economic policy terms, not a question of good will but rather a difficult optimization task ("achieving ecologically necessary objectives in an economically efficient manner"), something which can have substantial consequences for economic, employment and income structures.

(4) In July 1990 the Federal Ministry of Economics awarded Prognos AG a contract to carry out a fundamental study on the "Identification and Incorporation of External Costs Associated with Energy Use" as a contribution towards a necessary and inevitable research process. The formulation of the topics in question illustrate the two main problems connected with an economic and energy policy having a strong ecological orientation:

1. Sufficient information on the factors involved in connection with the extent, effects and consequences of the energy-use situation (identification)

2. Assessment, structuration and organization of adequate measures for the reduction/ avoidance of external costs (incorporation)

(5) The project was concluded in the summer of 1992. The main body of the Prognos report will be published in September by Schäffer-Poeschel-Verlag, Stuttgart, Germany. Also available are numerous individual studies compiled by various working groups within the overall project. This summary indicates the main results of the project.

2   Main Results

2.1   Identification

1. If external energy-use costs are not "incorporated" in the decision-making strategies of energy consumers in the energy sector as well as in the energy-policy sector, the function of

prices as the main steering mechanism of a market economy ceases to be effective. Massive misallocations occur in virtually all areas of the economy.

2. Although a few energy-use-related external costs have been known to us for quite some time, they have often been ignored, their importance played down or, indeed, contested.

The energy sector, including the transport sector, has come to be recognized as the main cause of considerable environmental damage.

3. In addition to "simple" external costs which can be identified, quantified and monetarized with relative rapidity, there are numerous effects that can be described in quantitative terms, but which do not lend themselves easily to precise quantification or monetary evaluation. All areas of the energy sector are - albeit to varying extents - involved in causing these effects. They are, however, in part the result of extremely complex causal chains.

4. On the basis of the synergetic and accumulative effects of a wide range of different emissions from the energy sector, revealing highly complex processes when examined in detail, it is probable that many effects have not yet been discovered or will not become evident until some time in the future ("ecological time bombs" caused, for instance, by the accumulation of heavy metals in the soil or the man-made greenhouse effect).

5. Despite a lack of knowledge in specific instances it has been firmly established that external effects are being caused and will continue to be caused to an unacceptable extent. There is a risk that processes could be triggered, the damage potential of which would be immeasurable.

6. The **"classical" pollutants** resulting from energy use, i.e. sulphur dioxide, nitrous oxides, various types of dust, carbon monoxide and hydrocarbons cause a wide range of different types of material damage to buildings and infrastructures. They constitute a hazard to human health and life and result in damage to or destruction of the natural environment.

7. Road traffic will be the main cause for this in the future, since in other areas of energy use environmental regulations and energy-saving measures will result in a strong reduction of "classical" emissions. Unchecked growth of the transport sector continues to be a source of concern.

8. An **impending climate disaster** as a result of the accumulation of trace gases due to energy use (particularly CO_2 and methane) is one of the most critical external effects of the

energy system. The existence of the greenhouse effect has been scientifically established. The **consequences** are still speculative, but immeasurable damage cannot be ruled out. Indeed, the ability of mankind to survive on this planet may be threatened by **irreversible processes**. Even very conservative scenarios assume disastrous droughts, mass population movements involving millions of people and devastating weather anomalies in many regions of the world.

9. The use of nuclear energy is seen by many as a way out of the CO_2 dilemma. However, it could cause considerable external costs of a different kind. The primary damage category in this case would be an unpredictably high death toll as a result of a major accident involving a **reactor-core meltdown**. The probability of an accident of this kind occurring is a subject of considerable controversy (probability estimates range from 1:3.300 to 1:280.000 per reactor year). However, calculations with these figures tend to make us forget that a nuclear accident is something that could happen any day. This results in a new external effect, the fact that the risk of an accident of this kind is imposed on people without its being clear whether or not this is based on voluntary acceptance. The imposition of a risk of death without the existence of voluntary acceptance is a serious external effect of our current energy system.

10. **Regenerative energies** result in comparatively low levels of external costs.

11. **Overall,** a fairly good qualitative **identification** of the complex and multilayered dimensions involved with the external effects of energy use in the current energy system is possible. With regard to quantification, on the other hand, deficits are quite evident.

12. This casts an obvious shadow on any attempts at **monetarization.** The immediate and basically trivial calculation of costs involved, for instance, in connection with repairing damage or, in general, in connection with "factor input for repairs", covers only a small part of the actual damage caused, be this as a result of the fact that the damage in question is simply irreparable or as a result of the fact that non-material values are involved in the broadest sense of the term, which are "compensatable" but are located in the general sphere of human benefits and preferences.

2.2 Incorporation

13. In view of the **possible** dimensions of the external effects involved, the objective of any attempt to incorporate the external effects of energy use in an appropriate manner must be the extensive avoidance of these effects.

14. Numerous **instruments for the incorporation** of external costs are available to government in the form of regulations, taxes, certificates as well as economically oriented provisions of liability law.

15. In connection with the quantitative structure of the instruments involved, the government sector will need to overcome its continued obsession with the acquisition of legitimating details (given the fact that the information situation will continue to be uncertain in the future in any event) and to employ carefully assessed **pragmatic strategies.**

16. The **economic effects** of using incorporation instruments would be noticeable first and foremost in the energy sector itself, resulting in a notable **decline of business volume** and (depending on the pollutants targeted by the instruments) also in a more or less strong trend towards fuel substitution.

The creation of fuel- or pollutant-specific instruments, something the authors generally view critically, would appear to make sense only with regard to the necessary reduction of CO_2 emissions. All other areas can be covered by a general tax on energy consumption. No significant threat to the competitiveness of German industry is to be expected. Structural change will be accelerated.

17. The monetary assessment of identified external effects will frequently have to concentrate on the determination of preference structures. The latter are general indicators for a **societal** process of consensus making. Economic incorporation instruments are part of this process and definitely not the least important of them. However, this process involves considerably more than this, since it necessarily includes the numerous interdependencies that exist between reciprocal technical/physical, economic and social factors associated with external effects.

18. The social incorporation of the external effects of energy use implies to a strong extent the need for a complete revision of established lifestyles and prosperity-related ideals, which can be motivated by economic or liability-related regulations, but which over the long term cannot be implemented on the basis of these instruments alone. The international dimension, the degree of interdependency and integration that exists in the global economy, as well as the size of the development-related differentials that exist between the various regions of the earth, accompanied by extensive ecological dependencies, will make general changes in living habits necessary based on the principle of sustainable development.

19. As a recommendation to the government sector it should be emphasized that it will be of crucial importance to act now wherever external effects ar known to exist or are becoming evident and not to wait until the last bit of physical evidence for external costs has been gathered. In connection with improving the information base in the sense of a **permanent learning process** three focuses emerge:

- insistent promotion of activities that contribute towards a better understanding of the growing external effects of energy use.

- concentration of empirical research on analysis of the possible consequences of climate change

- promotion of research on methods aimed at identifying societal preferences and risk assessment.

Processes of this kind constitute an initial step which should be institutionally supported by forming appropriate research and information pools with the necessary financial backing.

With regard to the incorporation debate it is recommended that the available incorporation instruments be used on a broad front and in various mixes and that this be done immediately. This will also require a change of attitude among policy makers.

3   The Different Dimensions of the External Effects of Energy Use

(1) In the first part of the study an attempt is undertaken to examine the full spectrum of external effects at least in qualitative terms.

Since a complete quantification and monetarization of these effects can be ruled out from the beginning due to their large number and multi-dimensionality, the various levels of external effects are indicated in the study on the basis of seven highly indicative examples. They are:

- material damage as a result of energy-related air pollution

- soil pollution, damaged forests and endangerment of biodiversity as a result of energy-related air pollution

- damage to health as a result of energy-related air pollution

- external costs of climate change as a consequence of energy use

- external costs of a nuclear accident

- external costs of regenerative energy technologies

- discussion of intertemporal misallocation of exhaustible energy sources.

On the one hand, growing potentials for damage are indicated in stages ranging from "limited" damage to buildings on down to extreme damage in connection with a climate-related disaster or a major nuclear accident. The increasing uncertainty in assessing external effects and, consequently, the increasing difficulty in assessing risks can be very clearly illustrated from one stage to the next.

The **identification,** i.e. the qualitative description of possible external effects is possible in all the areas studied. Many effects are observable and it is possible in principle to describe a number of causal relationships. In the health sector or in connection with the threat posed to biodiversity no precise qualitative description is available at the present time. However, this in no way detracts from the reliability of the statement that extensive damage is probable. This applies even more clearly with regard to impending global disaster as a result of large-scale climate change or the risk of a nuclear disaster.

(2) Identification gaps limit possibilities for **quantification.** What is still unknown cannot be measured. The study shows that additional conceptional or empirical problems exist with regard to technical/physical **measurement** of external effects identifiable in qualitative terms.

In connection with material damage or damage to health there is at least in principle the possibility of establishing **quantitative** relationships between certain causes and negative effects, e.g. on the basis of dosage-effect analyses and/or epidemiological studies. The fact that this has not yet been done to a sufficient extent is a financial and/or organizational problem, but not a fundamental problem.

Deficits are still unmistakably evident in other areas. The damage being caused to the plant and animal world is extremely difficult to **measure** in scientific terms. The dosage-effect values of radiation with regard to cancer-related deaths has to be based in empirical terms on the meagre flow of information on the Chernobyl disaster. Similarly, the quantitative effects of an impending climate disaster are still unclear.

(3) A wide range of different approaches are taken in the study with regard to the **monetarization** of external effects of energy use. Direct damage assessment strategies, compensation schemes, polls and abatement cost concepts are analyzed and discussed in detail in a special report following the second main section of the study. The examples dealt

with show that for the most part "simple" (direct) damage assessments have been attempted in practice thus far, but that these procedures are totally inadequate. Consequently, the study consciously avoids any attempt to make statements such as: "The external costs of energy use amount to a total of X billion DM". However, even the few tangible and quantifiable effects illustrate the dimensions involved:

- Emission-related material damage may amount to DM 10 billion per year.

- A number of authors estimate damage to forests at an annual amount of between DM 200 and DM 500 million. Other authors, who also take into account the ecological and recreational function of forests, speak of up to DM 9 billion per year.

- Damage to health (in connection with a few selected pollutants and a few selected health-related effects) is estimated at about DM 3 billion per year (of which DM 1.5 billion is energy-use-related) -- interpreted as costs deriving from work lost. Double-digit figures in the billions were estimated, on the other hand, on the basis of polls taken to determine willingness to pay and preferences.

- Attempts to monetarize the consequences of a major nuclear accident are extremely controversial. The amount of damage involved could go into the trillions. Then however, these amounts are evaluated on the basis of probability theory. For example, an occurrence probability of 1 to 100,000 per reactor year "reduces" the amount of damage by the factor 10^5. However, as this occurrence probability is also highly controversial, results of monetarization vary considerably from author to author.

- The costs of a climate disaster could be infinite.

(4) The direct monetary assessment of damage cost caused by eliminating the damage covers only a very small part of actual damage, be this since it is unknown or irreparable, or since non-material values are involved, for which "compensation" is possible but which are located in the general **sphere of human benefits and preferences**. In connection with the **possibility** of extremely large amounts of damage, for instance as a consequence of a nuclear accident, there is also the problem of "accurate" **risk assessment** on the part of those affected.

4   The Incorporation of External Costs of Energy Use

(1) The incorporation of external effects is for the most part a social process involving decision-making and regulatory effects. In this context the process of **economic** incorporation is an important but, as a result of the information deficits described in the first part of the study and the complex multi-dimensionality of the problem, limited part of the overall

system. The complexity of the incorporation debate is addressed in the second part of the study in that it connects both economic and non-economic incorporation strategies.

Economic incorporation approaches are discussed extensively in the study. But, in this short report, the presentation will concentrate on non-economic incorporation aspects:

The incorporation of external costs in connection with energy use on the basis of economic analysis is, of course, based solely on the satisfaction of needs through market processes. However, the **parallel existence of different forms of need satisfaction** is observed in reality, reflecting the fact that human needs differ and cannot always be formulated in economic terms.

(2) Economic incorporation strategies thus need to be included in an overall complex of **extra-economic processes**. Their integration in a complex of this kind is by no means in contradiction with the economic debate. On the contrary, it is a prerequisite for incorporation.

A "political incorporation" of external costs connected with energy use will only be possible if there is a reorientation of the value structures that have dominated thus far. This will make it necessary to:

- **accumulate additional information** on scientific and social factors

- increase public **willingness to learn** and adjust

- **gain time** and

- develop new and ecologically based conditions for more appropriately **adjusted value and regulatory structures.**

An important contribution is provided here by the **concept of sustainable development.** It is based on the observation that the type of industrial growth that has characterized the past 200 years has already led to an overuse of ecological resources in the industrialized countries and for this reason can no longer serve as a model for a future world economic order.

(3) The **processes of communication** in society are equally important for the way industry and society in general deal with the external costs associated with energy use. There are a number of factors involved:

- **Language:** Non-experts not acquainted with scientific usages easily associate misleading and in some cases "opposite" meanings with certain terms (e.g. "residual risk") with the result that not enough importance is attributed to statements of this kind.

- The **existence of communicable standards**: Often there are no easily understandable indicators such as the incidence of deaths or illnesses, the amount of costs caused by this, or votes lost.

- Established **organizational structures**: Existing institutions (administrations, labor unions, associations, political parties, clubs etc.) concentrate for the most part on their traditional range of functions.

- **Communication barriers**: The communication process is characterized by massive conflicts of interests and a desire to preserve what has been acquired in each case.

The formation and recognition of language, standards and institutions necessary for the discussion of external costs will make it possible to expedite the communication processes indispensable for promoting willingness to learn and adjust.

The mere awareness of the problems involved must evolve from an initially **unpolitical discussion** into an open political controversy before less conflictive forms of decision-making and, ultimately, broad acceptance of new lifestyles are achieved

(4)   With this in mind, the discussion of the **"ethics of pluralism"** takes on importance. The emphasis of pluralistic structures in a society has a high price. The clarity of processes is necessarily diminished and with this often their political and democratic controllability.

And, the establishment of the "pluralism ethic" as a principle for the **development** of new objectives and instruments takes time. The cases dealt with in the first part of the study (e.g. climate change or the risk of a major nuclear accident) show that the time available is short and the available latitude for adjustment is growing narrow.

(5)   **Over the long term** it will be necessary be incorporate environment-friendly value orientations and structures. However, this implies the development of new lifestyles and thus, ultimately, new economic and social structures.

Ideals of this kind can, however, not simply be invented. They can only develop in an evolutionary process within the framework of today's global competition of cultural ideas and values and, as such (as has been the case for thousands of years) in an interplay of factors involving the formation of human values, natural conditions and regulatory structures in society.

Key factors are:

- a system based on **sustainable development** and
- ethics of **pluralism.**

While concepts on sustainable development attempt to formulate the ecological framework in which nature has placed man and within which he will permanently be constrained, the ethics of pluralism provide important indications regarding the way in which instruments should be employed, the prerequisites for increasing creativity and raising the level of error-friendliness in an evolutionary process.

5   Conclusions and Recommendations for Action

(1)   The study clearly shows the sectoral and overall economic importance of external costs in connection with energy use. The current state of knowledge is far from sufficient to be able to identify, quantify and monetarize the amount of these costs as well as their economic and energy-sector effects on the basis of uniform standards. Nonetheless, a purely qualitative identification of the external effects currently being observed provides a large number of indications that these effects possess an extraordinarily high level of relevance and, indeed, that they may even pose a threat to life. Because of the enormous complexity of the interrelationships involved, the quality of information decreases considerably in connection with attempts to quantify external effects. The process of monetarization, necessarily based on quantification, is possible only in a few areas.

(2)   Based on the results of the study, the amount of external costs associated with energy use quantified and monetarized is sufficient to signal an **urgent need to act**. This applies all the more when it is taken into account that the costs that can be indicated are merely the tip of an iceberg. A defeatist attitude and insistence that we still do not have enough evidence is not an acceptable option in this connection.

Based on the analyses carried out, options for action exist in two central areas:

- the gathering and perception of additional information and, at the same time,
- the employment of available instruments for incorporation.

The Information Problem

(3) The information problem has two different dimensions:

- improvement of the given state of information and knowledge with regard to external effects, their quantitative importance and potential monetary assessment, as well as

- the permanent diffusion of this knowledge as a prerequisite for political decision-making.

(4) With regard to the current state of information and knowledge three particularly critical problem areas were identified in the study:

- improvement of information on the effect of the so called diffuse external effects in connection with pollutant emissions generated by the burning of fossil fuels,

- detailed analyses aimed at assessing the effects of the greenhouse effect,

- further development of methods for the determination of social preferences and for risk assessment.

(5) In connection with pollutant emissions generated by the burning of fossil fuels, for years now the main focus of attention has been on the **"classical" pollutants**, i.e. sulphur dioxide, nitrous oxides, carbon monoxide and various types of dust. We can expect to see a strong decline of these emissions in many areas of the energy sector, particularly in the electricity-generating industry, in the course of the next two decades (gradual effect of environmental policy regulations). Although, as a result of this, many external effects will decline in importance and, as such, any significantly large amounts of new empirical research on these areas of pollution would seem dispensable, it certainly cannot be said that the danger has passed.

(6) In all areas of the energy sector based on fossil fuels there are numerous other effects which cannot be quantified to the usual extent, in connection with which there is a danger that damage potentials may be much greater than has been observed in connection with the mass emissions of the "classical" pollutants thus far. Particular attention needs to be given to the wide range of different carcinogenic and toxic heavy metal compounds, to as yet unknown synergetic and accumulative processes in the biosphere, in the food chain and in the human body and, finally, to any and all interreactions of substances which we do not yet understand or which are simply unknown to us.

(7)   Here a picture is drawn of diffuse pollution processes which, primarily on the basis of gradual accumulation and enrichment, have long "maturation times", often extending over decades. **The analysis of the processes involved in diffuse pollution** of this kind, should be first an focus aimed at improving the politically relevant information base.

Because of the nature of these pollution processes no spectacular information can be expected right away (e.g. volumes of damage on a scale comparable to nuclear accidents). However, persevering and government-supported efforts to acquire new knowledge of damage potentials in an ongoing process will be indispensable, since as a result of the vague nature of the damage involved there is a considerable danger that public awareness will decline as a result of recurrent reports of success in connection with the "classical" pollutants.

(8)   In a second critical area of external effects - **climate change** as a result of energy-related accumulation of trace gases in the atmosphere - this danger does not exist to the same extent. The damaging influence of human activities on the global climate can be considered proven. The damage-related risk is enormous since the earth's general regenerative and regulatory capacities are under threat. A lack of clarity exists "merely" in connection with the consequences of the climate change, in particular for the agricultural sector, the regional distribution of damage and its effects on global social consensus. The creation and preservation of fundamental knowledge on these consequences is a second information-related focus for political defensive strategies.

(9)   Information gaps of an entirely different kind are found in a third complex of external effects. They concern the approaches taken to the risk of a **major nuclear accident.** The "Deutsche Risikostudie Kernkrafterke Phase B" (German Risk Study Nuclear Power Plants -- Phase B) has revealed a considerable number of uncertainties in the scientific/ engineering sector in connection with the analysis of potential scenarios. The "Gesellschaft für Reaktorsicherheit" (Society for Reactor Safety) notes that further research and development work will be indispensable in this area.

(10) Deficiencies have also shown up in the **assessment procedures for** nuclear-accident **risks.** This concerns in particular the problem in assessing the value of human life in general. Here, the extremely crude approach taken thus far using the "human capital method" should be replaced by other assessment approaches making use of the "contingent valuation method". In Germany research on this subject is still in the initial phases.

(11) The reference to the importance of **risk assessment,** particularly with regard to nuclear accidents (enormous damage potential with a relatively low occurrence probability), draws

our attention to a phenomenon which is a recurrent theme throughout the descriptions of the different dimensions of external effects. The deficiencies with regard to the ability to quantify and monetarize numerous effects does not permit "easy" assessment procedures, i.e. simple determination and monetary calibration of objectively occurring damage costs.

However, since this does not eliminate the external effects it will be necessary to determine in an appropriate manner how much a reduction or complete elimination of the qualitatively (and to a limited extent also quantitatively) describable **risk itself** is worth to individual persons or to society as a whole.

(12) The information problem in connection with awareness of external effects points to the problem concerning **diffusion of information** in societal decision making processes. It is a direct prerequisite for the discussion of any questions concerning potential incorporation strategies and, at the same time, a question of the institutionalization not just of the gathering and financing of information, but also the public discussion necessary for political decision-making.

(13) Institutionalization aimed at improving the availability of information and the diffusion of this information will need to be based on the following considerations:

- in Germany a sufficiently large scientific potential should be created and maintained for the analysis and observation of the external effects associated with energy consumption, with the indicated focal areas of activity.
- This scientific potential should not be centrally organized.
- The results of different studies should be evaluated regularly by a body not responsible for awarding funds, but capable of gaining public attention, a "public perception committee", and then be submitted, for example, to the cabinet or parliament where reply would be mandatory.

In the study possible organizational structures are suggested for satisfying these requirements.

The Incorporation Problem

(14) The primary objective of any incorporation strategy must be that of avoiding external effects to the largest possible extent, as soon as there is a risk that these effects could become intolerably large or constitute a threat to the basis of life on earth.

The second part of the study which deals with incorporation instruments shows that a large number of incorporation options are available to society and the political institutions. The **joint effect** of four different policy elements should be emphasized in this context:

- economically oriented strategies
- liability-related regulations
- social conflict strategies
- changes in human life styles.

(15) In public debate on these issues the first two elements indicated are the primary focus of interest due to the fact that their structural characteristics are easy to understand. As was shown, both are based in a very similar manner on economic incentive or penalization systems. The fact that they must always be integrated in a consciousness-raising and learning process which will probably involve our entire system of values is clearly stressed in the study.

In weighing the advantages and disadvantages of the different economic instruments what is important in the final analysis is whether

- external damage is to be "merely" mitigated,
- uncertainties exist with regard to the occurrence and extent of damage or
- damage is to be definitely prevented.

Standards or certificates reliably satisfy the last requirement, while in connection with the imposition of taxes it is (initially) doubtful whether the desired effect, e.g. the use of certain energy-saving technologies to achieve a reduction in emissions, will actually occur. Taxes move very slowly towards the achievement of the goal in question - at least as long as radical increases are not involved. In other words, there is a considerable time factor to be taken into consideration.

(16) To defend against the external effects of "classical" pollutants the set of instruments constituted by standards has been used with considerable success thus far. Although this instrument shows disadvantages with regard to optimization of economic efficiency, it should not be given up easily in the areas in which it has already been introduced and is successful.

(17) In connection with the effects of a wide range of different heavy metal and hydrocarbon compounds referred to above as "diffuse effects" and taking into account the still largely unclarified synergetic and accumulative processes, a standards-approach as well as a certificates-approach would not be very effective. A key prerequisite is missing for the use of these instruments, i.e. awareness of pollutant-specific quantities. Here, only taxes will

contribute towards better goal attainment, i.e. reducing **potential** effects on a **broad front.** The diffuse character of the large number of potential effects makes it impossible and nonsensical to impose fuel-specific taxes. Here, the only available instrument is that of imposing a general tax on energy consumption.

(18) While the tax proposed here - ultimately fair in terms of the polluter-pays principle - relates to all areas of the energy sector equally, additional strategies will need to be developed in the **transport sector.**

- The majority of nitrous oxides, hydrocarbon compounds and carbon monoxide emissions are caused by the transport sector. The external costs are very high.

- In contrast to the other energy-consumption areas, in which there will be at least a trend towards autonomous technological progress in the direction of energy savings in the course of the next twenty years (e.g. modern heating installations, building insulation etc.), the transport sector continues to show an unbroken record of growth, a factor that will make this sector a very critical problem area in the future.

In view of the very considerable status value attached to ownership of luxury vehicles and the **extremely high level of value attached to unlimited mobility** it is more than doubtful that a moderate energy tax will show any effect at all in this sector. Here, very high taxes or regulations (e.g. limitation of average fleet consumption) would need to be introduced. Consideration should also be given to certificate systems (e.g. limitation of available fuel on the basis of commercially available fuel certificates). A mix involving all of the instrument packages would probably be the best approach.

(19) A **third** element in the incorporation strategy is that of fending off an impending **climate disaster.** The "availability" of individual economic incorporation instruments for certain types of damage described above would make a certificate system the best approach to take here, since there is a need to ensure reliably that damage will not occur. Complications do arise, however, as a result of the fact that the accumulation of climate-relevant gases in the atmosphere is a global process and for that reason **quantity limitations,** e.g. for CO_2 emissions, must be imposed **worldwide.**

(20) Despite the need to limit pollutant emissions worldwide, taxes can be imposed at the national level with a view to initiating an energy-saving process. The mix proposed by the European Community consisting of a CO_2 tax and a general energy tax is -- aside from the much too low level and the fact, viewed critically by the authors, that there is a strong focus on CO_2 alone (e.g. methane problem) -- the right approach, since a "general energy tax"

would also promote the necessary adjustment processes in the other damage categories indicated above.

(21) The external effects of **nuclear energy use** and here, in particular the "residual risk" of a major nuclear accident, extends beyond the limits of the classical economic incorporation debate. Since incorporation on the basis of liability-law or insurance-related strategies can be ruled out (non-insurability of a nuclear accident), this "incorporation" can ultimately only be brought about on the basis of risk limits and a public discourse regarding the accepted height of these limits.

(22) The example regarding nuclear energy use makes the limitations of the economic and liability-law-related incorporation efforts discussed thus far quite clear. There is a need to take into account that even these measures can only be taken if a sufficiently large political majority can be found for this and the measures involved are granted broad public acceptance. Incorporation measures will recurrently encounter the determined resistance of interest groups or collide with established power structures. If resistance of this kind cannot be overcome, corresponding initiatives will either fail to have an effect or simply not be undertaken.

(23) For this reason, the incorporation of the external costs associated with energy consumption is a process, that has to be started much earlier than the planning of specific economic or liability-law-related measures. Here, there is a direct link between the information and the incorporation problem above and beyond the questions referred to concerning quantification and calibration of individual measures.

(24) The incorporation measures to be taken in each case must necessarily be structured in such a manner that established (and accepted) attitudes towards energy consumption are strongly questioned. As such, incorporation is a **critical sociopolitical process**.

The more **established behavior patterns and prosperity objectives** in society prove to be unsustainable and to constitute a threat to life over the short or long term,

- the more painful the necessary interventions will be, and

- the greater the danger will be that potentially effective proposals for measures and changes will be watered down or torpedoed in connection with conflicting interests and, as a result, not implemented at all.

(25) As such, a strong integration of the potential external effects in the values, lifestyles and prosperity objectives of the general public are an indispensable prerequisite for successful

incorporation. Aside from specific incorporation measures, the political authorities will need to strengthen public understanding of the **inevitability of a change of course**. This will be required not just with regard to the general public and industry, but also in the ranks of government and public administration.

It cannot be ruled out that this may, in part, involve going to the limits of the controllability of pluralistically organized societies. The attempt to find sustainable solutions will never lead to a great design, but rather to a persevering trial-and-error process to a searching and learning process constantly confronted with new insights, contradictions and resistance.

The political will to engage in a permanent learning process of this kind is at least as important as the implementation of the proposals indicated above for improving the information base and for economic incorporation. They are always to be viewed "only" as initial steps in the right direction, not as a final solution to the problem constituted by external costs associated with energy use. The famous Polish renegade Kolakowski once said: "Where you are going is not so important, the road that takes you there is what counts." Transferred to our problem this means that the direction in which the government sector needs to go to reduce or to avoid external costs associated with energy use is known. Even if it is still uncertain how long the road may be, it is in any event the right thing to do to set out on this road here and now.

9. Measuring the External Costs of Fuel Cycles in Developing Countries[1]

A. Markandya
Harvard Institute for International Development
Cambridge MA 02138, USA

1 Estimating Environmental Damages

In estimating environmental damages of fuel cycles, one would ideally like to start with the basic relationship between the energy outputs and the environmental impacts, as shown in Figure 1. In doing this there may be one or more stages, as for example is the case with air pollutants, where the energy source generates emissions and these emissions in turn impact on the environment depending on how they are deposited spatially. The method of estimating the energy-environment relationship is referred to as the dose-response function, which can be fairly simple (linking the pollutant - eg SO_2 to the measured variable of interest - eg. no of work days lost work day); or very complex, with the dependent variable being the probability of losing a work day, as a function of age, occupation, pollution levels of more than one pollutant, relative humidity etc. The next step is to value the impacts, using price and market data and this stage can also range from the simple direct valuation to a very complex model that allows for all the inter-linkages between markets and changes in prices.

An alternative to the three stage valuation procedure is to go directly from the energy source or pollutants to a valuation of these pollutants. This can be done by looking at the actual loss in property values attributable to the pollutant (the hedonic price method), or by asking individuals what their willingness to pay (WTP) for a reduction in the pollutant or their WTA for an increase in the pollutant is. This method is referred to as the contingent valuation method (CVM). For details of the application of these methods see Braden and Kolstad (1991).

---

[1] This paper is based on ongoing research undertaken by the author as part of projects with the European Commission (DGXII), and with the World Bank (ESMAP). Financial support from both institutions is gratefully acknowledged, without of course implying the approval of either of these institutions with the contents of this paper. The author would also like to thank P. Barnes of the Oxford Institute of Energy Studies and J. Homer of the World Bank for useful comments; and B. Rhodes for research assistance.

Ideally this is how each energy source should have its external effects valued, with the precise choice of method depending on what is being valued and on what the underlying information base is. Note that the estimates obtained will be in terms of damages in $ per unit of pollutant, and should refer to the marginal damages. Converting back to damages per unit of energy can generally be done, and will be required if one is to calculate the Marginal Social Cost (or MSC).

Figure 1:   Site Specific Damage Estimation Paths

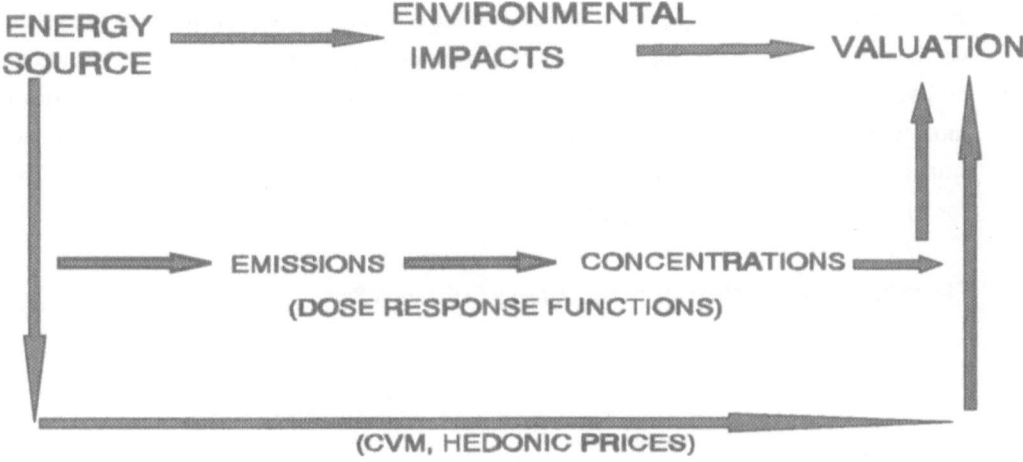

Unfortunately such a procedure is very cumbersome, and the estimates are very site specific. Damages cannot easily be transferred from one situation to another. The issue of transferability is discussed further below, but at this stage it is simply noted that in many cases the process of obtaining a damage per unit of energy or per unit of pollutant associated with that energy source starts from an overall value of damages that is based on much broader `macro' assumptions. The broad structure is illustrated in Figure 2. For example, a dose response model might have been used to estimate the total national health damages associated with SO_2. The exercise of interest would then consist of attributing the share of those damages that arise from a particular source - eg. coal fired electricity - and then calculating the damages per unit of emissions or per unit of energy from that source. Such a method inevitably provides only a rough estimate of the damages, and one that is based on average rather than marginal damages. If, as is normally the case, marginal damages are higher than average ones, the process will result in an underestimate of the desired marginal external cost (MEC).

In a large number of cases neither of the above procedures can be carried out and one has no estimate of the damage in monetary terms. This is typically the case when dealing with hydro power, where a lack of basic data often prevents monetary estimation taking place. It also applies to situations where the impact-valuation link is good but the energy-impact cannot be easily quantified. This is the case, for example, with noise impacts of energy. The data on the environmental costs of noise in $/dBA is generally good, but it cannot be translated into costs in terms of $/kWh of electricity generated. The attribution referred to above cannot be made with the existing data base.

It follows from this discussion that: (a) some of the estimates of the damages will be based on detailed site-specific data dealing directly with the energy source in question, (b) some will be based on an attribution of the damages from a wider estimate and (c) in some cases no monetary quantification will be available. The task for the analyst is to use the estimates from the first two sources as best s/he can, and identify and document those physical impacts for which no valuation has been possible.

Figure 2:   Broader "Macro" based Damage Estimation Paths

1.1   Transferability

Inevitably the information on environmental damages will never be available for exactly the areas and pollutants that are of interest. Hence numbers will have to be taken from other sources and transferred to the situation of interest. The transferability of estimates raises a number of issues which need to be addressed. The most obvious one is, what makes an estimate of damage transferable and what prevents it from being so? The list below illustrates the modifications that have to be made to the transferred information, in increasing order of complexity or difficulty in transference.

The order in which transferability becomes increasingly difficult or less valid is as follows:

(a) the most easily transferred data is the dose response function, relating environmental impacts adjusted for population. Thus numbers in the form $0.8 \times 10_{-6}$ excess deaths per $\mu g/M_3$ would be transferable across studies as long as adjustments to the other variables in the dose response function were made (eg. relative humidity). The additional local information that is required to use such data is simply local market conditions, costs and prices;

(b) the next ones in order of difficulty are monetary estimates of damages per unit of pollutant by concentration. Results are reported, eg, in $\$/\mu gM_3$, or in $\$/Km/person$ of lost visibility. Estimates may vary according to population affected, in which case an analysis of such variations would be desirable. Other socio-economic variables that could be of relevance are income level of the affected population, age, background environmental variables such as rainfall etc. If enough studies are available a `meta-analysis' can be performed, in which the mean estimated value is regressed against these variables. Then the relevant adjustment to the estimates are made, given the local values of the explanatory variables. No additional local variables should be required;

(c) similar to (b) above are estimates of monetary damages in terms of emissions or units of energy produced. In such cases one needs all the information listed above, plus details of how the emissions or energy units relate to the concentrations or whatever impacts are responsible for the damages. For example, damages may be quoted as $\$x/kWh$ for coal. The relevance of this estimate to a different situation will depend on how the kWh is related to emissions and how the emissions are converted into concentrations in the area where the impacts were measured, plus the variables with which the relationship between concentrations and damages vary. Thus most work will have to be done in these cases.

Transferability depends on being able to use a large body of data from different studies and estimating the systematic factors that would results in variations in the estimates. Unfortunately in most cases the range of studies available are relatively few. More can be done to carry out the meta analysis of the type indicated but it will take time. The best practice in the meantime is to use estimates from (a) in preference to (b), in preference to (c). Where even limited data on factors that could explain differences in estimates is available it should be used. That is what has been done in Section 3 of this paper, where national differences in international per capita income have been used to derive international differences in damage estimates.

1.2   Discount Rates

The choice of discount rate is of considerable importance in determining the relative costs of different technologies and different fuels. The higher the discount rate, the less will projects with high `up front' capital costs be preferred in comparison to investments which involve a sequence of investments over a period of time.   How this impacts on the environment, however, is unclear. A high discount rate would, for example act against expensive hydro projects, and favour a sequence of thermal plants. The environmental implications would depend on the relative environmental impacts of these two forms of energy.

A number of environmentalists have argued that high discount rates act against the environment, on the grounds that many conservation projects, eg. reforestation, have long gestation periods which result in low present values of benefits at high rates (see Markandya and Pearce (1991) for details).   While there is some merit in this argument, it is not completely valid. At least part of the problem in many `environmental' projects is that the future benefits are undervalued. If proper values were attached to them, the rate of return could be as high as that derived from more conventional projects. In particular, increasing values of environmental benefits as development takes place and real incomes rise have to be allowed for.

In terms of fossil fuels, a high discount rate also implies a smaller user cost (MUC) or depletion factor, which would tend to result in a faster depletion of the mineral resources of a country. Again, conservationists argue that a lower discount rate is environmentally desirable on these grounds.

In developing countries, the present World Bank practice is to use a real rate (ie adjusted for inflation) of 10% and over, reflecting the opportunity cost of capital in the country concerned. The justification is that if the marginal projects can achieve these returns, it should be a requirement for any public sector projects financed by the Bank as well. Unfortunately the discount rate is being made to play two roles here: that of valuing present versus future benefits and costs, and that of allocating a scarce amount of capital at the present time. There is something to be said for selecting the desired mix of projects, and depletion of natural resources, on the basis of a socially desirable discount rate, and then applying a capital rationing device to allocate scarce funds. These issues have been discussed more fully in Markandya and Pearce, (1989), in which the question of what is a `socially desirable discount rate' is addressed. It turns out that for most developing countries looked at, this rate would be around 5-7%.

For the purposes of this analysis, it is suggested that two rates be examined: a conventional rate at which the Bank would normally appraise projects in the country concerned, and a social rate which, for the sake of argument could be taken as 6%. The difference between the two in terms of energy strategy should be informative and revealing.

2   External Costs of Air Pollutants Generated by Fossil Fuels

Following the methodology set out in the previous sections, it is possible to estimate the damages arising from the use of fossil fuels. There are two major studies in which the direct environmental costs have been reported on a $/kg basis for the main air pollutants (SO_2, NO_x, and Particulates). One is the PACE study carried out in the US, and based on a review of the damage estimates from a large number of studies, also largely in the US. The other study is the one carried out for Poland, (Coopers, 1990), where a large amount of evidence was reviewed to arrive at a set of estimates of damages per kg for all the emissions listed above. In this section the results from these studies are summarised and their transferability to other countries discussed.

2.1   The PACE Study and Results From Other Developed Country Studies

The four categories under which estimates were obtained for damages from electricity generation were health (morbidity and mortality), materials, crops and visibility. There are other areas of potential damages, namely ecosystems and historical monuments, and other pollutants, particularly acid rain, but no specific numbers were reported for these on grounds of a lack of coherent data. Details of the studies on which the estimates are based can be found in PACE, 1991. In the figures reported in Table 1 below, the final summary figures from PACE have been modified somewhat, for reasons that are made clear in the discussion below. All damages are given in 1991 US dollars.

2.2   Health Costs

The health costs refer to the costs arising from the increased risk of death (mortality costs) and from the increased expenditures plus pain and suffering resulting from illness (morbidity costs). In neither case is it assumed that increased risks of death or illness among the workers employed in the generation of the fossil fuels (including the mining and transportation) are relevant. The argument here is that such costs are internalised in the wages paid to such workers, a view that is generally supported by the evidence. In fact a large part of the valuation of mortality costs is based on exploiting exactly such a relationship, ie. estimating the willingness to accept (WTA) a higher risk of death from the differential in wages between workers with different risk levels. These and other studies carried out in the US are reported in terms of a `value of life'. For example, if a worker needs an increase in salary of $30 per month to accept a risk of death at work that is 1/100000 higher than in a safe profession (one in a hundred thousand people will, on average, die every month from work-related risks in that occupation), the value of life is reported as $30x100000, which equals $3mn[2]. Other methods of estimating mortality costs include a questionnaire approach, in which individuals are asked what their willingness to accept an increase in risk of death in a well defined situation is. Finally there is evidence on the willingness to pay (WTP) to avoid risk by studying the behaviour of individuals who buy safer products at a higher price (eg. air bags in cars). The results from these studies have been summarised in PACE and Fisher et al (1989) and they appear to centre around a value of $4.3mn for the value of a statistical life in the US.

Table 1:   US Environmental Costs of Major Air Pollutants: $1991/kg

	SO$_2$	NO$_x$/Ozone	TSP
Health:			
Morbidity	4.084	0.807	0.784
Mortality	0.119	0.689	0.071
TOTAL	4.203	1.496	0.855
Materials	0.540	0.135	0.000
Agriculture	0.012	0.024	0.000
Visibility	0.332	0.404	1.971
TOTAL	5.087	2.059	2.826

Source: PACE, 1991, and own calculations (See text).

---

[2] This approach based on WTA is quite different from a valuation based on the loss of earnings, which was used in earlier studies. It is now accepted that the latter is not appropriate to valuing the risk of death.

The main concern about using this estimate for valuing increased risk of death from energy use is that it is largely derived from an analysis of the WTA a increased risk when that risk is voluntarily borne. Evidence from survey and market data suggests that involuntary risk may be valued anything from 10 to 100 times as highly (Starr, (1976), Litai, (1980)). Hence the valuation of $4mn can be regarded as a lower bound.

As with mortality, morbidity estimates are concerned to value the WTP to reduce the risk of illness. These include the costs of the illness in terms of hospital costs plus loss of earnings but are not limited to them. In addition there are expenditures incurred to avoid the illness, and a sum to account for general pain and suffering, both of the individual and of others affected by a person's illness. Estimates of these are harder to come by, but it is becoming clear that a simple costs of illness approach is not satisfactory and that, ideally, what is required is an estimate of the WTP to reduce the risk of illness in much the same way as was done for mortality. The PACE estimate in Table 1 is based on an arbitrary but slightly high, cost of $430,000 for a `morbidity unit'[3], but needs much more work.

Once the mortality and morbidity costs reported above have been obtained, the PACE report calculates the environmental costs of different pollutants on the basis of a relationship between concentrations of the pollutants and expected increases in mortality and morbidity, as evidenced in various studies. The estimated damages in terms of expected deaths and illnesses are then multiplied by the cost figures given above, to obtain an estimate of the cost of pollutant concentrations. These are then converted in $/kg of emissions by dividing the total damages by the amounts of emissions. For example, one of the base line data points used in the PACE study is the ECO report on a coal fired power plant in the North West of the US close to a large city. It is estimated from basic dose response relationships that an excess of 20.7 deaths and 5.9 morbidity units. Multiplying these by the value of life, and costs of a morbidity in 1991 prices gives a total figure of $91.7mn. At the same time the estimated total emissions of 21.82mn kg, which gives a cost per kg of $4.203, as reported in Table 1.

It is clear that the estimates are dependent on the particular plant and its site characteristics. Hence it would not be appropriate to use these figures to value the damages from a particular plant. However, it is reasonable to use them to obtain average or indicative values for the benefits of reducing particular pollutants from an unspecified plant. If possible, one should use the dose response functions directly on concentrations data, and then apply local costs of

---

[3] A morbidity unit is defined as 600 days of illness, which gives a value of $718 per illness day, which is on the high side, compared to other US studies, eg. Cropper and Freeman (1991).

mortality and morbidity, but where this cannot be done, as a first step one may use the emissions estimates.

2.3 Agricultural and Forest Costs

Estimates of crop damages are well surveyed, and in general quite a high level of reliance can be placed on them. In general the view is that acid deposition, ozone, NO_2, and SO_2 have impacts on crops[4]. It will be noted, however, that no estimates for the damages associated with long distance depositions of sulphates and nitrates (commonly known as acid rain) are provided. Although there are some estimates of such damages, the PACE study concluded that they could not be used to derive a meaningful generic figure. The same study further concluded that there were no significant damages from SO_2. However, that is not the view taken in this report. A number of dose response studies have shown such damages to be relevant (van Eerden et al (1987), Mendelsohn (1980)), and there are several estimates of damages per kg available. In order to keep the figures in Table 1 exclusively from the US (for reasons that will be clear later), the central estimate from the PACE study has been replaced by the figures from the Mendelsohn study[5].

The estimated damages range in the sophistication with which they treat market behaviour. When levels of pollution rise producers will shift away from crops that are more seriously affected, and will use inputs such as fertilizer to limit the damage caused. With changes in pollution levels, one would expect the relative output of different crops to change, resulting in new output prices, as well as changes in other inputs. Simply multiplying the change in yield by the current price and subtracting the current costs of production can therefore result in an underestimate of the benefits of the shift. However, in the light of all the other uncertainties involved this one is not likely to seriously mislead the calculation. Hence it is recommended that where concentrations data are available, direct estimates of crop losses should be made and local information of prices and costs used. But as a first approximation, estimates of damages per kg of emissions may be employed.

Forest related damages are quite controversial. Most of them are related to acidic deposition and ozone. The NAPAP study for Canada and the US (NAPAP (1991)) has concluded that

---

[4] The figures of NOx and ozone cannot be separated. It is difficult to identify the separate impact of ozone from those of NOx and HC, since it is produced as a chemical reaction between these two substances.

[5] The EC/US Fuel cycle study has been reviewing the more recent literature on morbidity costs and has come up with updated figures. Unfortunately these cannot be reported at this stage as the studies are not yet completed.

forest damage is not a serious issue, with much of the observed impacts being a result of natural factors, and some small amount of damage being attributable to ozone. In contrast forest damages estimated for Europe by IIASA (IIASA (1991)) have been very large, with as much as $30bn being accounted for by acid deposition for all of Europe. Even if one accepts the figures from IIASA, they are very specific to Europe, and cannot be transferred to other developing countries. Hence, for the purposes of this exercise they have been excluded. If the case study is to be carried out in a country where acid deposition is a major concern, the estimates of damages should be reviewed separately.

2.4 Materials Damage

Estimates of material damages are based on dose response functions applied to inventories of materials in the affected areas. Hence it is again necessary to model the link between concentrations and emissions. Damages levels in physical terms are a function of the concentrations but also of relative humidity. The PACE estimates range from some based on national data, to others which make use of the ECO model of a specific power station referred to above. Since the specific model will provide results that are dependent on the special conditions prevailing in the neighbourhood of that station, it is preferable for the purposes of this study to make more use of national level estimates. The three US studies quoted in PACE cite figures of between $0.28/kg to $0.81/kg of SO_2. Independent studies in Europe yield estimates of $0.31/kg (Denmark), $1.27/kg (Netherlands), and $2.91/kg (Germany) (See Pearce et al (1992)). The higher European figures would indicate that the US estimates may be on the low side. It is accepted that these estimates include the impacts of acid rain, which cannot be distinguished from those of SO_2, but that is also true of the US estimates. On the basis of these figures, the recommended value of $0.285/kg of SO_2 in PACE seems to be too low. Instead it is suggested that the average of the US national level studies be taken (ie. 0.540/kg of SO_2). The reason for taking the US figure is that all the other estimates in Table 3.1 are based on that data and it would be better to employ a set of figures reflecting the particular biases of that one country.

As far as NO_x and ozone is concerned, the US national level estimates range indicate that damages from these sources are approximately 20% of total damages, the other 80% being due to SO_2 and particulate. Since the best guess of direct particulate damage on materials is close to zero, the corresponding central estimate for NO_x/ozone damages would be 0.135$/kg.

As with crop damages, the estimates quoted vary in the way they treat avertive measures and actions. For example, damages can be reduced if more resistant building materials are used. Estimates of damages taken from areas where such materials are employed will show low coefficients on the dose response functions. In an area where less robust materials are employed, the damage could be greater. Ideally one should use the dose response functions on concentrations data, with separate functions being employed for steel and zinc sheets, metals, concrete, stone and paint. However, as an order of magnitude these figures are probably acceptable.

2.5   Visibility

Data on the impacts of the environmental damages caused by restricted visibility are based on studies of changes in property prices in areas of low visibility (the hedonic method, (Markandya and Pearce 1989)) and on the questionnaire approach where individuals are asked what their WTP for an improvement in visibility is. The PACE report reviews various studies and concludes that the average WTP for improved visibility is around $11.7 per lost km of visibility per person per year. (The range around this estimate is $1.2 to $108, in 1991 prices). It is this set of estimates that is the most transferable and the PACE study has allocated the damages from a typical plant to the three pollutants that impact to visibility to arrive at the figures quoted in Table 1.

Unfortunately the estimates obtained from the US are not supported in the one study completed outside the country. This was completed in Norway (Hylland and Strand (1983)) and showed that of the WTP for improved visibility was to a large extent a response to the WTP for the improved health conditions that would follow. There is certainly a difficulty in separating out the benefits associated with improved visibility from the other impacts of air pollution.

2.6   The Polish Study Results

The only less developed country for which damages data are available is Poland. The Coopers study quotes a number of national level estimates, from which it draws a selected range of damages per kg of pollutant. In Table 2 central estimates from the various Polish studies are presented. These do not correspond exactly to the Coopers figures, with reasons for varying from their conclusions being given below. National damages estimates do not break down the damages by pollutant, but one can apply relative factors from other studies to arrive at a decomposition of the damages. The Coopers study took an arbitrary relative weighting based on a priori judgments, but in Table 2 the relative weights have been taken from this report's review of the US studies -ie. from Table 1.

Table 2:   Polish Environmental Costs of Major Air Pollutants: $1991/kg

	SO_2	NO_x/Ozone	TSP
Health	0.308	0.110	0.063
Materials	0.107	0.027	0.000
Agriculture	0.027	0.054	0.000
Visibility	n.a.	n.a.	n.a.
TOTAL	0.442	0.191	0.063

2.7   Health

The health figures reported in Table 2 are based on a health damage estimate of $450mn, taken from Symonowicz (1990) and reported in Coopers. This estimate is based on costs of illness and loss of earning alone and is therefore likely to be an underestimate[6]. The resulting figure is multiplied by a further factor of 2.5 to account of the difference between domestic and border prices of related goods (see Coopers, (1991) and US Office of Energy, (1990)). The resulting figure gives the costs per kg of emissions as reported above.

---

[6]   The extent to which it is an underestimate for that reason, however is not clear. Certainly morbidity costs are undervalued but they are a very small part of the total costs. Mortality costs are also undervalued when using a productivity approach, but by how much in a less developed country context is not clear.

2.8   Materials

The materials estimates emerging from the Polish studies have been rightly dismissed as being too high by Coopers, who have assumed a damage level equal to 0.8% of GDP. However, even this figure would appear to be much too high. The German studies quoted above, which are among the highest, yield estimates of damages of around 0.2% of GDP. In view of this, the estimate for materials damage has been scaled down by a factor of 4 from the Coopers estimates. This yields an overall figure of $375mn (after adjusting for the gap between border and domestic prices) and yields the damages per kg emitted that are quoted.

2.9   Agriculture

The damage estimates are based on a total damage figure of $400mn, as obtained from Symonowicz (1990) and used in the Coopers Report. In addition there is an estimate of damages to forests of $225mn, which has not been included, on the grounds that these are mainly the result of acid rain effects, which are not being excluded in so far as it is possible to do so.

2.10   Comparison Between the Polish and US Estimates

The following points emerge from the comparison between damages per kg in the two countries:

(a)   the US health costs are 13.6 times higher than those of Poland.

(b)   damages to materials are 5.1 times higher than in the US;

(c)   the agricultural damages are higher in Poland than in the US by a factor of 2.3;

(d)   no damages are reported for visibility in Poland.

The higher health costs in the US are what one would expect, given the higher real costs of medical treatment, and the fact that per-capita income is greater. In 1987 (the year of the estimates), income per capita in Poland was estimated at $1930 and that for the US was estimated at $18530. In terms of an `elasticity' of damage costs with respect to per capita income, this yields a figure of 1.15, which would not appear to be unreasonable. The damages figures for materials in Poland are based on the German ratio of damages to GDP to the European ones, and imply an elasticity of around 0.72. In that case, however, the Polish figures may well be on the high side. The agricultural damages being lower in Poland simply reflects the fact that the US figures are an underestimate of national level damages. They are based on the Mendelsohn study which looked at an area with little affected agriculture. Hence in this case the Polish figures are more likely to be closer to the correct ones. Finally the lack of concern with visibility in Poland is not surprising, given that there are more urgent items of expenditure that people would place higher on their list of priorities.

2.11 Implications for Transferring Estimates to Other Countries

The above analysis suggests that, as far as health damages are concerned , an elasticity of 1.15 with respect to per capita income would not be unreasonable. This would give the figures for health damages per kg of pollutant emitted as a function of per-capita income presented in Table 3. For materials damages, the elasticity that emerges from the comparison appears to be too low, because the Polish estimates are probably overvalued. Nevertheless as it is the best guess that is available, it has been used to report the figures for material damages as a function of per-capita income given in Table 3. Finally there are the agricultural damage estimates. As stated above, the Polish figures are probably closer to the correct ones for countries at that level of development. As real incomes rise, value added in agriculture should rise, resulting in larger losses. In the absence of any estimated elasticity it is proposed that a value of one be taken for that parameter. This gives the figures cited in Table 3.

For the reasons given earlier, no external damage is associated to loss of visibility. The other main item that has been excluded is that of forest damages. Where the analysis is being carried out in a country with significant forest areas that are likely to suffer from acid deposition, it is suggested that a separate damage estimate be calculated for that item. As an indication, one might use the figures from Poland, although the actual values will depend very much on the prevailing meteorology.

The damages cited above would apply for the current year in the case of a country with a given level of per capita income. It is important, however, to increase this figure over time to reflect an increasing value of environmental improvements. This can be done by simply using values corresponding to higher levels of per capita income. Values between points in the table can either be interpolated, or calculated using the appropriate formula.

Finally there is the question of uncertainty. As a guide to how much the estimates matter, a sensitivity analysis can be carried out using damage estimates that are one half of those given,

Table 3:   Estimated Damages from Air Pollutants by Level of Per Capita Income

PER CAPITA INCOME ($)	HEALTH DAMAGES $1991/KG			MATERIALS DAMAGES $1991/KG		AGRIC. DAMAGES $1991/KG		TOTAL BY POLLUTANT $1991/KG		
	SO2	NOX	TSP	SO2	NOX	SO2	NOX	SO2	NOX	TSP
300	0.036	0.013	0.007	0.028	0.007	0.004	0.008	0.068	0.028	0.007
400	0.050	0.018	0.010	0.035	0.009	0.006	0.011	0.090	0.038	0.010
500	0.065	0.023	0.013	0.041	0.010	0.007	0.014	0.112	0.047	0.013
600	0.080	0.029	0.016	0.046	0.012	0.008	0.017	0.135	0.057	0.016
700	0.096	0.034	0.020	0.052	0.013	0.010	0.020	0.157	0.067	0.020
800	0.111	0.040	0.023	0.057	0.014	0.011	0.022	0.180	0.076	0.023
900	0.128	0.046	0.026	0.062	0.015	0.013	0.025	0.202	0.086	0.026
1000	0.144	0.052	0.029	0.067	0.017	0.014	0.028	0.225	0.096	0.029
1200	0.178	0.064	0.036	0.076	0.019	0.017	0.034	0.271	0.116	0.036
1400	0.213	0.076	0.043	0.085	0.021	0.020	0.039	0.317	0.136	0.043
1600	0.248	0.089	0.051	0.094	0.023	0.022	0.045	0.364	0.157	0.051
1800	0.284	0.101	0.058	0.102	0.025	0.025	0.050	0.411	0.177	0.058
2000	0.321	0.115	0.066	0.110	0.027	0.028	0.056	0.459	0.198	0.066
2400	0.396	0.141	0.081	0.125	0.031	0.034	0.067	0.555	0.240	0.081
2800	0.473	0.169	0.097	0.140	0.035	0.039	0.078	0.652	0.282	0.097
3200	0.552	0.197	0.113	0.154	0.038	0.045	0.090	0.750	0.325	0.113
3600	0.632	0.226	0.129	0.167	0.042	0.050	0.101	0.850	0.368	0.129
4000	0.714	0.255	0.146	0.180	0.045	0.056	0.112	0.950	0.412	0.146
4500	0.818	0.292	0.167	0.196	0.049	0.063	0.126	1.077	0.467	0.167
5000	0.924	0.330	0.189	0.211	0.053	0.070	0.140	1.205	0.523	0.189
5500	1.031	0.368	0.211	0.226	0.057	0.077	0.154	1.335	0.579	0.211
6000	1.140	0.407	0.233	0.241	0.060	0.084	0.168	1.465	0.635	0.233
6500	1.251	0.447	0.256	0.255	0.064	0.091	0.182	1.597	0.692	0.256
7000	1.362	0.487	0.279	0.269	0.067	0.098	0.196	1.729	0.750	0.279

Sources: See text

Notes: 1 Health figures are based on an elasticity of 1.15
             2 Materials figures are based on an elasticity of 0.72
             3 Agricultural figures are based on an elasticity of 1.00

131

and twice those given in Table 3. There is no scientific basis for this, but it does provide an indication of the robustness of the results.

References

J. Braden, and C. Kolstad. 1991. *Measuring the Demand for Environmental Quality*, North-Holland-Elsevier, Amsterdam.

Coopers & Lybrand Deloitte. 1991. *Environmental Assessment of the Gas Development Plan for Poland*, The World Bank, Washington DC.

M.L. Cropper and R. Freeman. 1991. *Environmental Health Effects* in J. Braden, and C. Kolstad (op. cit.).

A. Fisher, L.G. Chestnut and D.M. Violette. 1989. `The Value of Reducing Risks of Death', Journal of Environmental Economics and Management, 8,1, pp 88-100.

A. Hylland and J. Strand. 1983. `Valuation of Reduced Air Pollution in the Greenland Area' Working Paper 12-83, Department of Economics, University of Oslo, Norway.

IIASA. 1991. *European Forest Decline: The Effects of Air Pollutants and Suggested Remedial Policies*, Laxenberg, Austria.

D. Litai. 1980. A Risk Comparison Methodology for the Assessment of Acceptable Risk, PhD Thesis, MIT, Cambridge, MA.

A. Markandya and D.W. Pearce. 1987. `Environmental Considerations and the Choice of the Discount Rate in Developing Countries', The World Bank, Environment Department, Washington DC.

A. Markandya and D.W. Pearce. 1989. *Environmental Policy Benefits: Monetary Valuation*, OECD, Paris.

A. Markandya and D.W. Pearce. 1989. `Development, Environment and the Rate of Discount', World Bank Research Observer, Washington DC.

R. Mendelsohn. 1980. `An Economic Analysis of Air Pollution from Coal Fired Power Plants', Journal of Environmental Economics and Management, 7, pp 30-43.

NAPAP. 1991. (US National Acid Precipitation Assessment Program), `Acid Deposition: State of Science and Technology', Washington DC.

PACE [1991], *Environmental Costs of Electricity*, Oceana Publications inc., New York.

D.W. Pearce et al. 1992. `The Social Costs of Fuel Cycles', Report for the UK Department of Energy, London, UK.

C. Starr. 1976. `General Philosophy of Risk-Benefit Analysis', in H. Ashley et al eds., *Energy and the Environment: A Risk Benefit Approach*, Pergamon, Oxford.

J. Symonowicz [1990], See Coopers & Lybrand Deloitte (op. cit.)

US OFFICE OF ENERGY [1990], `Poland: An Energy and Environmental Overview', Office of Energy, USAID, Washington DC.

van der Eerden et al [1987], `Economische Schade Door Luchtverontreiniging aan de Gewasteelt in Nederland', Publikatiereeks 65, Ministry of Public Housing, Physical Planning and Environmental Management, Leidshendam, Netherlands.

10. Evaluation of the External Costs of a UK Coal Fired Power Station on Agricultural Crops

M.R. Holland and N.J. Eyre
Energy Technology Support Unit (ETSU), B156, Harwell Laboratory
Oxfordshire OX11 ORA, UK

Introduction

The work described in this paper forms part of the collaborative project between Directorate General XII of the Commission of the European Community and the United States Department of Energy to assess the external costs of fuel cycles. A partial evaluation of the damages to agricultural crops arising from the operation of a new coal fired power station is presented to demonstrate the project methodology. This involves the use of a damage function approach to allow impacts to be expressed on a marginal basis, suitable for use in energy planning analysis. The methodology is thus different to that used in other studies in which externalities have been calculated from estimates of national damages and aggregated measures of polluting activities[1] [2].

A consequence of adopting this approach is that many of the results will be site specific. Special care is thus required, particularly in the following areas:

1. Definition of the reference technologies and locations;
2. Definition of reference environments;
3. Choice of models of atmospheric transport and chemistry;
4. Choice of dose - response functions.

The concept of the reference environment for this study is complex. It should cover the full area which may be affected by fuel cycle activities. In the case of fossil fuel cycles this takes in the entire planet as a consequence of CO_2 emissions and the resulting increase in global warming potential. For other impacts the area of interest can be much smaller: occupational health effects on miners are clearly restricted to those working, or who have worked, in the mine; impacts of power station construction on biodiversity may only affect the area on which the plant and associated infrastructure are built. Long range transport of pollutants poses a particular problem in this study, as the amount of material carried to a point several hundred kilometres from the plant under analysis, and hence the likely impact, will both be small when considered on an individual site basis. However, they may be important when summed over the total area affected, and when added to other apparently trivial damages. Artificial truncation of the reference environment is thus to be avoided in order to reduce error.

There are few problems with valuation of the impacts described in this paper because they concern changes in the annual production of a marketable commodity. Long term changes caused by the effects of pollution on agriculture are not expected because farmers can take mitigating action, such as increasing soil liming rates (which should, accordingly, be included in damage estimation). Valuation of damages to forests and other ecosystems is more difficult as their benefits are often not marketable and impacts may persist for many years.

This paper presents the methodology used to estimate the impacts of the direct effects of SO_2 and O_3 produced by a single power station in the UK on agricultural crops. O_3 is well established as a serious pollutant for agricultural crops. The results of the NCLAN project suggested that associated annual losses are about \$3 billion in the USA[3]. The effect of SO_2 is less certain. Recent reductions in levels in western Europe, and the virtual elimination of point source problems have reduced atmospheric concentrations to the point where direct effects may be negligible. The report by UK TERG[4] concluded that:

"Major agricultural crops in the UK are unlikely to be damaged directly by current rural concentrations of SO_2 and NO_2. Recent evidence suggests that interactions between pollutant stresses and other stresses such as pests, may be extremely important in influencing crop yields."

A similar conclusion was reached by NAPAP for the USA[5].

However, van der Eerden et al.[6] estimated that air pollution reduced total crop volume in the Netherlands by 5%, 3.4% of which was caused by O_3, 1.2% by SO_2 and 0.4% by HF. This equated to damage costing roughly \$77 million for the direct effects of SO_2. This study did not include any estimation of interactions between different pollutants, climate or pests, which would almost certainly have increased estimates further. If it is assumed that van der Eerden's calculations were correct, it seems reasonable to consider the figure of \$77 million as the lower bound for the economic impact from SO_2 induced crop damage in the Netherlands each year.

Methods and Sources of Data

Definition of the Impact Analysed

An impact pathway has been described (figure 1) to show the progression of effect from power station emissions to economic damage. The pathway is designed to be as comprehensive as possible; all known impacts are included, whether they are quantifiable given current knowledge or not. Depiction of pathways in this manner enables the analyst to place those effects that can be estimated into context with those that cannot. It also allows the identification of subjects that require further research.

The analysis presented here is only concerned with the direct and separate impacts of O_3 and SO_2 on yield of wheat and barley. The sequence of processes required in the evaluation of the direct impacts of these pollutants is shown in figure 2. The following effects are not included in this example:

1)   Impacts on other crop species;
2)   Interactions between pollutants and pests;
3)   Effects of acidic deposition to soils;
4)   Interactions between pollutants and climate;
5)   Interactions between pollutants, including NO_x;
6)   Interactions between pollutants and pathogens.

The first 3 of these are being assessed within the project. Impacts 4, 5 and 6 will not be studied further because insufficient data are available in the literature at the present time. This does not reflect on the potential importance of these impacts.

Description of the Power Station

The West Burton 'B' coal-fired power station, situated by the River Trent in the East Midlands of England was selected[7]. It is the only unbuilt coal burning plant for which planning permission currently exists in the UK. The design is in accordance with the Large Combustion Plant Directive of the European Community. It would be fitted with pulverised fuel boilers, 99.7% effective electrostatic precipitators, 90% effective flue gas desulphurization plant and low NO_x burners. Capacity would be 1800 MW gross, with 1710 MW sent out, based on 2 sub-critical 900 MW units. It would have a load factor of 76% and a projected life-span of 40 years. The thermal efficiency of the plant is estimated at 38%.

Figure 1: Impact pathway for the effects of photo-oxidants and acidic deposition on crops. Effects are capable of extensive interaction with one another. There are also numerous points at which feedbacks occur. To improve clarity and establish some degree of uniformity between pathways for different receptors, the pathway has been split into 7 pre-defined stages from (I), burden imposed by a given fuel cycle activity to (VII), economic valuation.

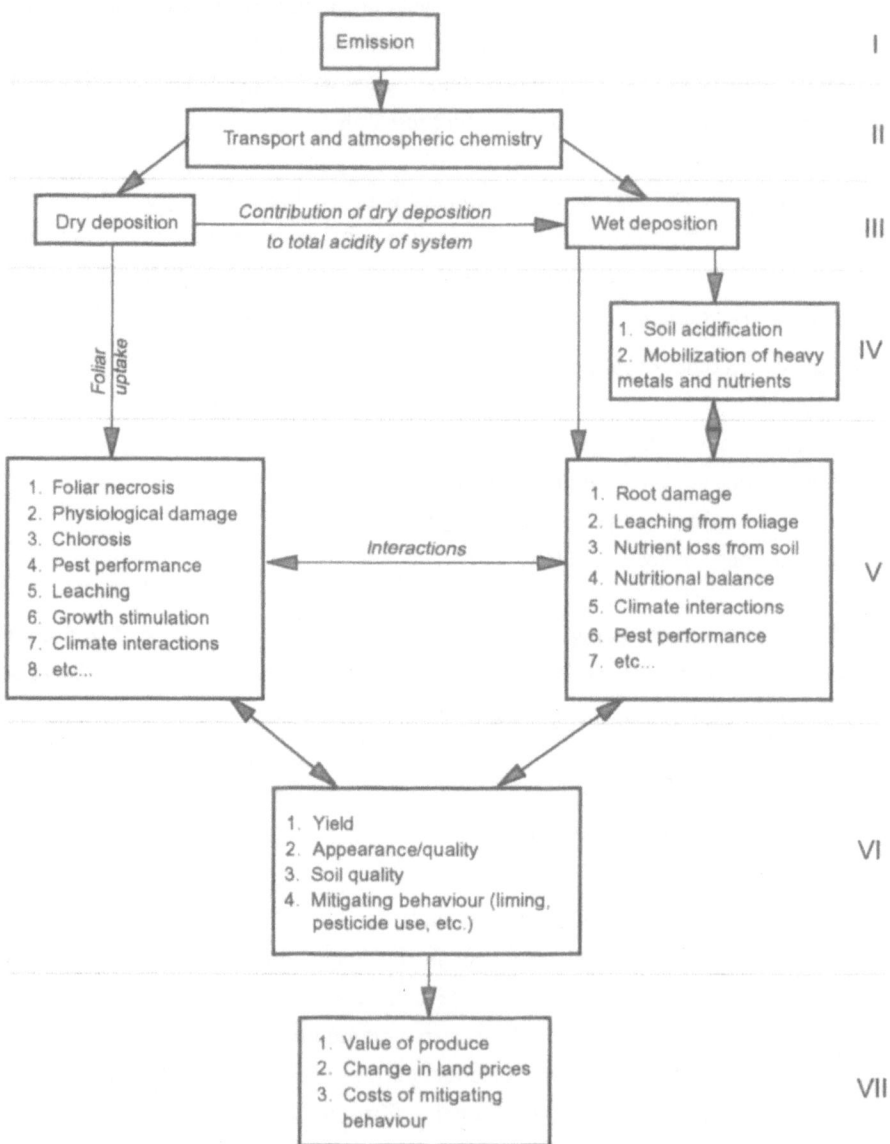

Figure 2: **The sequence of processes required during the analysis of the direct impacts of ozone and sulphur dioxide on crops.**

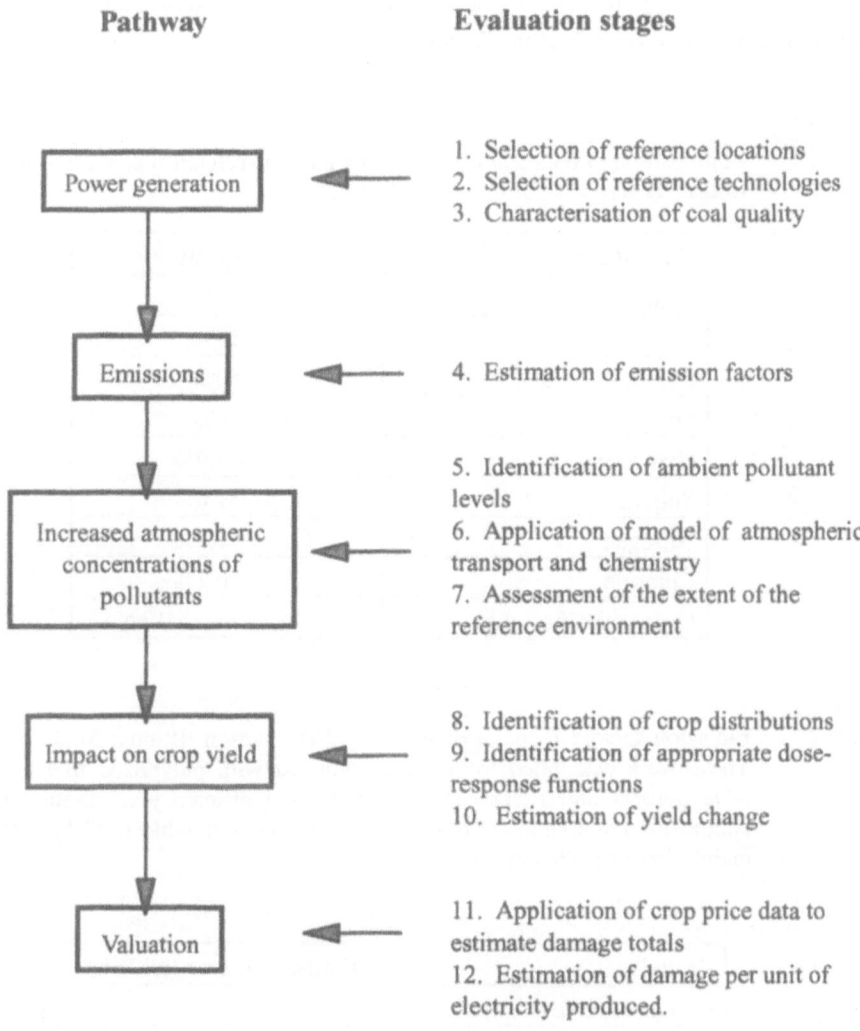

Emissions from coal fired power stations are dependent not only on the technologies employed, but also on the composition of the coal that they use. It is assumed that West Burton 'B' would burn locally mined coal. The underground mine at Maltby in South Yorkshire, 45 km away, was selected as representative of mines that could be used by West Burton 'B'. Typical figures for the composition of coal from this area are shown in table 1. Emission factors for West Burton 'B' were calculated from these data and are shown in table 2.

Table 1:   Composition of coal from the area in which the reference coal mine for the West Burton 'B' power station is located.

Constituent	% composition
Water	12%
Ash	15%
Carbon	60%
Oxygen	6%
Hydrogen	3.9%
Sulphur	1.6%
Nitrogen	1.3%
Chlorine	0.2%
Gross calorific value	24.5 MJ/kg

Table 2:   Emission factors associated with the West Burton 'B' coal-fired power plant. These were calculated for a plant equipped with pulverised fuel boilers, low NO_x burners and FGD and with a thermal efficiency of about 38%. The composition of coal in the reference area is shown in table 1. CH_4 emissions are mainly from the coal mine.

Emission	Emission factor (g/kWh)
Carbon dioxide (CO_2)	880
Methane (CH_4)	2.9
Nitrous oxide (N_2O)	0.06
Particulates	0.16
Sulphur dioxide (SO_2)	1.1
Oxides of nitrogen (NO_x)	2.2

Modelling Atmospheric Transport and Chemistry

Atmospheric transport and chemistry of SO_2 released from West Burton 'B' were modelled using a receptor orientated Lagrangian model[8][9]. This type of model is better suited to estimating the fate of emissions from power station stacks than either Eulerian grid or statistical type models[10]. Preliminary inspection of results demonstrated that the incremental increase in annual mean SO_2 concentrations caused by operation of West Burton 'B' was greatest in areas close to the plant, and was negligible (<0.01 ppb) at distances more than about 400 km away. At such distances most SO_2 has either been deposited to the ground or converted to sulphate. Our reference environment for direct impacts of SO_2 can thus be restricted to the UK.

The atmospheric chemistry of O_3 is highly complex. It may be formed or destroyed by a variety of reactions involving different precursors (principally NO_x and hydrocarbons). It has been shown that power station emissions are capable of reducing ambient O_3 levels for several hours after their release[11] in consequence of a high $NO:NO_2$ ratio within the plume. Net O_3 production will only commence once the appropriate conditions are attained within the plume. In extreme cases O_3 deficits are still present after more than 500 km of travel[12]. Results from a plume model have been used in this study (10) to estimate 7 hour (09:00 to 16:00 daily) mean values over the growing season. As the model is still being developed the results used here must be considered as preliminary and are subject to change. Inspection of the results suggested that the reference environment for impacts of O_3, like SO_2, could be restricted to the UK.

Description of the Reference Environment

A grid system was used to describe the distribution of wheat and barley crops within the reference environment. Inspection of the results of the models of atmospheric transport and chemistry suggested that it was appropriate to analyse impacts at 2 different scales for SO_2; within a 10 x 10 km grid at distances of up to 100km from the power station inside which SO_2 increments are greatest, and within a 100 x 100 km grid for the rest of the UK. The 100 x 100 km grid was used throughout for O_3. These grid systems are shown in figure 3.

The most authoritative data on crop distribution in the UK is collected by the Ministry of Agriculture, Fisheries and Food (MAFF[13]). However, this is not available in the format required for this work. The ITE (Institute of Terrestrial Ecology) Land Classification and Database[14] was used instead to provide the distribution of wheat and barley crops by area sown. The accuracy of data in the ITE database was checked against the results of the MAFF dataset and found to be in close agreement for the crops of interest to this paper.

Figure 3: Grid systems used to define the UK reference environment for the effects of sulphur dioxide and ozone on wheat and barley. Ozone effects were analysed at the full 100 x 100 km grid size throughout. Sulphur dioxide effects were estimated for a 10 x 10 km grid around the reference power station, and the 100 x 100 km grid for the rest of the country.

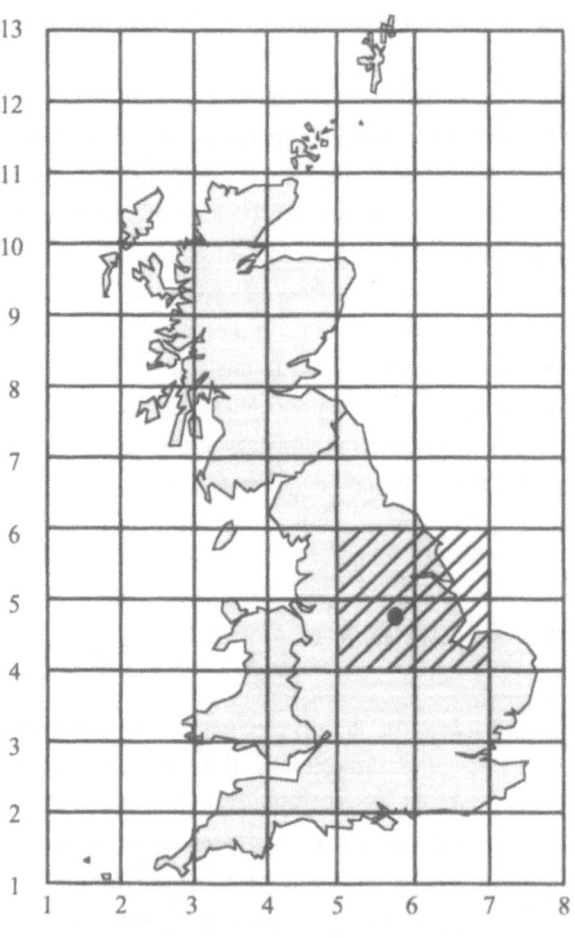

● Site of the West Burton 'B' power station

▨ Area for analysis in 10 x 10 km grid cells

The dose-response relationships to be used estimate % yield loss of crops in response to SO_2 and O_3. Yields were calculated by multiplying areas assigned to each crop within each grid cell by average yield for the UK over a 5 year period to account for variation in weather conditions between years (13).

Selection of Dose - Response Relationships

The NCLAN study in the US is the most extensive assessment so far of the sensitivity of crops to O_3. Response of a range of crops and varieties were found to be best described by a Weibull function (figure 4):

$$y = a \exp \{ - (x / s)^c \} \qquad (1)$$

in which y = observed yield,

a = hypothetical yield at 0 ppm O_3 (usually normalised to 1),

x = 7 hour (09:00 to 16:00) seasonal mean O_3 concentration (ppm),

s = O_3 concentration when y = 0.37a,

c = a dimensionless shape parameter.

For winter wheat NCLAN provided estimates for s and c of 137 and 2.34 respectively[15]. The figures for spring wheat were 186 and 3.20[15].

These functions were used because they are currently the best available to describe the impacts of O_3 on wheat. However, the results obtained must be treated with some care for the following reasons:

1) Conditions in the UK are very different to those in some of the areas in which the NCLAN study was performed. Variation in climate (heat and moisture) affects stomatal opening and will thus alter the rate at which a plant can take up a pollutant. Hence plants exposed to the same atmospheric concentration of O_3 may be subjected to different doses. Those experiencing more favourable conditions with temperatures closer to the optimum for growth and less water stress will be subjected to higher doses. On the other hand such plants may be better able to repair O_3 induced damage than more stressed individuals.

2) Varieties used in the NCLAN project are not grown in the UK. Differences in sensitivity between varieties were noted by NCLAN.

Figure 4:   The Weibull function used to describe the response of crop yield to ozone. The result for cotton (a highly sensitive species) is taken from the NCLAN study (15), and used to demonstrate the shape of the function. It should be noted that large differences in sensitivity were found between different crops by NCLAN.

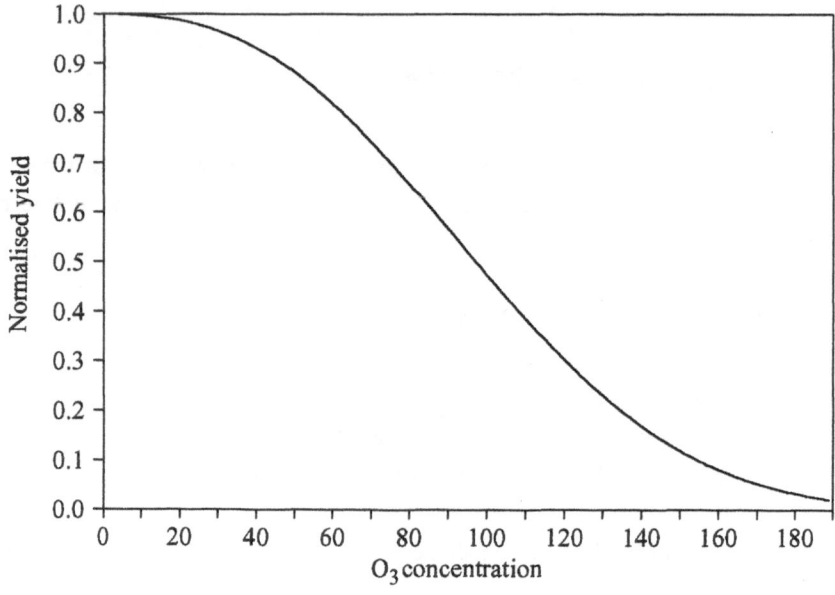

$$Yield = a \exp(-[x/s]^c)$$

x = ozone concentration
a = hypothetical maximum yield at zero ozone
s = ozone concentration when yield = 0.37 (max)
c = exponential loss function

Within NCLAN, work on barley (cv. Paco) in California suggested that this species is relatively insensitive to O_3 (3). However, other work has found barley to be more sensitive to O_3 than wheat[16]. For this assessment it has been assumed that the 2 species are equally sensitive, on the basis that they are quite closely related and that the harvested product (grain) is similar. Although there is uncertainty in the application of the NCLAN data to the European situation it is expected that the final results will be accurate to well within an order of magnitude. The parameters for winter and spring wheat were thus used to estimate high and low values respectively for yield loss of both wheat and barley. The mean of these 2 figures for each crop was taken as the best estimate.

Previous estimates of SO_2 induced yield loss (6) have relied on a function first proposed in the early 1980's:

$$\% \text{ Yield Loss } = \frac{100 \exp (y)}{1 + \exp (y)} \qquad (2)$$

Where $y = (3.8 \log (C - 5.8) - 9.2)$

and $\quad C$ = annual 7 hr (1000 to 1700) daily mean SO_2 concentration in μg m^{-3}.

This was derived from work on the pasture grass, *Lolium perenne*. The attraction of this species is that its sensitivity to SO_2 has been studied more extensively than that of any other crop. Van der Eerden *et al* adapted this function for use with other species through the introduction of what they admitted was a somewhat arbitrary sensitivity index, dependent on the manner in which each crop was used. Subsequent work[17] in which all data for this species were assessed, however, suggested that the original relationship provided a poor description of the response of *L. perenne*. Instead, a linear function that accounted for more of the observed variation in response was provided:

$$\% \text{ Yield Loss} = 2.75 - 0.18(z) \qquad (3)$$

Where z = annual mean SO_2 concentration, ppb.

This relationship estimates significantly smaller yield reductions for any given SO_2 level than equation 2.

For this case study the most appropriate dose - response relationship derived from work on cereal crops is that of Baker et al.[18]. SO_2 related yield loss was observed in a series of open air fumigation experiments on barley (cv. Igri) in the UK over 3 growing seasons. It was described by the equation:

$$\% \text{ Yield loss} = -0.69z + 9.35 \qquad (4)$$

Unfortunately neither equations 3 or 4 were calculated using data below 16 ppb. This corresponds with the highest rural levels observed in the UK. Accordingly, extrapolation was required to apply this function to areas where annual mean SO_2 levels are less. This is not straightforward. At low concentrations it is now well accepted that some growth stimulation occurs[19]. 2 approaches have been used to extrapolate the function to 0 ppb SO_2;

i) Direct extrapolation of functions to 0 ppb, assuming that no growth stimulation occurs. This provides a maximum estimate of crop loss from direct SO_2 effects.

ii) Extrapolation of functions to 0 ppb and 0% yield change, assuming a non-linear relationship and growth stimulation between 0 ppb and the SO_2 level at which equations 3 and 4 predict 0% yield change. Maximum yield was assumed to occur midway between these points. The precise form of the curve was adjusted to the tangent formed by the existing slope in each case.

Figure 5: Illustration of the methods used to extrapolate dose response functions for sulphur dioxide, using the results of Baker et al (18) as the example. The straight line shows the published function (equation 5b), and is extrapolated back from the point at which 0 % yield change was predicted to 0 ppb sulphur dioxide to provide the high estimate of crop losses. The curve shown here (equation 5a) is drawn as an arc to the published function at the point predicting 0 % yield loss, and also passes through the origin. It attains its maximum value midway between these 2 points. The 'best estimate' uses the function describing this curve between the points at which 0 % yield change is predicted, and the published function for higher sulphur dioxide levels. Extrapolation in this manner is based on the observation that low levels of sulphur dioxide can stimulate crop growth (19). Although not ideal, this approach does at least allow the estimation of damages to an accuracy of within an order of magnitude.

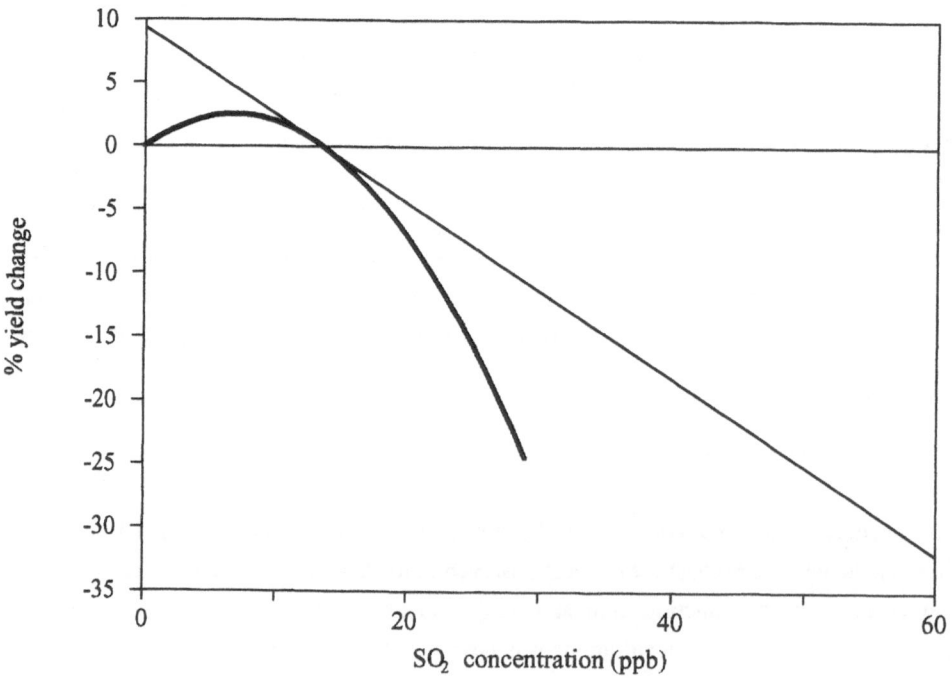

This approach gives 4 dose - response relationships, as follows, in which z = the mean concentration of SO_2 expressed in ppb (see also figure 5):

Baker 1:   %yield loss = -0.69z + 9.35

Baker 2:	%yield loss = $0.74z - 0.055z^2$	(from 0 to 13.6 ppb)	(5a)
	%yield loss = -0.69z + 9.35	(above 13.6 ppb)	(5b)

Roberts 1:   % yield loss = -0.18z + 2.75

Roberts 2:	%yield loss = $0.20z - 0.013z^2$	(from 0 to 15.3 ppb)	(6a)
	%yield loss = -0.18z + 2.75	(above 15.3 ppb)	(6b)

For SO_2 the best estimate is provided by equations 5a and 5b (Baker 2), the high estimate by equation 4 (Baker 1) and the low estimate by equations 6a and 6b (Roberts 2).

Damages were finally calculated using the 1989 prices of 157.8 ecu/tonne of wheat and 154.3 ecu/tonne of barley on the UK market. These prices are about 50% higher than those quoted by the International Wheat Council[20].

Results

Estimated high, low and best estimate figures of lost yield and damages for both pollutants are presented in table 3.   As expected from the literature, losses caused by O_3 are greater than those from SO_2.   However, given that the results of the model describing the atmospheric transport and chemistry of O_3 are preliminary at this stage, firm conclusions on the relative importance of the emissions of SO_2 and O_3 from a single power station in the UK should not be drawn from these results.   The difference between high and low estimates reflects the uncertainty associated with calculations of this type.

To convert total damages (as calculated here) to the marginal damage per unit of electricity, costs are divided by the total output of the West Burton 'B' power station over a 1 year period. This gives figures of 7.8 E-5 ecu/kWh for O_3 damage to wheat and barley and 3.7 E-5 ecu/kWh for SO_2 damage.   It is important to appreciate the fact that, as stated above, these figures represent only a partial evaluation of the impacts of the reference plant to agriculture. To make a full assessment all other crops grown in the UK would need to be considered, along with interactions between crops and climate, pests, pathogens and soil quality.

Table 3:   Lost yield of wheat and barley and associated economic damages estimated for the effects of SO_2 and O_3 produced as a consequence of the generation of electricity from the West Burton 'B' power station in the UK.

	O_3 Lost yield (tonnes)			SO_2 Lost yield (tonnes)		
	High	Low	Best estimate	High	Low	Best estimate
Wheat	4065	582	2324	2042	308	1471
Barley	5877	850	3363	1925	257	1258
	O_3 related damages (000 ecu)			SO_2 related damages (000 ecu)		
	High	Low	Best estimate	High	Low	Best estimate
Wheat	642	92	367	322	49	232
Barley	907	131	520	297	40	194

Conclusions

The methodology of the CEC/US Fuel Cycles Study has been demonstrated through the estimation of damages to wheat and barley crops associated with the direct effects of SO_2 and O_3 produced as a result of the generation of electricity from the West Burton 'B' power station. It is stressed that the evaluation given is only partial and hence the final figures for damage per unit of electricity should not be regarded as fully representative of the external costs of electricity production on these 2 major crops. A number of interactions, with pests, pathogens, climate and between pollutants, were not examined in this analysis. It is possible that some of these effects are implictly included in the dose-response functions used, particularly the one provided by Baker (18) to describe the effect of SO_2 on the yield of barley, as this experiment was conducted under field conditions.

Greatest increments in SO_2 from West Burton 'B' were in the area of the UK that has the highest annual mean rural SO_2 levels and the highest proportion of land used for cereal growing. Given these factors and the shape of the dose - response curves, the figures obtained may be considered a maximum for this type of power station in the UK. Had the power station been built in an area such as SW England, where SO_2 levels are very low, it is likely that the net effect on wheat and barley crops would have been positive. Conversely, O_3 levels are relatively low around West Burton, and the model results did not predict an overall increase in concentration until the plume had travelled some distance. Further analysis is necessary before we can say whether the level of damages reported for O_3 is typical of the UK or not. The potential for variation in results depending on where the reference power plant is sited illustrates the fact that selection of reference locations requires careful consideration. In the case of West Burton 'B' we selected the only site in the UK for which planning permission to build a coal fired power station exists.

The following basic research requirements were noted during implementation of this pathway:

1. The need to produce dose-response data for realistic concentrations of SO_2.

2. The need to derive dose-response functions for O_3 for the European situation.

3. The development of models that will allow the integration of the impacts of mixtures of different stresses (other pollutants, climate, pests etc.).

The example presented in this paper illustrates the fact that many of the impacts of interest to this study are highly complex and require analysis at a multi-disciplinary level. As a result of this a wide variety of groups are involved. They include experts from the fields of economics, energy technology, computer science, atmospheric and aquatic transport modelling, medicine, forestry, agronomy, terrestrial and aquatic ecology and materials science. This level of expertise is essential if the external costs of fuel cycles are to be assessed on a marginal basis, using the best information and models available at the current time.

References

[1] Hohmeyer, O. (1988) Social Costs of Energy Consumption. Springer-Verlag, Berlin.

[2] Ottinger, R.L., Wooley, D.R., Robinson, N.A., Hodas, D.R. and Babb, S.E. (1990) Environmental Costs of Electricity. Oceana Publications Inc, New York.

[3] Heck, W.W. (1989) Crop losses from air pollutants in the United States. In "Air Pollution's Toll on Forests and Crops" by J.J. MacKenzie and M.T. El-Ashry (Eds.), pp. 235-315. Yale University Press, New Haven.

[4] UK TERG (United Kingdom Terrestrial Effects Review Group) (1988) First Report, prepared at the request of the Department of the Environment. HMSO, London.

[5] Shriner, D.S. (1991) Report 18 - Response of vegetation to atmospheric deposition and air pollution. From "Acid Deposition: State of Science and Technology; Summary Report of the U.S. National Acid Precipitation Assessment Program". P. M. Irving (Ed.), US Government Printing Office, Washington D.C., pp 143-151.

[6] van der Eerden, L.J., Tonneijck, A.E.G. and Wijnands, J.H.M. (1988) Crop loss due to air pollution in the Netherlands. Environ. Pollut. **53**, 365-376.

[7] CEGB (1988) Proposed West Burton 'B' Coal Fired Power Station. Environmental Statement, Central Electricity Generating Board.

[8] Derwent, R.G. and Nodop, K. (1986) Long range transport and deposition of acidic nitrogen species in north-west Europe. Nature, **324**, 356-358.

[9] Derwent, R.G., Dollard, G.J. and Metcalfe, S.E. (1988) On the nitrogen budget for the United Kingdom and north-west Europe. Q.J.R. Meteorol. Soc., **114**, 1127-1152.

[10] C.E. Johnson (1992) Modelling and Assessments Group, Harwell Laboratory, *personal communication*.

[11] Altshuller, A.P. (1986) The role of nitrogen oxides in non urban ozone formation in the planetary boundary layer over N. America, W. Europe and adjacent areas of ocean. Atmospheric Environment, **20**, 245-268.

[12] Clarke, P.A., Fletcher, I.S., Kallend, A.S., McElroy, W.J., Marsh, A.R.W. and Webb, A.H. (1984) Observations of cloud chemistry during long range transport of power plant plumes. Atmospheric Environment, **18**, 1849-1858.

[13] MAFF (1990) Agricultural Statistics United Kingdom 1988. HMSO, London.

[14] Institute of Terrestrial Ecology Land Classification and Database (1991), ITE Merlewood, Cumbria, UK.

[15] Adams, R.M. (1989) Estimated economic consequences of ozone on agriculture: some evidence from the US. In "Atmospheric Ozone Research and its Policy Implications" by T. Schneider *et al* (Eds.), pp 869-879. Elsevier Science Publishers B.V., Amsterdam.

[16] Taylor, H.J., Ashmore, M.R. and Bell, J.N.B. (1987) Air pollution Injury to Vegetation (a guidance manual commissioned by HM Industrial Air Pollution Inspectorate of the Health and Safety Executive). IEHO, London.

[17] Roberts, T.M. (1984) Long term effects of sulphur dioxide on crops: an analysis of dose - response relations. Phil. Trans. Roy. Soc. London **B 305**, 299-316.

[18] Baker, C.K., Colls, J.J., Fullwood, A.E. and Seaton, G.G.R. (1986) Depression of growth and yield in winter barley exposed to sulphur dioxide in the field. New Phytologist **104**, 233-241.

[19] Murray, F. and Wilson, S. (1990) Growth responses of barley exposed to SO_2. New Phytologist **114**, 537-541.

[20] B. Rhodes (1992) Dept. of Economics, University College London, *personal communication*.

Acknowledgements

The authors would like to thank Dr. David Howard and Prof. Mike Hornung of ITE for providing the data on crop distributions and Jeremy Colls of the University of Nottingham's Department of Agriculture for providing data on dose - response relationships of barley from the experiments conducted at Sutton Bonnington.

11.   Economics of Nuclear Risks - A German Study

Hans-Jürgen Ewers, Klaus Rennings
Institut für Verkehrswissenschaft an der Universität Münster
Am Stadtgraben 9, 4400 Münster

1   Introduction

Estimating the risk of a nuclear meltdown is one of the most delicate problems associated with the monetary valuation of external effects. The relationship between the incredible damage potential of such an accident and its exceedingly small probability is extreme. Experts differ greatly in their evaluation of nuclear energy rating it between being the safest or the most dangerous form of energy production depending upon whether the emphasis is put upon the probability of an accident or upon the the damage potential.

Further, a nuclear meltdown can cause effects that can only inadequately be measured by standard methods of monetary valuation of environmental damages, the most problem ridden of which are fatalities, consequences for following generations and irreversible environmental damages. The purpose of this paper is to discuss possibilities of dealing with these and other risks of nuclear energy.

In Germany damages from reactor accidents are only insured to a maximum of 500 million DM. All costs for damages that go beyond that would be passed on to the general public. Therefore, electricity from nuclear power is possibly being offered at prices lower than its real costs.

The main purpose of the Atomic Liability Law (Atomhaftungrecht) is the protection of victims.[1] Legally, however, this doesn't allow setting the insured amount so astronomically high that running a nuclear power plant, in effect, would become impossible. "Particularly because the liable holder of a nuclear plant, even when not at fault, is liable only for damages that the government explicitly allows in an official licensing process"[2]. From a legal viewpoint it appears that victim protection is guaranteed inspite of the possibly inadequate coverage of damages because, in the case of a catastrophe, the state would quickly jump in with unbureaucratic aid. From an economic point of view, however, it is necessary to examine if passing the costs of nuclear power risks on to the public doesn't eliminate the incentives for reducing these risks.

2   Methods and Problems of the Monetary Valuation of Health Risks

The loss of human life represents the largest category of damages in valuating the risks of nuclear accidents - almost all research ascertains its proportion to be over 80% of all damages. Depending upon the valuation method used, however, there are great differences in the estimations. The results of the hedonic price-analysis, widespread in Anglo-Saxon research, are much higher than the results of Human-Capital method used in Germany. In order to understand these differences the different methods for valuating health risks will now be discussed and evaluated.

2.1   Basic Principles

The first problem in valuating health risks is that fatalities, regardless of the monetary value placed on life, cannot be recompensated. This is a problem that all methods have. When valuating health risks the problem isn't how to compensate for loss of life after the fact but to estimate the statistical risks before the damages take place - it isn't known which individuals will actually be damaged but it can be ascertained to what extent damages can be expected. In other words, the problem isn't, to determine ways of carrying out the (impossible) ex post compensation (after the fact) but how to achieve an ex ante compensation (by reducing the risks). The economic value of a health risk is here the amount an individual is willing to pay to avoid a risk, or the amount for which the individual would be willing to accept the risks.[3]

A second problem in calculating the health risks is concerned with questioning the value of using "risk indicators". Conventionally, the socalled final risk of nuclear power plants is calculated as the product of the accident probability and amount of damage. In the case of a reactor accident both factors take on extreme opposit values. The amount of damage is extraordinarily high and the probability is extraordinarily low. The extremely high amount of damage is then concealed by using expectation values.[4]

A third problem in calculating these risks can be seen in the assumption that the risks of nuclear power are voluntarily accepted. A reactor meltdown, however, is a involuntary risk that no one can avoid. Therefore, high standards must be set in connection with the acceptance of the final risks of nuclear power plants. One possibility for setting these standards can be found in John Rawls' "A Theorie of Justice".[5] According to Rawls, in a developed society with only mild scarcity of resources, basic human rights like the right to protection against bodily harm cannot be compensated for through material goods. Cost- Benefit analyses are only then justified when the basic human rights aren't impaired.

Thus, operating a nuclear power plant is only then legitimate when the risks do not impair basic human rights. In such cases Rawls demands that every citizen "under the veil of uncertainty" - whether he profits from, or is a victim of, nuclear energy - must be in agreement with the use of nuclear energy. This means that the use of nuclear energy is only legitimate if unanimity in the case of a societal contract can be expected.

2.2   Methods for Valuating Health Risks

Environmental damages can be calculated in two different ways depending on the compensation concept upon which they are based.[6] The first method, the analysis of willingness to pay (WTP), asks how much citizens are willing to pay to achieve improvements in the environmental situation. The second method, the analysis of the willingness to accept compensation (WTA), asks to what extent citizens would demand recompensation in order to accept environmental damage.

The willingness to pay or the compensation demands by the public can either be directly ascertained through surveys or indirectly by plausible estimations of the costs of environmental damages.

2.2.1 Direct Methods

The direct methods of ascertaining the willingness to pay can be categorized under the term Contingent Valuation Methods (CVM). Hoevenagel[7] describes the advantages and disadvantages of CVM as follows:

Advantages of CVM

- widespread fields of application (for many environmental goods it is the only method available);
- measurement of non user-values;
- generation of its own data;
- direct measurement of the Hicksian measures of welfare changes.

Surveys have the basic advantage that the willingness to pay for the avoidance of intangible damages (for example psycho-social costs) can also be assessed. Therefore, based on individual or subjective preferences, Contingent Valuation methods are the most capable of extensively measuring changes in benefits or welfare levels.

Disadvantages of CVM

- they are based soley on reported behavior or willingness to pay;
- results are very sensitive to the type of research design;
- therefore a diffentiated design is necessary (this makes CVM very costly);
- they demand theoretical decisions that most people have never made before.

The possibility of strategical answering by individuals are often overestimated. This can be controlled by a differentiated (and therefore costly) survey design. Thus, the danger of manipulation by the researcher is greater than by the persons being interviewed. Nevertheless, because of their theoretical and practical advantages CVM are valued as the best methods for estimating environmental damages.[8]

The CVM is already being widely used in valuating health risks. The results differ depending upon research questions and the type of risk being investigated. In an overview of the different older studies Violette/Chestnut give a range of the estimates of the value of fatal health risks - from 1986 on - between 315,000 US \$/fatality and 7.4 mio. US \$/fatality.[9]

2.2.2 Indirect Methods

Because of the problems (above all because of the costs) of direct questionnaires research in the valuation of human life in the past often resorted to indirect methods. These methods assessed individual preferences on the basis of plausible cost estimates.[10] The most important of these methods, the Human- Capital and the Hedonic Price Analysis (HPA), will now be discussed and evaluated.

Hedonic Price Analyses estimate the prices of environmental goods indirectly by comparing the prices of goods that have differing environmental quality. For example, the price differences of homes in different housing areas can indicate how much economic subjects are willing to pay for environmental qualities such as peace and quiet, and clean air.[11] HPA has already been widely used in the USA in the evaluation of health risks.[12] For example, empirical studies compare wages for occupations carried out under perilous conditions with wages for occupations carried out without these risks. The assessed difference is then interpreted as the wage increase for the increased risk of accidents. As a compensation payment the wage difference represents the economic (market) value of the risk associated with the activity.

This can then be used to help valuate the risk of fatal accidents. Ottinger calculated the average out of 10 previously conducted studies and came to the value of 4 million US-$/fatality.

The advantage to the HPA is that the results are based on real (and not on verbally reported) behavior. The disadvantages are to be found in the assumptions upon which the methods are based. These are as follows:

- the risks are voluntarily accepted;
- the qualities of the risks are transferable to other types of risks;
- the wage differences are based only upon the different risks associated with the tasks.[13]

The last problem can be partially solved by making an econometric analysis of the factors that influence wages. The other two disadvantages are more serious. Does a wage increase represent compensation for voluntarily working under hazardous conditions? It is questionable if the wage increase of a construction worker due to hazardous conditions can be used to valuate the risks of nuclear energy. In the case of nuclear energy several factors are neglected. One factor is that the assumed voluntariness of the activity is missing. The other neglected factor is the psychological phenomenon that the risk aversion of the population increases with the absolute increase of possible damage associated with the an accident.[14] Binswanger called this phenomenon of risk aversion the psychological risk.[15]

The socalled Human-Capital model calculates the production deficit that occurs through an accident. The most useful method consists of calculating the expected income of the victim from the present time and then discounting the interest. The resulting value can be used to estimate the lost benefits to the individual caused by his early death. This value can be supplemented by including factors like suffering, sorrow and other loss of benefits of third persons resulting from the fatality.

In Germany the Human-Capital Model has been almost exclusively used in calculating health risks. Using these methods the value for a traffic fatality was calculated at about 1.32 million DM for the year 1989.[16]

A certain plausibility and the ease of calculation speak for these methods. For the last 20 years, however, this model has been continuously challenged in the scientific community and is considered as being theoretically ungrounded[17] because it offers no information on the subjective preferences for avoiding risks or the subjective loss of benefits as risks increase.

The more modern methods of HPA and CVM have only achieved usable results in the area of fatal health risks (mortality). Studies in the valuation of non-fatal health risks (morbidity) are so scarce[18] that in this area the Human-Capital Model must be used.

3   Quantification and Monetarisation

At this point in time there are a series of studies which attempt to quantify the economic value of the risks of nuclear meltdowns. A discussion of some issues will now be given. These issues are:

- different estimations of the potential risks (quantification);
- different methods for valuating these risks (monetarisation).

Based upon this discussion we will present a monetary valuation of the consequences of a nuclear accident as could be expected after a nuclear meltdown in Germany.

3.1   Quantification

The potential risks of nuclear meltdowns are determined by the probability of its occurrence and the extent of possible damage.

Regarding the probability of the occurrence of a nuclear accident the estimates vary greatly. An anylysis of the external cost caused by nuclear risks in the USA by Richard Ottinger (PACE-University) assumed the propability of a meltdown to be at 1 accident per 3,333 reactor operating years. This estimate, made by the Nuclear Regulatory Commission (NRC), states that there is a probability of 45% that within 20 years one of 109 US nuclear reactors will have a serious accident.[19]

According to the "Deutsche Risikostudio Kernkraftwerke", Phase B (abbrieviated as DRS Phase B) - a study of German nuclear power plant risks - a meltdown in a reactor of the type Biblis B is only to be expected every 33,000 years.[20] In this case, however, plant internal emergency measures, as part of an accident management plan, are not considered in the calculations because there is as of yet no evidence that these measures contribute to reducing the risk of accidents.[21] The DRS Phase B assumes that these plant internal emergency measures would have a large rate of success. Thus, were these measures taken into consideration the probability of a reactor meltdown would decrease to 1:250,000.

Because studies for other German nuclear power plants do not exist, we will assume that the probability of 1:33,000 years - as calculated by DRS Phase B (without accident management) - is representative for all German reactors.

Regarding the extent of potential damage the question is whether a catastrophe of Chernobyl dimensions is possible in German reactors. Using the data of the old american risk study WASH-1400 Alfred Voß conducted a study which estimates that relatively little radioactivity would be set free in the case of an accident.[22] Since the publication of DRS Phase B, however, it can be assumed that a meltdown in Germany can release more radioactivity than in Chernobyl (for our purposes we will then double the Chernobyl values)[23]. Also, because the population density in Germany is 7 times greater than that of the areas affected by Chernobyl, we estimate a radioacitve immission rate (measured in person-rem) 14 times greater than the official rate at Chernobyl.

One indicator that can be used to measure health risks is the Radioactive Cancer Risk (Strahlenkrebsrisiko). This is usually measured in units of Tumors per million person-rem. The International Commission for Radioactive Protection (ICRP) assumes that per million person-rem immission, 500 cancer fatalities, 100 non-fatal cases of cancer and 130 cases of extreme genetic damage can be expected.[24]

In addition to the health risks most studies take material damage into consideration when valuating the consequences of a reactor meltdown. Voß, for example, included the cost of evacuation, agricultural losses and the costs of long-term protection measures. Ottinger considered agricultural production losses in his calculations. We calculated the total loss of production in areas with radioactive contamination in our Biblis Study.

3.2   Monetarisation

When valuating health risks it is necessary to differentiate between fatalities caused by radio-activity (mortality) and other non-fatal damages to health.

Using the Human-Capital Model to estimate Health damages in his Study "Social Costs of energy consumption", Olav Hohmeyer calculated the production losses for cancer fatalities at one million DM/fatality.[25] He then reworked these calculations using the Hedonic Price Analysis and came up with a corrected estimate of between three and six million DM/Fatality.[26] Ottinger analysed eight studies which used the HPA to estimate health risks and came up with an estimate of four million US-$/fatality.

In the following section we will use results gathered using HPA in calculating mortality risks, whereas in the estimation of morbidity risks we must resort to the use of Human-Capital Method as no HPA studies for ascertaining the WTA or WTP exist. For the estimation of fatalities using the HPA the value of four million US-$/fatality is seen as representative so that the amount of six million DM/fatality has been set. For non-fatal damages the value of 0.5 million DM has been set - this corresponds to the loss of income over 10 years.

By calculating the monetary health risks of a nuclear reactor meltdown using a "tariff" of 6 million DM and pro fatality and 0.5 million per case of non-fatal or serious genetic damage one comes up with the figure of 10,466.4 billion DM. The monetary value of the mortality risks amounts to 10,080.00 billion DM (480 mil. pers.-rem x 7 x 500 fatalities/million pers.-rem x 6 mil. DM = 10,080.00 bil.). The monetary value for the morbidity risks amounts to 386,4 billion (480 mil. pers.-rem x 7 x 230 cases/mil. pers.-rem x 0,5 mil. DM = 386,4 bil. DM).

The material damage risks of such an accident, even though extraordinarily high, fade into the background when compared with the magnitude of the health risks, which amount to many times that of the annual German GNP. We assume that the material damages estimated in our Biblis study[27] represent an extreme value because of the relatively dense population near the Biblis plant. Therefore, we reduced the values of material damage to an average that is 55% of the Biblis figures. This results in a value of between 231 billion DM for the average material damage to be expected from a reactor meltdown in the FRG (compared to the 420 billion DM that can be expected from a Biblis accident). The estimated value of total damages from a nuclear reactor meltdown in the FRG can be seen in the following table.

Table 1:   Total damages of a reactor meltdown in Germany (in bil. DM)

Damage to persons:	
Mortality	10,080
Morbidity	386
Subtotal	10,466
Material Damage	231
Total (persons and material)	10,697

4 Dealing with the Meltdown Risk: Internalization or an Ecological Framework?

There are several different possibilities for dealing with the damages of a reactor meltdown. One of these consists of integrating the social costs of a nuclear accident in the market price through price corrections and thus internalizing the costs. One variant of this internalization is the socalled Pigou-Tax. This transfers the estimated monetary value of potential damages to the price of nuclear electricity. Another possibility consists of setting risk limits which are put on the energy producers as restrictions. This strategy aims at achieving an "ecological frame-work" for the market economy, within which the price mechanism would still provide for the optimal allocation of resources.[28]

4.1 Internalizing the Costs of a Reactor Meltdown

In order to internalize environmental damages they must first be monetarized. The cost of a reactor meltdown in the FRG was calculated above at 10,697 billion DM. According to DRS Phase B the probability of a meltdown is 1:33,000 years (not including accident management). By dividing the probability of an accident over the 20 German reactors operating in 1989 a reactor meltdown can be expected every 1,666 years. The yearly costs of these damages are then 6,42 billion DM. In 1989 the 20 German reactors produced 537,8 petajoules of energy (or 149.4 TWh).[29] The resulting external monetary costs per energy unit amount to 1.2 Pf/kJ (or 4.3 Pf/kWh).

These results correspond approximately to the estimates made by Ottinger (2,3 cents/kWh), are near those of the Hohmeyer study (1,2 - 12 Pf/kWh), are about 3 times lower than the values of Ferguson (5 pence/kWh)[30] and about 60 times higher than the estimates made by Voß (0.008 - 0.07 Pf/kWh).

The resulting costs could be raised in the form of a duty. A viable alternative would be an insurance plan. The necessary insurance premium should, however, go much beyond the estimated social costs because a private insurance company certainly won't be able to afford a saving period of 30,000 years.

If a postulation were to be so formulated, that the risks in a market economy can only be allowed when they can be privately insured, the polluter-pays-principle would then be so established that the cost of an accident must be payed for by the polluter in the form of premiums. The postulation of private insurance would give incentives to:

- raise liability limits and the coverage amounts for nuclear accidents so that they correspond to the expected costs of damages;

- reduce the potential risks so that they become insurable;

The Federal Republic of Germany, Japan and Switzerland are the three nations where operators of nuclear power plants (NPP) are liable for all damages even when without fault. Operators of NPP in Germany must be insured to the amount of 0.5 billion DM, next to the USA the highest coverage.[31] Compared to possible damages, however, these amounts appear negligible. Raising the coverage amounts to the level of 10 billion DM as is now being discussed in Germany would cross the limits of what is insurable by private insurers. The US Atomic Liablility Law currently covers nuclear damages to the amount of 7,8 bil. US-$: 200 million through liability insurance, 7.607 billion US-$ through a reinsurance system in which the owners of NPPs must subsequently pay certain amounts in the case of damages.[32] Privately insuring full potential meltdown damages - which, as we have shown in the Biblis scenario, could amount to several trillion DM - appears to be hopeless.

Postulating the private insurability of the risks would then result in the reduction of the risks or the construction of plants that absolutely exclude accidents of Chernobyl dimensions. This kind of reactor could be the socalled "inherently secure reactor". A nuclear facility can be considered inherently secure when the security of the facility isn't dependent upon the intervention of personnel or the functioning of electromagnetic components but upon the unchangeable laws of physics and chemistry.[33]

The problem that all internalization strategies have is that they attempt to monetarily recompensate damages that have already occurred. These attempts meet their limits in the dimension of potential damages of a reactor meltdown, the consequences of which endanger the health of persons of whole regions and future generations. The prerequisite for the compensation of environmental damages is that the damages can be recompensated. In the case of a serious nuclear accident this is only imaginable with a very restrictive interpretation of Rawls' "societal contract" in which the people unanimously accept the risks of nuclear energy.

4.2   Limiting the Risks: The Ecological Framework

Because an optimal internalization of environmental risk often cannot be realised, environmental economists are being increasingly pressured to set environmental quality goals that must be adhered to by producers and consumers in the economic process. Standards for dealing with possible reactor meltdowns can be imagined on several levels:

- quantitative risk-limits;
- additional security criteria;
- standardized methods for evaluating risks.

When formulating environmental policy standards it must also be examined to what extent intergenerational justice can be achieved by building in operational sustainability rules. Without such rules it can be feared that in the political process environmental dangers will be systematically passed on to future generations. General approaches for such rules have been developed by supporters of the concept of sustainable development.[34]

4.2.1 Quantitative Risk-Limits

Risk-limit values for NPP concerning catastrophic risks in the FRG have, until now, not been specified. General consensus in the construction of an ecological framework consists solely in the point that the environmental subsistence level must remain under complete protection.[35] The ecological subsistence level can be so described, that the environment will be sufficiently protected, maintained and preserved for future generations[36]. Further, the questions must be asked, how the postulation of an ecological subsistence level can be more concretely defined and quantified, and if this subsistence level is endangered through the use of nuclear energy.

Internationally experience has been made with risk-limit values in the area of nuclear energy. 1986 in the USA the policy statement of the NRC went into effect under the title "Safety Goals for the Operation of Nuclear Power Plants". The quantified "Safety Goals" made there set the limits for individual risk and for the occurrence frequency. Binding safety goals in the form of maximum allowed probabilities for nuclear accidents are also existent in Great Britain. France and Canada have non-binding risk-limits.[37]

A weakness of these international risk-limit values is that the risk-figures do not account for the magnitude of potential damages: "The formulation of the safety goals allow for arbitrarily large maximum damages as long as the probability of the occurrence is correspondingly small"[38]. Further development in quantifying risk-limits could, here, consist of setting additional absolute limits for possible amounts of damage, because the amount of potential damage can be so large that, regardless of the probability of their occurrence, these risks are not acceptable.

Monetary units are not useful as a dimension for risk-limits. Monetary damage amounts can, however, offer orientation in setting standards in that they can indicate what environmental costs remain within the set limits and also indicate the magnitude of costs which can be avoided by setting these limits.

4.2.2 Additional Security Criteria

Uncertainty in the valuation of environmental damages has hindered the use of quantitative risk-limits. The existence of uncertainty on one hand and the goal of achieving ecological sustainability on the other, suggest the necessity of risk aversive behavior. When in doubt a very low limit should be set. When setting risk-limits it is helpful to keep certain characteristics of risks in mind. These criteria could be:

- the irreversibility of the risk;
- the non-voluntary aspect;
- the consequences for future generations.

If any one of these characteristics is present it makes sense to make the limits particularly risk aversive or to completely forbid the risk.

4.2.3 Standardized Methods for Evaluating Risks

Quantified limits require supplementation and concretization through standardized methods which apply these limits to specific cases. When quantified risk-limits do not exist or when these limits allow too much room for discretion or interpretation, the attempt should be made to achieve some sort of consensus on the procedure. These procedures could take on the form of an environmental impact statement along with the participation of the public or public interest groups.[39]

5   Conclusion

We have now presented and discussed three possibilities for dealing with the risks of reactor meltdown: The internalization through a Pigou-Tax, a private insurance solution as well as improved standards in the form of risk-limit values. In view of the risk of irreversible damages, long-term consequences over generations and its non-voluntary aspects, we are of the opinion that, basically, compensatory solutions for managing the risks of reactor meltdown are inacceptable. At best, it is imaginable that they may be applied under the very restrictive assumptions of Rawls' social contract in which the public unaminously decides to accept the risks of a reactor meltdown. Better, however, would be the development of improved risk-standards, in the sense of an ecological framework for the market economy, that would not endanger the survival chances of current and coming generations. Such a concept of sustainable development would contain more emphasis for the aspect of security. In any case, the risk of a reactor meltdown of the magnitude of Tschernobyl would certainly not be consistent with any definition of an "ecological subsistence level".

References and Endnotes

[1] Pelzer, Norbert (1991): Überlegung zur Novellierung des atomrechtlichen Haftungs- und Deckungsrechts in den 90er Jahren. in: Lukes, Rudolf (ed.): Reformüberlegungen zum Atomrecht. Köln, pp. 455 - 503, see p. 45.

[2] a.a.O: P. 460 (Translated by author)

[3] Ottinger, Richard L. et. al. (1990): Environmental Costs of Electricity. New York, London, Rom, p. 56.

[4] Ottinger, Richard L., et al: a.a.O., p. 79; Binswanger, Hans-Christoph (1990): Neue Dimensionen des Risikos, in: Zeitschrift für Umweltpolitik und Umweltrecht, 13. Jg., pp. 103-118.

[5] Rawls, John (1979): Eine Theorie der Gerechtigkeit, Frankfurt a.M.; Bogai, Dieter (1989): Technikfolgen, Ökonomie und Ethik. München; Ewers, Hans-Jürgen, Klaus Rennings (1991): Die volkswirtschaftlichen Schäden eines Super-GAU's in Biblis, in: Zeitschrift für Umweltpolitik und Umweltrecht (ZfU), 4/91, pp. 379 - 396, see here. pp. 391.

[6] The concepts of Compensating Variation and Equivalent Variation can be used. See Keppler, Jan (1991): Wieviel Geld für wieviel Umwelt? Entschädigungskonzepte und ihre normativen Grundlagen, in: Zeitschrift für Umweltpolitik und Umweltrecht. 4/91, pp. 397 - 410.

[7] Hoevenagel, R. (1991): An Assessment of the Contingent Valuation Method, Institute for Environmental Studies, Free University Amsterdam, unpublished, pp. 13; Hoevenagel, R.

(1991): A comparison of Economic Valuation Techniques, Institute for Environmental Studies, Free University Amsterdeam, unpublished, p. 20.

8 Buchanan, Shepard C. (1991): Contingent Valuation Study of the Environmental Costs of Electricity Generating Technologies, in: Olav Hohmeyer, Richard Ottinger (eds.): External Environmental Costs of Electric Power, Berlin, Heidelberg, New York, pp. 159-167; Hoevenagel, R.: An Assessment of the Contingent Valuation Method, a.a.O., pp. 55; Hoevenagel, R.: A comparison of Economic Valuation Techniques, a.a.O., pp. 21; Ewers, Hans-Jürgen (1986): Kosten der Umweltverschmutzung - Probleme iher Erfassung, Quantifizierung und Bewertung, in: Umweltbundesamt: Kosten der Umweltverschmutzung, Berlin, pp. 9-19, see here pp.16; Mishan, E.J. (1971): Evaluation of Life and Limb: a Theoretical Approach, in: Journal of Political Economy, Vol. 79, pp. 687-705, see here pp. 703.

9 McDaniels, Timothy L. (1988): Comparing Epressed and Revealed Pfeferences for Risk Reduction: Different Hazards and Question Frames, in: Risk Analysis, vo. 8, No. 4, pp. 593-604, see here p. 599.

10 Mishan, E.J. (1971): Evaluation of Life and Limb: a Theoretical Approach, a.a.O. pp. 689.; Hanusch, Horst (1987): Kosten-Nutzen-Analyse, München, pp. 90.

11 Pearce, David W., R. Kerry Turner (1990): Economics of natural Resources and the Environment; New York, London, pp. 143.

12 Ottinger, Richard L. et al.: Environmental Costs of Electricity, a.a.O., pp. 96.

13 Hoevnagel, R.: A comparison of Economic Valuation Techniques, a.a.O., p. 18.

14 Ottinger, Richard L. et al.: Environmental Costs of Electricity, a.a.O. p. 81.

15 Binswanger, Hans-Christoph: Neue Dimensionen des Risikos, a.a.O., pp. 103

16 Bundesanstalt für Straßenwesen (BAST) (1989): Gesamtwirtschaftliche Unfallkosten für das Jahr 1989, Bergisch Gladbach, unpublished.

17 The judgement that from a welfare economics position the Human-Capital Model is unsustainable has been most crassly formulated by Shulze/Kneese"Unfortunately, earlier attempts at measuring the value of safety programms have given economists a `black eye'

for supposedly advocating that individual human lives could be valued as the lost economic prductivitiy associated with a shortened life span.... On economic theoretical grounds,
all of these calculations have shown to be nonsense." Schulze, William D., Allen V.
Kneese (1981): Risk in Cost-Benefit-Analysis, in: Risk analysis, Vol. 1, pp. 81- 88, see
here pp. 88. See also Conley, Brian C. (1976): The Value of Human Life in the Demand
for Safety, in American Economic Review, Vol. 66, pp. 45-55, see here p. 45; Mishan,
E.J. (1971): Evaluation of Life and Limb: a theoretical approach, a.a.O., p. 704.

[18] Ottinger, Richard L. et al.: Environmental Costs of Electricity, a.a.O. p. 100; Baumann,
Angelika, Robert Hill (1991): External Costs/Benefits of Energy Technologies - Development of a Methodology. Newcastle upon Tyne Polytechnic, p. 38)

[19] Ottinger, Richard L. et al.: Environmental Costs of Electricity, a.a.O., p. 379.

[20] Gesellschaft für Reaktorsicherheit (GRS) mbH: Deutsche Risikostudie Kernkraftwerke
Phase B - eine zusammenfassende Darstellung, a.a.O., pp. 87.

[21] Hahn, Lothar, M. Sailer (1987): Charakterisierung von Sicherheitsphilosophien in der
Kerntechnik, Öko-Institut Darmstadt; Fischer, B., L. Hahn, M. Sailer (1989): Bewertung
der Ergebnisse der Phase B der Deutschen Risikostudie Kernkraftwerke, Öko-Institut,
Darmstadt, pp. 137.

[22] Voß, A,; R. Friedrich, et al (1990): Externe Kosten der Stromerzeugung, 2. Auflage,
Frankfurt, pp. 58.

[23] For the justification of this see; Ewers, Hans-Jürgen, Klaus Rennings: Die volkswirtschaftlichen Schäden eines Super-GAU's in Biblis, a.a.O., p. 392.

[24] International Commission on Radiological Protection (ICRP)(1990): 1990 Recommendations of the International Commission on Radiological Protection, Oxford, New York, p.
70 and p. 135.; Schmidt, Mario (1991): Der Diskussionsstand zum Strahlenkrebsrisiko
und notwendige Konsequenzen für den Strahlenschutz, in: WSI-Mitteilungen 12/1990,
pp. 769-777.

[25] Hohmeyer, Olav (1989): Soziale Kosten des Energieverbrauchs, 2. Auflage, Berlin, Heidelberg, p. 64.

26 Hohmeyer, Olav (1990): Latest Results of the International Discussion on the Social Costs of Energy - How does Wind Compare Today?, Paper read at the EC Wind Energy Conference in Madrid, 1990, unpublished, p. 5.

27 Ewers, Hans-Jürgen, Klaus Rennings (1991): Die Volkswirtschaftlichen Schäden eines Super-GAU's in Biblis, a.a.O., p. 393.

28 Hansmeyer, Karl-Heinrich, Hans Karl Schneider (1990:) Umweltpolitik - Ihre Fortentwicklung unter marktsteuerneden Aspekten, Göttingen, pp. 12-15; Kemper, Manfred (1989): Das Umweltproblem in der Marktwirtschaft, Berlin, pp. 10-15.

29 Brecht, Christoph, et al (eds.) (1990): Jahrbuch Bergbau, Öl und Gas, Elektrizität, Chemie 1989/90, Essen, see here Table 96.

30 The results from Ferguson are hypothetical willingness to pay that have not yet been empirically examined. Ferguson, Ross (1991): Environmental Costs of Energy Technologies - Accidental Radiological Impacts of Nuclear Power, unpublished outline.

31 Pelzer, Norbert: Überlegungen zur Novellierung des atomrechtlichen Haftungs- und Deckungsrechts in den 90er Jahren, a.a.O., pp. 460; Smets, H. (1987): Compensation for Exceptional Environmental Damage Caused by Industrial Activities, in: Kleindorfer, P.R.; H.C. Kunreuther (eds.): Insuring and Managing Hazardous Risks: From Seveso to Bhopal and Beyond, Berlin, Heidelberg, New York, pp. 79-138, see here p. 110.

32 Pelzer, Norbert: Überlegungen zur Novellierung des atomrechtlichen Haftungs- und Deckungsrechts in den 90er Jahren, a.a.O., see here p. 462.

33 Hahn, Lothar, Michael Sailer: Charakterisierung von Sicherheitsphilosphien in der Kerntechnik, a.a.O., p. 111.

34 Pearce, David W.; R. Kerry Turner: Economics of Natural Resources and the Environment, a.a.O., pp. 43; Brenck, Andreas (1991): Moderne umweltpolitische Konzepte - Sustainable Development und ökologisch-soziale Marktwirtschaft, Diskussionspapier Nr. 3 des Instituts für Verkehrswissenschaft an der Universität Münster, November 1991, pp. 6.

35 Hansmeyer, Karl-Heinrich, Hans Karl Schneider: Umweltpolitik - Ihre Fortentwicklung unter marktsteuernden Aspekten, a.a.O., p. 20.

[36] Wicke, Lutz (1991): Umweltökonomie, 3 Auflage, München, p. 584.

[37] Hahn, Lothar, Maichael Sailer: Charakterisierung von Sicherheitsphilosophien in der Kerntechnik, a.a.O., pp. 93.

[38] Hahn, Lothar, Michael Sailer: Charakterisierung von Sicherheitsphilosophien in der Kerntechnik, a.a.O., p. 103.

[39] Ewers, Hans-Jürgen (1988): Möglichkeiten der Früherkennung von umweltbelastenden Technologien - Denkbare politische Strategien, in: Ewers, Hans-Jürgen; et al (eds.): Produktionsprozesse und Umweltverträglichkeit, Veröffentlichung der Akademie für Raumforschung und Landesplanung, Beiträge, Vol. 104, pp. 75-86.

12.   Environmental Impacts of Photovoltaics/Solar Energy

Angelika E. Baumann and Robert Hill
University of Northumbria
Newcastle upon Tyne, UK

Introduction

The generation of electricity is generally known as a major source of environmental pollution. This applies not only to conventional energy systems but also to some extent to renewable energy technologies, if one considers the complete life-cycle of these technologies. During the various stages of the life-cycle of each given technology, which include fuel extraction, preparation, transport, conversion, operation, distribution, utilisation, waste processing and disposal, emissions are generated and dispersed into the environment, thereby imposing a burden on living systems and items of value to human society. These burdens, in turn, have an impact on the physical and biological environment as well as on human health and life. As these impacts impose significant costs on society, decisions concerning energy planning and the selection of energy systems ought to be based on a comparative analysis of various energy technologies and their "upstream" and "downstream" activities.

Photovoltaics solar energy was identified in the early 1980s as a very promising renewable source of energy for the future. It is presently considered that it could be used for generating electricity at a cost comparable to that of fossil fuels early in the next century. To be able to compete with other energy technologies PV will still have to improve its performance and minimise its total costs. PV like any other energy system is not entirely free of environmental impacts, if its complete life-cycle is analysed. However, compared to conventional energy systems, its main environmental impacts do not occur at the operation stage where PV is almost entirely benign but are caused during production. Environmental and health problems are mainly associated with the manufacturing of PV-cells/modules. The materials used in PV-modules are not used-up during electricity generation and emitted into the environment but merely broken or degraded. As they can be recycled in principle, this minimises the problem of waste disposal.

Given further improvement in performance and a reduction in total costs from large scale production, PV technologies will be competitive with other technologies. At present, the production and operation of PV systems can not be compared on an even footing with those of existing conventional technologies as the latter have considerable advantage due to long term experience in developing and operating large and optimised power stations, while the production of solar cells and the operation of PV power stations is still undertaken on a significantly smaller scale. Hence, comparisons in quantitative terms under present conditions do not reflect the full potential and future feasibility of PV technologies.

Photovoltaic Technologies

There has been a transformation in the performance of solar cells and modules over the past 15 years. Conversion efficiency of PV cells has been improved, module costs reduced and the lifetime of PV modules increased to over 20 years with present technology. The technical advances made in PV modules which have been matched by advances in the balance-of-system technologies contributed to a reduction in total costs of PV systems and an increase in their reliability. At present, reduced costs and higher reliability of photovoltaic applications are mainly of relevance to consumer products and applications in remote areas, for telecommunications, water pumping, lighting devices etc.. A continuation of this development with regard to costs and performance of PV devices will be of particular importance in the future, as photovoltaic power systems are seen as a potential candidate for large-scale electricity production both within and outside Europe.

The major photovoltaic technology for power production is that based on flat plate silicon wafers, where the wafers are cut from either single crystal boules or from multicrystalline ingots.

"Single crystal silicon cells have been the work horses of PV systems designed for power generation"[1] with relatively high efficiencies, 24% efficiency of the best R&D cells, and a good record of reliability, with expected module lifetimes in practice of up to 25 years. Due to a number of innovations, mainly in the optical engineering of the cells, BP Solar have now begun to commercially produce cells of 18% efficiency by using semiconductor-grade silicon wafers. However, silicon wafers are not only expensive, but there are increasing supply problems as the PV industry expands.

Most recently, almost all PV companies are moving to the use of "solar-grade" silicon wafers to produce polycrystalline silicon cells. The wafers are multicrystalline, with grains about 1 cm or more in size, and form cells which are 1-2% less efficient than the single-crystal wafers, but at potentially lower cost. New technologies developed to avoid the costly, time-consuming and wasteful handling of material when cutting ingots or boules into wafers, promise cost reductions when they come into commercial production in the next few years[2].

An other technology in commercial production at the present time is based on a thin film amorphous silicon (a-Si). It is the only type of thin film PV technology adopted for mass production of consumer products such as solar calculators, watches and radios, outdoor lights for walks or patios, water pumps and many more. The development and acquisition of this niche market has provided a-Si manufacturers with the much needed revenue to improve the performance of a-Si cells and build towards the ultimate market - large scale power use.

By 1985 a-Si cells of 10% efficiency were produced, but these simple single-junction a-Si devices were unstable and lost efficiency when exposed to sunlight. Within a few months the initial efficiency of the cells would degrade by 20-50% until it would fall to about 4%. Since 1986 new research initiatives were started to replace the single-junction a-Si devices by more efficient and more stable multijunction a-Si devices. While improving efficiency of multijunction a-Si cells to 13.3% and minimising instability of the cell's performance, another factor influencing stability has been detected: "Light degraded a-Si cells can be heated under 100° C, and their original efficiencies can be restored. The same phenomenon affects modules outdoors. Operating temperatures in the range of 50° C are enough to cause a slow self-annealing of a-Si modules while they are outside. Especially during hot summer months, the self-annealing can raise efficiency of degraded a-Si cells toward their original value"[3].

Three other technologies using thin film semi-conductor materials are also under active investigation and are likely to become of commercial importance during the end of this decade: 1) Copper Indium Diselenide thin film cells, 2) Cadmium Telluride thin film cells, 3) Thin Film Silicon cells. Many appear to offer considerable improvements in efficiency and higher stability over long exposure to light. These technologies are presently at the commercial pre-production stage and as data on commercial scale practice are not yet available they will not be considered within this paper.

Environmental Impacts of Photovoltaics

Reduced and still falling total costs of PV systems, improved reliability and conversion efficiencies as well as a rapid increase in the volume of production for the commercial market have helped to improve the competitiveness of PV technologies and to ensure that PV is now regarded as a particularly important potential source of electricity generation in the near future. This is all the more important as environmental concerns are emerging as a powerful driving force behind the development of renewable energies meant to replace conventional energy systems in the future in order to help to alleviate and reduce their potential of polluting and damaging the environment. PV could make a major contribution to energy supply on an environmentally sustainable basis , as it is known as an energy source with comparatively little impacts on the environment and human health.

Environmental concerns will demand a full life-cycle assessment of all competing energy technologies, as only the identification of the "upstream" and downstream" activities of energy systems will allow a comparative evaluation of their environmental burdens, impacts and costs.

In order to perform a life-cycle analysis it is essential to establish the relevant environmental impacts associated with PV systems. The types of environmental impacts listed and discussed below refer to the various stages of the life-cycle of single crystalline and polycrystalline silicon cells and amorphous silicon cells.

- land use,
- visual impacts,
- influence on local climate,
- material input,
- energy input,
- emissions,
- waste disposal,
- accidents and health risks.

Land Use

Land-use impacts of installed PV system will differ for decentralised and centralised applications. Photovoltaic modules as small or intermediate, residential or commercial decentralised devices can be used as cladding on the south-facing walls or on roofs of buildings - commercial buildings in particular. No additional land is required. PV land requirements for large central electricity generation depend on several factors including insolation and system efficiency. Land requirement for ground-mounted PV systems can be substantial with 2ha/MWp or $3km^2/100MWp$[4]. However, PV arrays can be arranged and set up in a way that the area covered is available for farming or other alternative uses, and the land returns to a green-field site when the plant is decommissioned and removed.

Visual Impacts

Large PV arrays could certainly have adverse visual impacts, if installed in areas of natural beauty or other sites of scientific interest. As with land use impacts, visual impacts will vary by application type. The extent to which decentralised PV systems, e.g cladding of buildings, are perceived as acceptable or not depends on the way planners are able to integrate PV systems as part of the specific characteristics of the location. As with aesthetical values the general public's perception of new technologies reflects the values held at a particular time by society.

Impacts on Local Climate

Negative impacts on the local climate due to the use of PV technologies can only occur in connection with large central PV power stations when a large area covered with PV arrays would change the reflection of the incoming sunlight. In the Sahara, the same situation could influence the local climate in a positive way by helping to create a green area underneath the PV arrays.

Material Input

Each of the process steps have inputs of energy and materials and require capital equipment, and each also has potential hazards associated with it. Material input and its environmental and health impacts are very much dependent on the technology used. Table 1 and 2 indicate the potential material input used for the manufacturing of PV modules. As new developments in the manufacturing of PV cells aim at minimising costs and material input this will in the long run lead to a reduction in the use of materials and their associated problems.

The first process step in the production of any photovoltaic cell is a standard mining operation with associated hazards to the miners and inputs of diesel fuel and machinery. Metallurgical grade silicon is made in large quantities for the steel industry, with a small fraction going as input to the semiconductor industry. Its major emission is silica dust which can cause lung disease. During the manufacturing process silicon needs to undergo a purification process that can involve materials such as silane that are explosive and highly toxic. The doping of silicon involves toxic chemicals such as diborane, and phosphine, although only in small quantities diluted in inert gas. These materials are used in the microelectronics industry and they have proven to be capable of monitoring and controlling them safely despite their toxicity. Another supplementary component required for the manufacture of PV cells is glass. The fabrication of flat glass requires a substantial input of materials and energy depending on the thickness of the glass plates. The materials for construction of the complete PV system, other than the PV modules, are steel, aluminium, copper, concrete and electronic equipment, which are associated with the standard industrial hazards.

Table 1: Material input per MWh of generated electricity; assuming 1000 kWh/m^2 p. a. production rate with 10 5 efficiency and 20 years lifetime[5]

poly-Silicon		CIS		Cd/Te	
glass	2-5 kg/MWh				
frame	0.01-1 kg/MWh				
Si	1.5 kgMWh	CU	1.5 g/MWh	Cd	2.7 g/MWh
		In	2.5 g/MWh	Te	3 g/MWh
		Se	4.5 g/MWh	Sn, Zn, Ni	small amount
		Mo	8 g/MWh		

Table 2: Process materials per kWh of generated electricity for single crystalline silicon under same conditions as in Table 1

KOH	HNO$_3$	N$_2$	Phosphine	HF	O$_2$	Solvents
kg/MWH	kg/MWh	m^3/MWh	kg/MWh	kg/MWh	kg/MWh	kg/MWH
0.5	0.05	15	0.004	0.25	0.1	0.007

Energy Input

The energy used in manufacturing the PV modules and the other components of the PV system is derived from the fuel mix of society and is therefore associated with emissions of greenhouse gases and acidic gases. The energy content of PV modules using silicon wafers has been measured at the Photowatt plant by Palz and Zibetta[6] as 235 kWh (el)/m^2 for 1990 technology at 1.5 MWp/year production rate. The energy content for PV modules using amorphous silicon has been measured at the commercial manufacturing plant of Chronar by the same team as 68.8 kWh/m^2 for 1990 technology at 1.2 MWp/year production rate[6].

Table 3:   Carbon Dioxide Emitted in the Production of PV Modules

Cell Material	Production Scale	Efficiency (%)	Lifetime (years)	Carbon Dioxide Output (kTonnes/GWyr)
Crystalline silicon	Small	12	20	400
	Large	16	30	150
Multicrystal silicon	Small	10	20	400
	Large	15	30	100
Thin film silicon	Small	10	20	130
	Large	15	30	50
Thin film polycryst. materials	Small	10	20	100
	Large	14	30	40
Future (2020) multijunctions	Large	30	20	24

For comparison:- the most efficient coal plants emit 9 million tonnes of carbon dioxide per GWyr, whilst a BWR nuclear plant emits 75,000 tonnes of carbon dioxide per GWyr (plus an unknown amount for decommissioning and waste treatment).

The average solar input in Southern Europe is assumed to be 3kWh/m^2/day. If this is converted by a PV power plant into 400 Wh/m^2/day times 365 times 20 years lifetime the plant could generate a total energy output (el) of 2920 kWh/m^2. For 1 m^2 of crystalline and multicrystalline silicon modules a small scale 1990 manufacturing plant requires 235 kWh/m^2 energy input from conventional energy mix, which is equivalent to 200 kg CO_2. The output of PV modules thus has an equivalent emission rate for CO_2 of 0.07 kgCO_2/kWh. The CO_2 emissions estimated for the various current PV technologies are shown in Table 3, along with an estimate of values which might be typical of 2020 PV technology. The thin film polycrystalline materials have been studied by Hynes et. al.[7] and those data in Table 3 are taken from their work.

Table 4:   Hazardous Emissions from Photovoltaics

Material	Production	Operation	Disposal
Silicon	Silica dust Silanes Diborane Phosphene Solvents		
Copper Indium Diselenide	Hydrogen Selenide Cadmium Oxide Cadmium Dust Selenium Solvents	Cadmium Selenium (In a fire)	Cadmium Selenium (If not recycled)
Cadmium Telluride	Cadmium Oxide Cadmium dust Tellurium Solvents	Cadmium Tellurium (In a fire)	Cadmium Tellurium (If not recycled)
There is an energy input to all technologies, derived from the fuel mix of the nation producing the PV.			

Emissions

Air emissions from PV silicon systems are exclusively related to the construction of the plant and the emissions from steel, concrete, copper and aluminium plants that manufacture the raw materials and the plant for fabrication of PV cells. Hazardous gases can be released into the air if improperly handled and disposed . Quantities of pollutants produced at photovoltaic plants should be small, and properly designed pollution control systems such as multistage scrubbers can be 98% efficient and thermal incineration can remove 100% of the SiH_4, SiF_4, PH_3 and B_2H_6 emissions from effluent streams[8]. Table 4 shows the various PV technologies and the various stages were emissions could pose an environmental and health hazard, if not handled properly.

Accidental releases of large quantities of hazardous air pollutants may result either from leaking from storage, distribution or process systems or from the venting of process or control equipment during conditions such as fire, power failure etc. These releases could pose health risks to the employees and public alike. However, proper and careful management of stored gases as well as ventilated sheds with monitoring and fire-prevention devices can reduce the risks of accidental releases.

Waste Disposal

Potential problems of waste disposal arise during the extraction of raw materials and during the manufacturing stage. Used solutions such as acids and solvents can contaminate water sources if disposed without due care. Solid toxic wastes will require careful handling and disposal in controlled conditions. The decommissioning of PV modules can be seen as another problem to be considered although the recycling of the modules would solve the problem in an environmentally benign way.

Accidents and Health Risks

The most important hazard in PV cell manufacturing plants arises from the large variety of toxic gases and from electrical equipment if designed and used improperly.

Risk of fire and of burns due to contact with hot substrate materials or surfaces are another hazard associated with PV cell fabrication. But as with other hazards at this stage properly managed work places make the likelihood of occupational health risks very small.

The exposure to PV generated electricity could produce serious health effects. The risk of electrical shock could arise while installing, maintaining or removing roof-top PV systems. Fires associated with roof-top PV arrays caused e.g. by short circuits can pose another hazard though the probability of a fire caused under such circumstances has not yet been estimated[8].

Conclusion

It was the intention of this paper to identify the environmental impacts associated with two major and already commercially produced PV technologies used for power production, flat plate silicon and amorphous silicon cells. Both photovoltaic systems have been proven to be almost entirely benign in operation and environmental hazards mainly occur at the production and disposal stage. As the methods of monitoring and controlling these environmental hazards are well established and can be reduced to within international safety limits, the potential damage to the environment and human health can be seen to be small compared to the impacts of other electricity generating technologies.

References

[1] Energy Technology Support Unit (ETSU), 1990, "Review of Solar Energy Technology, Part III, Photovoltaic Power", ETSU, unpublished paper.

[2] Hill, Robert, 1991, "Solar Power Steps out of the Shadows", Physics World, Vol. 4, No 8, 30-34.

[3] Zweibel, Ken, 1990, "Harnessing Solar Power - The Photovoltaics Challenge", Plenum Press, New York/London.

[4] Fthenakis, , V.M., 1985, "Hazards from Radio-Frequency and Laser Equipment in the Manufacture of a-Si Photovoltaic Cells", BNL 518153, Brookhaven National Laboratory, Upton, New York.

[5] Zittel, W. and A.E. Baumann, 1992, "Ökologische Belastungen durch solare Stromerzeugung im Vergleich zu konventionellen Systemen", Proceedings for Intern. Sonnenforum, June, Berlin.

[6] Palz, W. and H. Zibetta, 1991, Intern. Journal of Solar Energy, 12, 3/4.

[7] Hynes, K.M., N.M. Pearsall and R. Hill, 1991, Proc. 10th EC Photovoltaics Solar Energy Conference , pp 461-464, Kluwer Acad. Publ., Dordrecht.

[8] OECD, 1988, "Environmental Impacts of Renewable Energy",OECD, Paris.

[9] A.E. Baumann and R. Hill, 1992, "Quantification of External Costs of Energy Technologies Applied to Photovoltaic Life-Cycle Analysis", paper presented at the OECD Workshop on Life-Cycle Analysis: Methods and Experience, 21-22 May 1992, Paris.

13. External Costs of Rational Use of Energy

Hermann Herz
Fraunhofer-Institut für Systemtechnik und Innovationsforschung (FhG/ISI)
Breslauer Str. 48, D-7500 Karlsruhe 1

1 Introduction

The evaluation and enumeration of external costs of energy consumption has gained increasing interest in recent years, since it has become obvious that a variety of environmental and societal damages are caused by the energy system. As long as these external costs are not included into the prices of energy, market mechanisms will never tend to reach the optimal structure of the energy system: offering the energy with minimum overall costs.

Investigations and enumeration of external costs and the recommendation of measures to incorporate these costs into the market oriented decision process in the energy sector have so far concentrated on the supply side only, and here mainly on the supply of electricity. This sector is characterized

- by its high flexibility in choosing between different options to generate electricity and in consequence the possibility to determine type and level of external costs, and

- that energy policy has significant opportunities to interfere: via rules and regulations public authorities in general have a direct influence in decisions about the type of power plants to be constructed and the prices (tariffs) of electricity.

It is one of the characteristics of the energy system that the marketed products (fuels such as electricity, gas, oil products) are not the final utility the consumer desires: it is the energy service the consumer is looking for, such as, for example

- comfortably temperated rooms and buildings or warm water,
- the lighting of rooms or streets,
- the transport of persons or merchandise,
- the heating, transformation or conversion of materials during manufacturing.

When including these energy services as the final demand into the systematic approach to find an optimal overall energy system, the variety of options increases significantly. Since there are numerous different technologies available to offer the required energy services via different fuels, the total costs of these services, including all their external costs, may vary broadly. Not taking into account these options with their respective internal and external costs intuitively will lead to a less desirable solution.

It has become common wisdom in the last years that investments or organizational measures to reduce the specific consumption of conventional fuels to meet a given demand for a specific energy service are often more convenient and cheaper than increasing the supply system of the respective fuel. This is often due to increasing marginal costs of the energy supply system (induced by e.g. limits to the economy of scale and security aspects, or reduced availability of resources such as hydropower) or the societal problems involved with the additional construction (e.g. nuclear energy). "Rational Use of Energy" and "Demand Side Management" therefore have become subjects with ever more increasing importance in energy planning[1].

2   Rational Use of Energy and the Concept of External Costs Analysis

It is an established concept when evaluating external costs, to follow the process cycle of a fuel - that is the whole way from extraction/production via transport and distribution until its final delivery to the consumer - and count for all corresponding emissions and burdens involved and which are not yet computed in the (internal) cost calculations. The assessment of the resulting impacts on society and their finally quantification leads to monetary values, serving as a guide-line for financial measures of their internalization.

It is obvious that the total costs of a fuel cycle grow with the increasing depths of the analysis and that the proportion of processing costs expands for account of the resource costs. Latter costs are generally fully internalized whereas the negative effects of their production, distribution, and consumption are the main components of external cost estimates. This implies that the proportion of external costs within total costs will increase with the incorporation of additional steps in the overall fuel cycle. It seems to be important to analyze, how total costs of measures of rational use of energy compare to their alternatives of expanding the supply.

3   The Importance of Rational Use of Energy

3.1   Definition

All energy demand induced by the need for energy services at the various stages of the conversion chain could be reduced in the future by a more rational use of energy (RUE). Here, the term RUE is understood in its technical sense and designates all forms of reduced direct energy use in an energy converting process, i.e. it comprises reduction of energy and exergy losses during the conversion process as well as the reduction of the specific demand for useful energy. The term "energy saving" is defined in a wider sense, and comprises beside the rational use of energy (or improved efficiency) the reduction of energy services, e.g. lowering the indoor temperature or reducing travelling speed.

3.2   The Future Potentials of RUE

Most recent energy demand projections and environmental analyses support the view that energy conservation will be one of the most important measures to restrict energy demand in industrialized countries. RUE will have to contribute to the solution of the most urgent energy problems in the future.

Many OECD countries have decreased their primary energy intensity, defined as the ratio of total primary energy consumed to gross domestic product, by 1.5 to 2.7 % per year between 1973 and 1988. Some 75 to 90 % of these changes are caused by improved technical energy efficiency[2]. In the latest report of the German Enquête-Commission "Preventive Measures to Protect the Earth's Atmosphere"[3] the importance of RUE measures is impressively proved, demonstrating the major contribution to the ambitious goal to reduce greenhouse gas emissions in Germany by 25 % until the year 2005. Figure 1 resumes the future technical RUE potentials, relative to the energy consumption of the year 1987 (the figures actually show CO_2 reduction potentials, which in general are almost identical to reductions in energy consumption). The largest potentials may be seen in the field of residential heating with 70-90 % reduction, depending on the type of building and heating system. Electricity consumption of appliances may be reduced by 30-70 %.

A similar effort to estimate the potentials for a future CO_2 reduction has been initiated by the CEC for all member States of the European Community: several options have been analyzed

Figure 1:   Technical potentials of efficient energy use in the FRG
             (Source: [3] Vol. 2, p. 163)

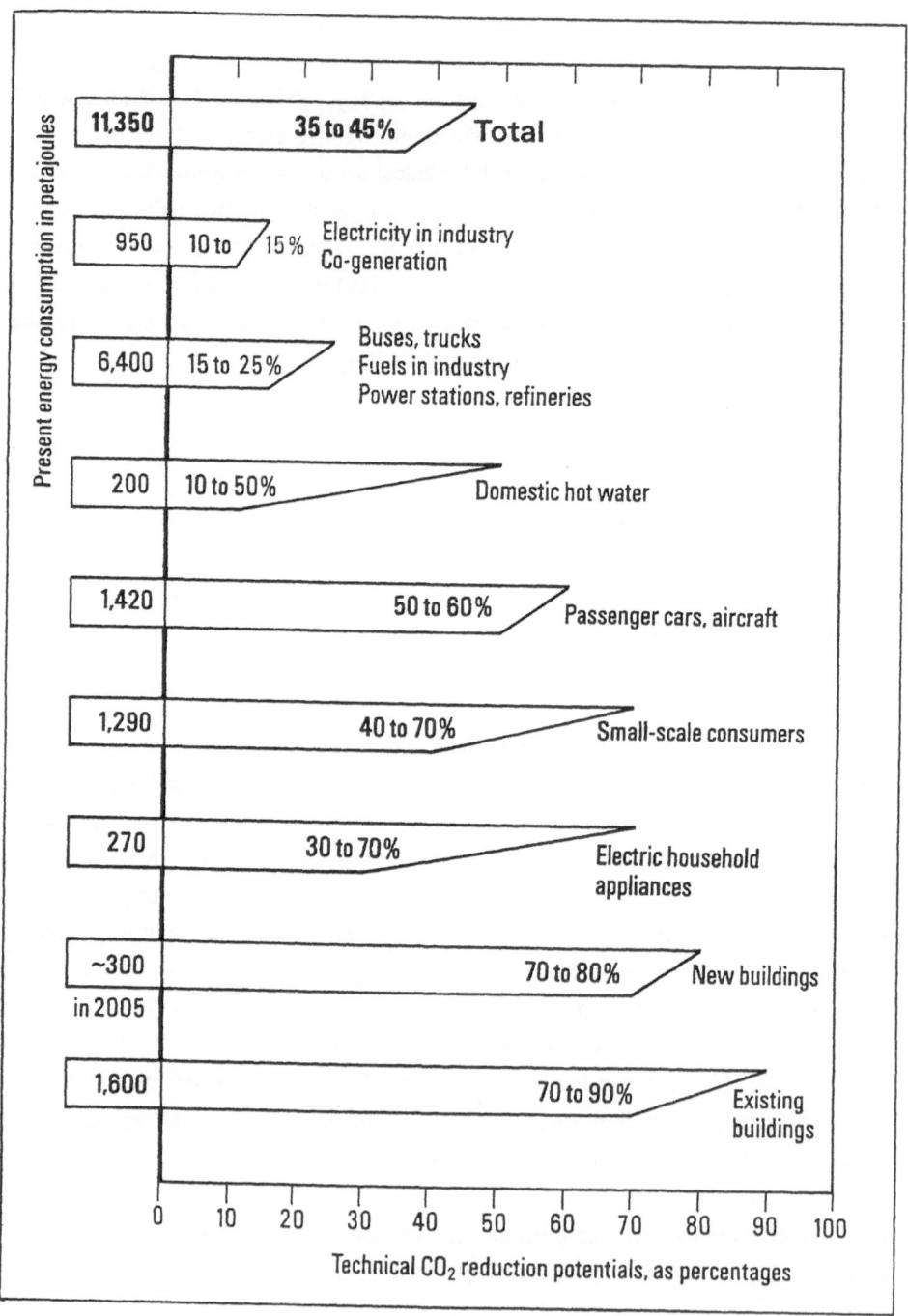

with respect to their cost-effectiveness to reduce CO_2-emissions[4]. The advantages of measures of RUE (MURE 1 and 2: MURE: Mesures d'Utilisation Rationelle de l'Energie) over

- measures of renewable technologies (DERE),
- high efficient fossil fuel generating technologies (FRET),
- fuel switching, and
- increased implementation of nuclear energy

have been demonstrated, not only with respect to the absolute amount of CO_2 emissions potentially reduced (Figure 2), but also with respect to the incorporated costs of CO_2 reduction (not including external costs).

The efforts to quantify the future technical and economic potentials of RUE measures are permanently continued on a national and international basis: the ongoing IKARUS project in Germany is evaluating more than 2 500 technologies with respect to their energy conservation potentials, their related costs, and reduced emissions[5]. Very similar activities of the CEC to compare such RUE measures for all Member States of the EC will be continued in the future and finally aim at a compound computerized data base.

The point has been made by several individuals in administration and industry that the potential of rational energy use may be more or less exhausted within the next 20 to 30 years. As has been demonstrated earlier, the actual potential of further energy efficiency improvements is estimated to be at least 40 % in western industrialized countries for the next few decades with major potentials in space heating, electrical appliances, transport and industry. Further great potentials are possible, if energy conservation concentrates on the improvement of exergetic efficiency and the reduction of the useful energy demand. The overall exergetic efficiency (the relation between availability or capability to work of useful energy and the availability of primary energy) for industrialized countries is estimated to be in the order of only 10 %[6][7] and theoretical potentials of improving energy efficiency in the next century have been estimated to be in the order of at least 80 %[8].

From the above mentioned arguments it is obvious that in the future the options of RUE will gain and increasing importance in the development of the energy system. Therefore it is imperative that RUE has to be integrated into the framework of accounting external costs,

- to include one of the most promising alternatives in the analysis of future energy options, and

- not to forget to evaluate the probable adverse effects and the related external costs of RUE, which are mainly in the field of manufacturing, installation, and operation.

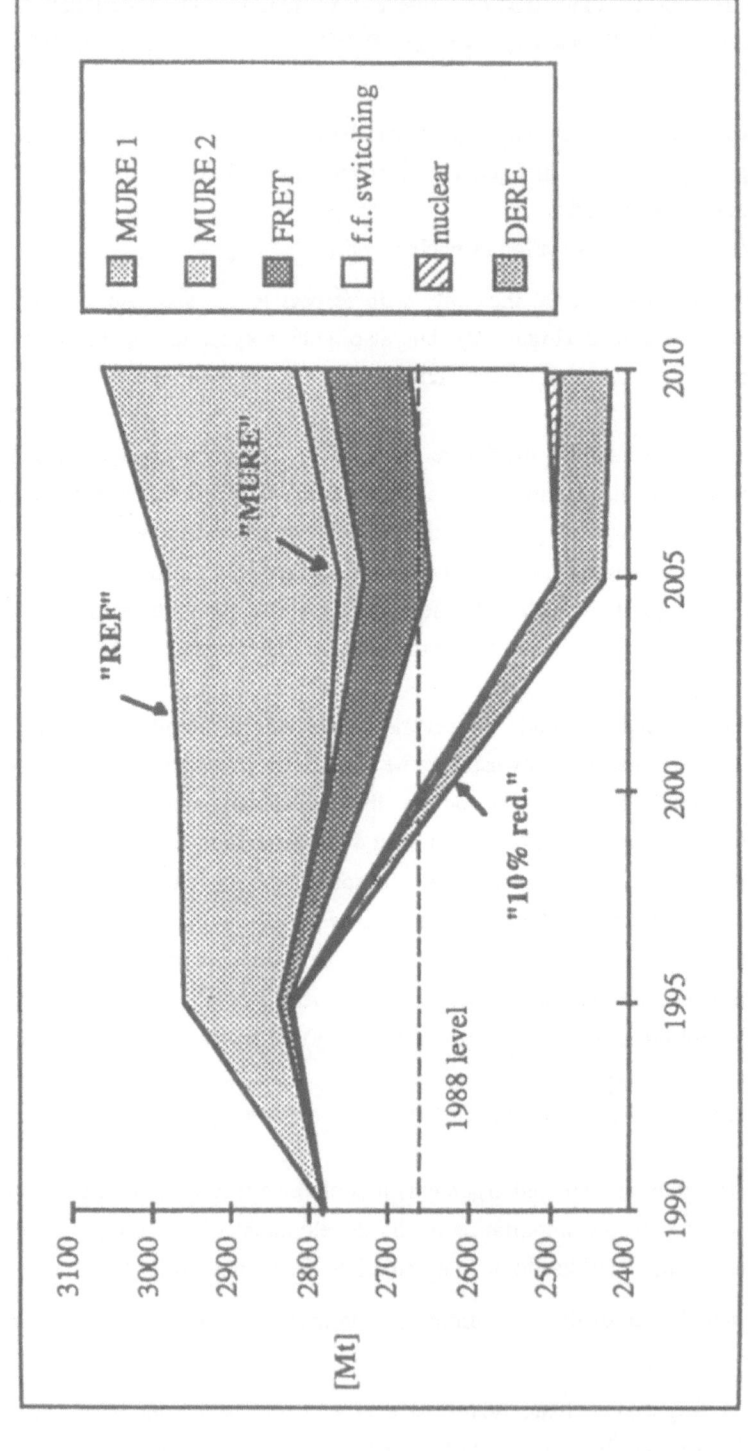

Figure 2: Contribution of technologies to the abatement of Community emissions (in Mt) (Source: [4] p. 28)

3.3 The Characteristics of RUE Technologies

The characteristics of RUE technologies vary from those of the conventional energy supply technologies. The economic activities of the supply system are concentrated in the extraction, transformation and transportation of fuels, generally concentrated in centralized production complexes. The analysis of the burdens of these activities concentrates up to now on the direct emissions and burdens induced by these activities, taking into account a regular operation and possible accidents.

RUE technologies are in most cases very decentralized measures, often distributed over millions of end consumers (house insulation, private cars, domestic appliances). Therefore the analysis of burdens of these technologies have to concentrate on

- the materials used and the variety of different processes involved in their production (e.g. CFC emissions in refrigerator production),

- their installation and operation (e.g. accidents during installation, harmful fibre emissions from mineral insulation materials[11]),

- and their final treatment after their lifetime (waste disposal, problems of toxic components;[9]

4   The Evaluation of External Costs: Two Approaches

Since in-depth investigations about external costs of RUE measures are just beginning to start, no detailed and coherent results may be presented until now. It is the purpose of this presentation to demonstrate two different components of a comprehensive investigation, which will be used in future investigations.

Taking the great potentials of RUE in the future, two different technologies have been roughly analyzed:

- The **insulation of private homes** to reduce energy consumption for space heating was chosen to demonstrate the involved external costs of insulation materials and their relation to the external costs of the equivalent energy service, provided by electrical space heating systems.

- The introduction of **compact fluorescent lamps** instead of conventional bulbs to reduce electricity consumption for lighting purposes. This example will demonstrate the utilization of an analytical instrument that concentrates on the evaluation of the burdens (emissions) of the manufacturing industries and electricity generation (Input-Output Technique).

Both methods will later be used for an integrated analysis of RUE technologies.

4.1 The Insulation of a 1-Family House

4.1.1 Technical Description

The potential of energy savings in residential heating has already been mentioned in chapter 3.2 and may be estimated to range between 70 to 90 %, depending on the type of building. A recent analysis calculated the energy savings of three different measures to insulate a 1-family house[10]:

(1) Putting a 15 cm mineral wool layer under the roof of a typical 1-family house (with a roof area of 81 m^3), reduces the k-value from 1.85 (W/m^3*K) to 0.3 (W/m^3*K), and therefore implies a reduction of the corresponding losses of 84 %.

(2) A 12 cm polystyrene hard-foam cover of the walls (113 m^3) reduces the k-value from 1.10 to 0.25 (W/m^3*K), representing a reduction of related losses by 77 %.

(3) A 12 cm polystyrene hard-foam layer on the ceiling of the basement implies a k-value reduction from 0.89 to 0.32 (W/m^3*K), equivalent to a 64 % reduction of losses through this area.

The absolute energy consumption figures (expressed in the localization their corresponding losses) and their potential amounts to be saved are shown in table 1:

Table 1:   Useful energy demand for space heating and the related energy savings by different RUE measures

Insulation measure	normal losses (kWh/a)	reduced losses (kWh/a)
roof: 15 cm mineral wool	7 550	1 224
wall: 12 cm polystyrene hard foam	3 861	877
basement: 12 cm polystyrene hard foam	3 460	1 242
total:	14 871	3 343
Total useful energy saved		**11 528**

Regarding to this example, the losses may be reduced by 78 %, resulting in an annual energy saving of 11 528 kWh.

4.1.2 The Induced Burdens (Emissions)

Hoffman analyzed the emissions of the insulation measure on the basis of a detailed process analysis of the materials polystyrene, hard foam, and mineral wool and concluded that significant emissions may only be found on the production side. He neglects the possible dangers of fibre emissions of mineral wool, a fact that has become the subject of recent adverse discussions[11]. Concluding the results of Hoffmann, the air emissions of the production are summarized in table 2.

Table 2:   Emission estimates of the insulation of a 1-family house

Activity	Annual *) air emissions (g/a)					
	SO_2	NO_x	CO	Part.	VOC	CO_2
Production of PS hard foam	354	161	40	57	13	76 109
Production of mineral wool	59	27	7	9	2	12 761
Installation of insulation			---	not evaluated	---	
Utilization	0	0	0	???	<<	0

*) Total emissions distributed over a life-time of 25 years

4.1.3 Additional Impacts

So far, investigations are concentrated on the damages of air pollutants on the environment, including human health, flora, fauna and material. More exhaustive analysis also takes into account known effects on e.g.

- **depletion (user) costs** of nonrenewable energy, which takes into account the limited nature of resources and a corresponding price mechanism that changes fuel values (and prices) with respect to their diminishing availability[12]. The results of the different models demonstrated that the depletion costs amount to substantial part of the present energy prices[13]

- costs of a future **climate change**, induced by the emissions of the greenhouse gases, mainly CO_2. Although these external costs have so far been monetized to a very small extend, they may become a central and dominant external costs element in the near future[14].

The measures of RUE are generally characterized by the fact that the energy intensity of an energy service is reduced by an unique investment (increase in capital intensity). In case of an economic measure, investment costs are more than compensated by saved fuel expenses. Detailed investigations of the net effects of insulation measures showed the macroeconomic advantages (net analysis compares all effects of alternatives to provide a given energy service, also taking into account the consumption patterns of saved budgets[15][16]). RUE measures in house insulation are

- resulting in an **increase in GDP** (here mainly higher income of entrepreneurs and employees) in the order up to DM_{87} 5 700 per PJ saved energy, and

- in times of severe unemployment additional social benefits of **increasing employment** occur, due to savings in social expenditures for the unemployed and increased income tax revenues. The employment effect has been estimated to be approximately 100 new jobs per PJ saved energy and their respective benefits are estimated to be DM_{87} 25 000 per year and additional job .

4.1.4 Monetized External Costs/Benefits

The numerical evaluation of the costs/benefits of a measure uses given relations between burdens (impacts) and their respective physical damages (dose-response functions) as well as monetary values of these damages.

Using the values for the monetization of environmental damages of the energy system from Hohmeyer[17], and actualized by Hoffmann in 1991, the total external costs/benefits of the insulation measure may be calculated and compared to the equivalent costs/benefits of the same energy service "space heating", provided by an electric heating system, using coal based electricity. The results of the enumeration are given in table 3.

What may be seen from the results is the fact that the RUE measure has an obvious advantage over the conventional technology in mainly two fields:

- the almost complete avoidance of air pollution and therefore absence of related external costs, and

- substantial macroeconomic net benefits, mainly due to substitution of energy imports by domestically produced and installed insulation material as well as additional consumption of the house owners.

In total, the investment in thermal insulation of homes has a comparative advantage in this example of more than 6 DPf/kWh.

Table 3:   Comparison of costs (+) and benefits (-) of house insulation versus (coal-based) electric space heating in DPf $_{87}$ / kWh useful energy

Cost Category	Thermal Insulation	Electric Space Heating
Environmental costs of air pollution		
- crops, timber	< 0.01	0.4 ... 0.5
- livestock	< 0.01	<0.01
- buildings & material	< 0.01	0.01
- human health	< 0.01	0.06 ... 0.23
total environm.:	~ 0.02	0.47 ... 0.74
Other external costs		
- climate change	> 0.02	> 2.55
- depletion costs	< 0.01	< 0.06
- net impact on GDP	-2.09	-
- employment (net)	-1.33	-
total others	-3.39	2.61
TOTAL:	**-3.37**	**3.08 ... 3.35**
Source: [9] p. 108		

4.2   Introduction of Compact Fluorescent Lamps

4.2.1 Technical Description

Using compact fluorescent lamps instead of conventional bulbs has demonstrated its technical and economic advantages: Although higher investment costs seem to impede a substitution, the longer lifetime and the cost savings for electricity result in advantages. A comparison of the characteristic of the two alternatives is given in table 4.

Taking the lifetime of 8 000 hours, the fluorescent lamp

- has higher investment costs of DM 29
- has a lower electricity consumption of 480 kWh,
- representing a value of DM 144 (with an electricity price of 0,3 DM/kWh).

These figures have been used to make an analysis of environmental net effects, using the methodology of an Input-Output Table, combined with a matrix of sectorial specific environmental effects. This analytical tool has been implemented recently, combining the 56 sectors Input-Output matrix of Germany (1988) with almost 60 different environmental burdens, including air and water pollution as well as waste treatment[18].

Table 4:   Comparison of the technical and economic characteristics of fluorescent lamps and conventional bulbs

Characteristics		Fluorescent Lamp	Conventional Bulb
Power	(W)	15	75
Lifetime	(h)	8 000	1 000
Electricity cons. (8000 h)	(kWh)	120	600
Investment costs	(DM)	45	2
Electricity price	(DPf/kWh)	30	30
Energy costs in 8000 h	(DM)	36	180
Investments 8000 h	(DM)	45	16
Source: [19] and own estimates			

This model is an ideal instrument to evaluate the direct emissions of a specific industrial sector as well as the indirect emissions induced by the inputs to the respective sector. Especially these indirect emissions are partially neglected in conventional process analysis but may significantly contribute to the overall emissions.

Using the economic data of the fluorescent lamp, the model was used to calculate the two alternatives:

- increase of the demand in the sector "electrical appliances", representing the additional investment costs of DM 29 for the production of the fluorescent lamp

- decrease of the demand for electricity by DM 144, representing the reduced electricity consumption during the lifetime.

The results for some basic air emissions of the alternatives are shown in table 5, comparing:

- the direct emissions of the production of the alternative,
- the emissions induced by the production of the direct inputs (1st level inputs), and
- the emissions of all other previous production steps.

Table 5: Comparison of direct and indirect emissions of fluorescent lamp production and electricity generation. All values normalized to the consumption of 480 kWh.

	SO_2 (g)	CO_2 (kg)	NO_x (g)	CO (g)	VOC (g)	Partic. (g)
Electricity production (480 kWh)						
direct emission	1 008	413	1 197	66	16	137
emissions 1st level inputs	94	42	125	60	17	25
all prior inputs	36	18	55	69	11	17
total	1 139	472	1 377	195	44	179
Fluorescent lamp production						
direct emissions	1.3	0.9	4.4	9.4	2.1	0.2
emissions 1st level inputs	5.0	2.3	8.7	15.9	2.1	1.9
all prior inputs	6.7	3.1	10.0	18.9	2.0	3.7
total	13.0	6.3	23.2	44.2	6.1	5.8

As may be seen from the table, the advantages of the fluorescent lamp are obvious: the overall emissions are significantly lower than those of the alternative electricity production. Greatest advantage may be seen in SO_2 emissions, where those of electricity generation are some 88 times higher than those of the lamp. The equivalent relation for CO_2 emissions is some 75 times and for CO and VOC some 5 to 7 times.

It may be further observed that these relations are very much dependant on the depth of the analysis: the direct SO_2 emissions of electricity production are some 87 % of total emissions, only 13 % are based on the inputs of this sector (see figure 3). In the production of the lamp

190

Figure 3: Share of direct emissions on total emissions of the alternatives

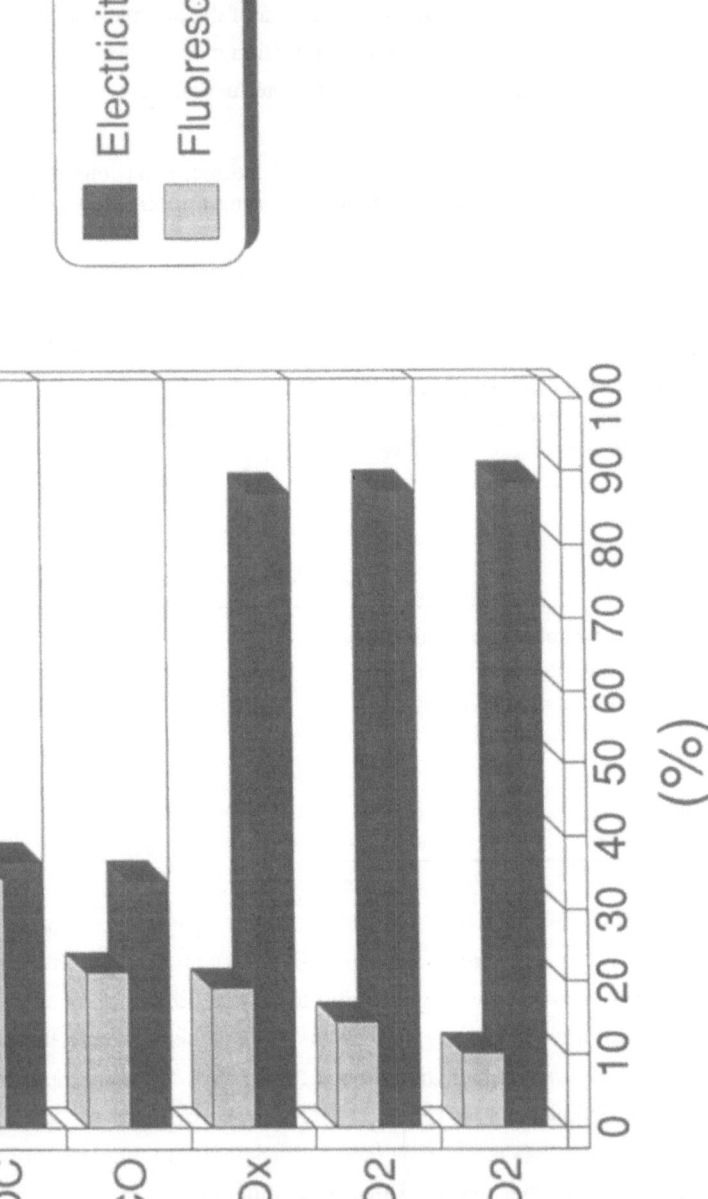

only a very small portion of the total SO_2 emissions result from the production, the major part comes from the inputs involved (e.g. electricity).

5   Conclusion

The rough estimates of the impacts of RUE measures have shown that they will have a great advantage compared to their respective alternatives of increasing the supply and therefore their incorporation in systematic analysis of energy systems is highly recommended. The true quantification of all advantages of RUE measures is important to assist the formulation of future energy policies.

Both methods used here demonstrated the importance of a detailed process analysis of the RUE technology, since major burdens may come from activities not yet taken into account in present frameworks of external costs estimation.

A new tool is readily available now that may assist in the estimation of a vast variety of environmental impacts of various industrial production processes and which will be extremely useful when performing detailed process analysis of RUE measures.

References

[1] IEA (International Energy Agency): Energy Efficiency and Environment. OECD, Paris 1991

[2] Morovic, T. et al.: Energy Conservation Indicators II. Springer. Berlin/Heidelberg/New York, 1989

[3] Enquête Commission of the German Bundestag: Protecting the Earth. Deutscher Bundestag, Bonn 1991

[4] COHERENCE: Cost-Effectiveness Analysis of CO_2 Reduction Options. Report for the CEC, DG XII. Brussels, 1991

[5] IKARUS: Instruments for Greenhouse Gas Reduction Strategies. Interim Summary Report for Project Phase 1, TFF, Forschungzentrum Jülich KFA, 1992

[6] Fredrikson, R.: The technological potential for more efficient use of energy in industry. STU-Report, Stockholm 1984, pp. 7 - 10

[7] Gyftopoulos, E. P.; Th. F. Widmer: Cost-Effective Waste Energy Utilization. *Ann. Rev. Energy* 7 (1982), pp. 293 - 327

[8] Jochem, E.: Long-term potentials of rational energy use - the unknown possibilities of reducing greenhouse gas emissions. *Energy and Environment* 2 (1991) 1, pp. 31 - 44

[9] FhG-ISI: Abfälle an mikroelektronischen Bauelementen und Leiterplatten in der Bundesrepublik. First Interim Report of an ongoing study at the FHG-ISI, Karlsruhe 1990

[10] Hoffmann, C.: Soziale Kosten des Heizenergieverbrauchs (Social Costs of Space Heating). Masters Thesis, supervised by O. Hohmeyer, FhG-ISI, 1991.

[11] Fischer, M.: Personal communication. Institut für Wasser-, Boden-, Lufthygiene, Berlin 1991

[12] Wagner, R.: Wie teuer sind uns die fossilen Brennstoffe ? *Zeitschrift für Energiewirtschaft*, 1/87, pp. 43-52

[13] Hohmeyer, O.: Beyond External Costs - A Simple Way to Achieve a Sustainalble Energy Future. Paper presented on the 2nd workshop, Racine, Wisc./USA

[14] Hohmeyer, O.; M. Gärtner: The Costs of Climate Change. Report to the Commission of the European Communities, DG XI. Karlsruhe, 1992

[15] Garnreiter, F. et al.: Auswirkungen verstärkter Maßnahmen zum rationellen Energieeinsatz auf Umwelt, Beschäftigung und Einkommen. In: Berichte 12/83 des Umweltbundesamtes, 1983

[16] Hohmeyer, O. et al.: Employment effects of energy conservation investments in EC countries. Luxembourg 1985

[17] Hohmeyer, O.: Soziale Kosten des Energieverbrauches. Springer Verlag: Berlin, 2. ed. 1989

[18] Hohmeyer, O (project leader) et al.: Methodenstudie zur Emittentenstruktur in der BRD (Methodological study on the structure of emissions in the FRG). FhG-ISI report, Karlsruhe 1992.

[19] Ebersperger, R.; W. Mauch: der kumulierte Energieaufwand von Glühlampen und Kompaktleuchtstofflampen. *HLH*, vol. 43 (1992), p. 109-114

14.   Economic Impacts of Electricity Supply Options

Ajay K. Sanghi
New York State Energy Office
Albany, New York 12223

Introduction

Public interest requires electric utilities to have a social obligation beyond the mere provision of electricity. As an example, the New York State Legislature recently passed (August 1992) an energy bill that establishes an integrated resource planning process to provide guidance to State electric utilities for procuring future electricity resources. It calls upon the State's investor owned utilities to consider all options and "select the source or sources which best serve the public interest, taking into consideration such factors as ... **preservation or creation of economic opportunities, ...**"[1] before purchasing power, investing in new plants, or repowering or extending the life of existing plants. To this extent, economic development has become an important consideration in the procurement of electricity resources.[2]

While the law requires consideration of economic development parameters, it does not provide guidance in terms of their trade-off with price. This paper presents estimates of employment and earnings impacts which would be created in New York State and the U.S. from construction and operation of select electricity generation and demand side management technologies. It also addresses the issue of how to consider economic development attributes in energy planning and decision making.

Methodology for Estimating Economic Impacts

The economic benefits which result from the construction and operation of electricity generation and demand side management resources are generally related to the earnings and employment impacts created by such expenditures. In comparing employment and earnings across alternative resources, impacts are normalized by an equal amount of expenditure made

---

1   New York State Energy Law, Section 6-106, Paragraph 5 [Senate Bill 4912-A].
2   Other factors specified include ratepayer impacts, system reliability, environmental impacts, conservation of energy resources, fuel efficiency, fuel availability and diversity, and public health and welfare.

on each technology (per million dollars of expenditure); and by an equal amount of energy service provided by each technology (per 10 Gwh).

Economic impacts per million dollars of expenditure are largely influenced by the amount of "economic leakage", or the percentage of total expenditure on imported goods from outside the regional economy. The higher the economic leakage the lower the economic impacts per million dollars of expenditure within the region. On the other hand, economic impacts based on an equivalent amount of energy service (per 10 Gwh) are largely determined by the cost of the technology. Thus, a relatively more expensive technology will generally create relatively more economic impacts on an equivalent energy service basis.

The estimates of employment and earnings impacts presented in this paper are obtained by using input-output modelling techniques. For this purpose the Economic Impact Evaluation System (EcIES) developed by the New York State Energy Office is used. EcIES estimates impacts at the state and national level resulting from an exogenous change in final demand. The primary source of data used in the EcIES is the Regional Industrial Multiplier System (RIMS II), developed by the Bureau of Economic Analysis of the U.S. Dept. of Commerce.[3]

The input/output multipliers capture employment and earnings impacts resulting from primary (direct) and secondary (indirect and induced) expenditures associated with the construction and operation of an electricity technology. Using earnings impacts as an example, the direct impact is comprised of the wages paid directly to workers who construct and operate an energy facility. The indirect earnings capture incremental wages paid to labor in the industries which provide the materials, equipment and services needed to construct and operate the facility. The induced earnings are the cumulative result of expansion which occurs as the direct and indirect earnings are spent in the economy.

The total cost of producing electricity from each technology alternative is distributed among five expenditure components: capital; operation and maintenance (O&M); fuel; finance and insurance, and taxes paid to federal and local authorities. Capital and O&M expenditures are further divided into expenditure categories (e.g., manpower and purchase of machinery and equipment).

---

[3]   Bureau of Economic Analysis, U.S. Department of Commerce. Regional Multipliers: A User Handbook for the Regional Input-Output Modelling System (RIMS II), 2nd Edition, May 1992.

The detailed breakdown by industrial sectors, termed a Bill of Goods (BOG), is used to allocate the total expenditure to the RIMS II industrial sectors.[4] Expenditures on each industrial sector are further adjusted by using location quotients to account for economic leakages.[5] The localized expenditures are then multiplied by the RIMS II multipliers to derive employment and earnings impacts.[6]

New York State Economic Impacts per Million Dollars of Expenditure

Employment and earnings impacts are presented for several electricity supply and demand side management technology alternatives. A listing of the technologies considered in this paper is presented in Table 1. Also presented are estimates of each technology's cost (real levelized 1991$) and the employment and earnings impacts per million dollars of expenditure.[7]

Technology costs reported in Table 1 are illustrated in Figure 1 for ease of comparison. As shown, the least cost alternatives are generally represented by the DSM technologies, whose costs range from about 0.8 ¢/Kwh to 2.9 ¢/Kwh (real levelized 1991$). The most expensive technologies tend to be the renewable resource alternatives (especially photovoltaics), with the cost of the fossil fuel-fired generation technologies falling somewhere between DSM and renewable resources technologies.

Employment Impacts per Million Dollars of Expenditure

The estimated employment impacts for each technology (reported in Table 1) are illustrated in Figure 2. The New York State employment estimates are strongly influenced by the composition of the total cost in terms of expenditures on fuel, O&M, capital, project financing and taxes. The percentage of each component cost to total costs (per million dollars

---

4   Beemiller, Richard. "Hybrid Approach to Estimating Economic Impacts Using Regional Input-Output Modelling System (RIMS II)" Transportation Research Record 1274.

5   The location quotient of an industrial sector is a ratio of two percentages: the percentage of the region's total employment which is in this industrial sector; and the percentage of the nation's total employment in this same industrial sector.

6   In the construction of the RIMS II multipliers, the Bureau of Economic Analysis of the U.S Dept. of Commerce uses location quotients to represent the proportion of responding in New York at each round of the multiplier process.

7   It is important to note that technologies are listing according to the magnitude of employment impacts in ascending order.

of expenditure) for each technology are presented in the upper part of Table 2. The lower portion of Table 2 shows employment impacts per million dollars of expenditure for each of the component costs. Variation in the proportion of total costs allocated to different factors, and variation in the ability of these cost factors to create economic impacts explains why the cost and job impacts from building and operating alternative electricity technologies are not well correlated. This fact is further highlighted by comparing Figures 1 and 2. There is no systematic relationship between costs (c/Kwh) of technologies and economic impacts from one million dollars of expenditure on these technologies.

Table 1: New York State Employment and Earnings Impacts of Selected Technologies per Million Dollars of Expenditure

	Cost[a]	Employment Impacts[b]		Earnings Impacts	
	(c/Kwh)	(Jobs)	(Rank)	(dollars)	(Rank)
Natural Gas Combined Cycle (NGCC)	5.81	5.28	1	144,161	1
Repower Oil/Gas with NGCC (RPNGC)	4.69	5.51	2	144,196	2
Photovoltaic (PV)	17.29	8.89	3	248,686	3
Small Hydroelectric (HYDRO)	9.17	9.81	4	289,639	4
Integrated Coal Gasif. Comb. Cyc. (IGCC)	4.49	11.90	5	332,551	6
Effic. Household Refrigerator	2.90	12.82	6	318,800	5
Circulating Fluidized Bed Coal (CFB)	5.30	13.27	7	362,105	8
Repower Coal with IGCC (RPIGC)	3.82	13.34	8	358,695	7
Advanced Pulverized Coal (APC)	4.84	13.59	9	369,550	10
Repower Coal with AFB (RPAFB)	4.29	13.66	10	363,837	9
Wind Turbine (WIND)	6.24	14.48	11	397,338	12
Press. Fluid. Bed Coal Comb. Cyc. (PFBCC)	5.53	14.60	12	395,741	11
Life Extend Conventional Coal (LECOAL)	3.39	16.4	13	424,985	13
T-8 Bulbs and Ballasts	1.13	16.44	14	437,500	14
Urban Wood Waste (WOOD)	5.96	17.47	15	470,892	15
Efficient Motors	0.81	21.50	16	559,776	16
Municipal Solid Waste (MSW)	6.82	21.74	17	590,074	18
Compact Fluorescent	1.60	21.87	18	571,976	17
Variable Speed Drives	2.33	24.96	19	661,383	20
Fiberglass Insulation	1.67	25.82	20	641,144	19

a/ Real levelized 1991 dollars
b/ Number of jobs created (full and part time)

Since fossil fuels are generally imported in New York State, the percentage of total costs which is comprised of fuel expenditures is the most important factor in explaining the variation in employment impacts across technology alternatives, as it represents economic leakage out of the State. The fuel component as a percentage of each technology's total cost (reported in Table 2) is further illustrated in Figure 3.[8]

---

8 Although the fuel cost for an MSW plant is shown as zero in Figure 3, it is generally a negative cost. This is because the MSW plant collects a tipping fee for its fuel (the removal of solid waste). Revenues from the tipping fee are estimated to reduce the net cost of electricity generated from an MSW plant by about 45 percent to 6.82 cents per Kwh. However, it is assumed that the tipping fee revenues are not reused, and therefore do not create additional economic impacts.

Figure 1:   Technology Costs (Real Levelized Cents / Kwh, 199 1$)

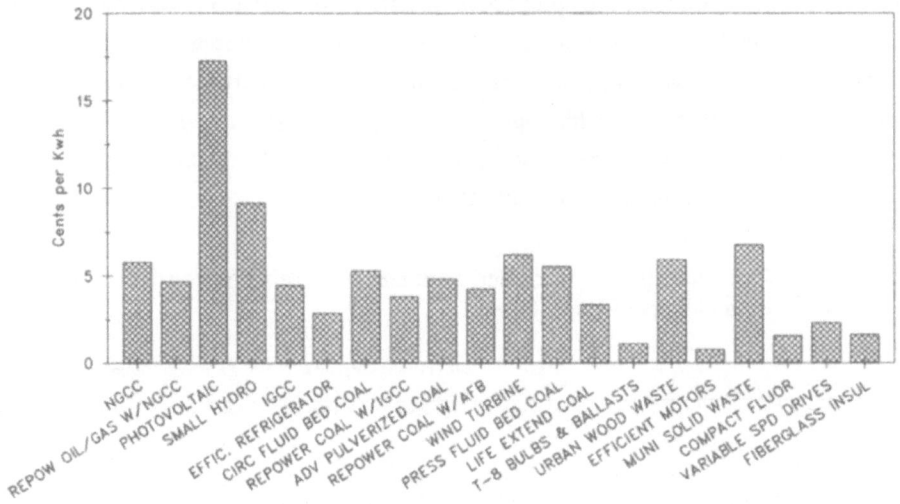

Figure 2:   New York State Employment Impacts Per Million Dollars of Expenditure

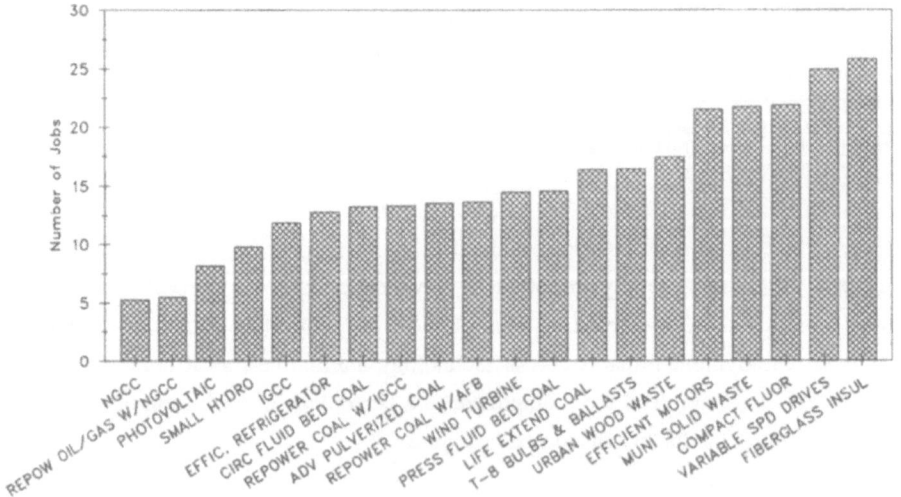

As shown in Table 2 and Figure 3, about 66 percent of the total cost of generation from a natural gas combined cycle plant consists of expenditure on gas. Since this represents the highest portion of the total cost attributed to fuel expense, the natural gas combined cycle plant creates the least amount of jobs in New York State per million dollars of expenditure.

On an equivalent expenditure basis, O&M costs generally create 40 to 50 percent more jobs than capital costs. This is because O&M expenditures tend to be more labor intensive than the other components. In addition, a large portion of the O&M expenditures consist of labor required to run the plant which is likely to be obtained locally.

Table 2:   Allocation of Total Cost by Expenditure Categories

Energy Resource Alternative	Capital	O&M	Fuel	Finance & Insurance	Taxes
		Component Cost Percentages			
Natural Gas Combined Cycle (NGCC)	3	11	66	11	10
Repower Oil/Gas with NGCC (RPNGC)	3	13	74	4	7
Photovoltaic (PV)	16	1	0	54	30
Small Hydroelectric (HYDRO)	14	7	0	51	28
Integrated Coal Gasif. Comb. Cyc. (IGCC)	11	18	25	33	13
Effic. Household Refrigerator	100	0	0	0	0
Circulating Fluidized Bed Coal (CFB)	9	27	24	28	13
Repower Coal with IGCC (RPIGC)	10	24	32	20	13
Advanced Pulverized Coal (APC)	9	25	24	29	13
Repower Coal with AFB (RPAFB)	8	30	29	18	15
Wind Turbine (WIND)	12	22	0	42	24
Press. Fluid. Bed Coal Comb. Cyc. (PFBCC)	9	28	20	29	13
Life Extend Conventional Coal (LECOAL)	10	37	36	8	9
T-8 Bulbs and Ballasts	100	0	0	0	0
Urban Wood Waste (WOOD)	13	30	5	34	18
Efficient Motors	100	0	0	0	0
Municipal Solid Waste (MSW)	20	34	0	41	5
Compact Fluorescent	100	0	0	0	0
Variable Speed Drives	100	0	0	0	0
Fiberglass Insulation	100	0	0	0	0
		Employment Impacts Per Million Dollars of Expenditure			
Natural Gas Combined Cycle (NGCC)	18.88	34.28	0.17	9.03	0.00
Repower Oil/Gas with NGCC (RPNGC)	21.27	34.28	0.17	9.03	0.00
Photovoltaic (PV)	23.92	46.05	0.00	7.70	0.00
Small Hydroelectric (HYDRO)	22.52	29.25	0.00	9.03	0.00
Integrated Coal Gasif. Comb. Cyc. (IGCC)	20.76	35.99	1.18	8.46	0.00
Effic. Household Refrigerator	12.82	0.00	0.00	0.00	0.00
Circulating Fluidized Bed Coal (CFB)	21.22	32.52	1.18	8.82	0.00
Repower Coal with IGCC (RPIGC)	24.25	35.99	1.18	8.46	0.00
Advanced Pulverized Coal (APC)	22.11	34.87	1.18	8.82	0.00
Repower Coal with AFB (RPAFB)	25.16	32.61	1.18	8.82	0.00
Wind Turbine (WIND)	21.18	37.23	0.00	8.72	0.00
Press. Fluid. Bed Coal Comb. Cyc. (PFBCC)	25.89	33.22	1.18	8.82	0.00
Life Extend Conventional Coal (LECOAL)	24.11	34.63	1.18	8.82	0.00
T-8 Bulbs and Ballasts	16.44	0.00	0.00	0.00	0.00
Urban Wood Waste (WOOD)	23.34	39.77	0.00	7.84	0.00
Efficient Motors	21.50	0.00	0.00	0.00	0.00
Municipal Solid Waste (MSW)	23.91	38.20	0.00	9.53	0.00
Compact Fluorescent	21.88	0.00	0.00	0.00	0.00
Variable Speed Drives	24.96	0.00	0.00	0.00	0.00
Fiberglass Insulation	25.83	0.00	0.00	0.00	0.00

Capital expenditures, on the other hand, are allocated among a variety of industrial sectors. Capital purchases from industrial sectors which are concentrated within the state will result in less economic leakage than purchases from sectors concentrated outside of the state. Thus, employment impacts associated with capital expenditures seem to vary more across alternative technologies than economic impacts associated with O&M expenditures, due to the economic leakages.

Figure 3:   Percentage of Fuel Cost in Total Expenditure

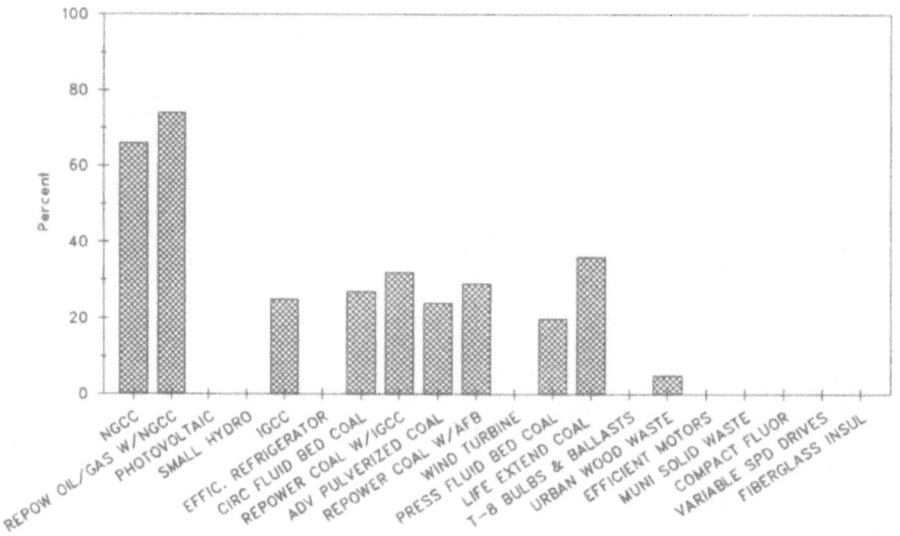

Taxes are treated as transfer payments. As such, there are no economic impacts associated with tax payments. Finance and insurance expenditures consist largely of financing costs which also act for the most part as transfer payments. It is assumed that only 5 percent of a project's financing costs induce economic impacts.

Figure 2 shows that there is substantial variation in the employment created across alternatives. As a group, the DSM and renewable resource technologies create more jobs per million dollars of expenditure than the fossil fuel-fired alternatives. Renewable technologies examined in this paper do not use fossil fuels. For these technologies, the jobs created tend to be on average 15 percent higher than the jobs created by fossil fuel-fired alternatives, as shown in Table 1.

The employment impacts per million dollars of expenditure across the group of renewable resources are of a similar order of magnitude. Differences among this group are largely explained by the percentage of financing costs and taxes in the total expenditure. As shown in Table 2, these components comprise 84 percent of the total expenditure for photovoltaic, 79 percent for small hydro, 66 percent for wind and 52 percent for urban wood waste. Since expenditures on finance generate lower employment impacts than expenditures on capital and O&M, photovoltaic and small hydro create a lower number of jobs per million dollars of expenditure than wind and urban wood waste.

The employment estimates do not indicate the intertemporal distribution of the jobs. However, the employment created by different components of the total expenditure can be used to approximate the sustainability of employment on an annual basis. Since capital expenditures generally occur only during the construction period, the resultant employment impacts are not likely to be sustained over time. On the other hand, O&M expenses are required over the entire useful life of the technology. Therefore, the proportion of total cost comprised of O&M expenditure indicates the sustainability of the economic impacts of that particular technology over its remaining life.

Earnings Impacts per Million Dollars of Expenditure

Figure 4 illustrates the estimated earnings impacts per million dollars of expenditure reported in Table 1. Earnings impacts are highly correlated with employment impacts. For example, the DSM technologies tend to generate higher than average employment and earnings impacts per million dollars of expenditure across alternative technologies. Fossil fuel generation technologies such as NGCC create very small amount of jobs as well as earnings impacts.

There are some exceptions to the correlation between earnings and employment impacts. For example, high efficiency refrigerators, compact fluorescent lighting, and fiberglass insulation technologies create relatively small earnings impacts in relation to their employment impacts. This is because a large proportion of the jobs created by these technologies involves installation activities using low cost labor.

New York State Employment and Earnings Impacts per 10 Gwh

In assessing the technology alternatives, it is also useful to evaluate potential employment and earnings impacts on an equivalent electricity service basis (10 Gwh). The estimated employment and earnings impacts per 10 Gwh for New York State are reported in Table 3 and illustrated in Figures 5 and 6, respectively. It is important to note that technologies continue to be listed according to the magnitude of their employment impacts per million dollars of expenditure (as in Figures 1, 2 and 3) for identification of alterations in ranking.

Figure 4:   New York State Earnings Impacts per Million Dollars of Expenditure

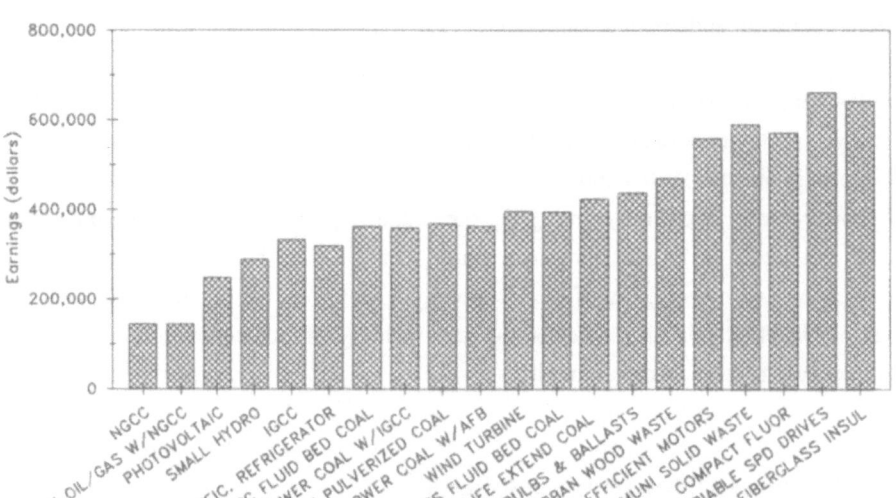

Relatively more expensive technologies generally have larger employment impacts per Gwh of electricity output. As shown in Table 3 and Figure 5, photovoltaics is the most expensive technology, and it creates the second highest employment impacts. On the other hand, the DSM technologies as a group create the fewest employment impacts per 10 Gwh across all alternatives because they are the least expensive technologies.

A big discrepancy is found in the ranking of technologies by employment impacts per 10 Gwh (Figure 5) when compared to employment impacts per million dollars of expenditure (Figure 2). For example, in terms of employment impacts per million dollars of expenditure, photovoltaics is the third lowest ranking technology. On the other hand, photovoltaics, is the nineteenth lowest ranking technology in terms of employment impacts per 10 Gwh.

Table 3: New York Employment and Earning Impacts of Selected Technologies per 10 GWH of Output

	Cost[a]	Employment Impacts[b]		Earnings Impacts	
	(c/Kwh)	(Jobs)	(Rank)	(dollars)	(Rank)
Natural Gas Combined Cycle (NGCC)	5.81	3.07	4	83,758	4
Repower Oil/Gas with NGCC (RPNGC)	4.69	2.59	3	67,628	3
Photovoltaic (PV)	17.29	14.16	19	429,978	19
Small Hydroelectric (HYDRO)	9.17	9.00	16	265,599	17
Integrated Coal Gasif. Comb. Cyc. (IGCC)	4.49	5.34	9	149,316	10
Effic. Household Refrigerator	2.90	3.72	6	92,452	6
Circulating Fluidized Bed Coal (CFB)	5.30	7.03	14	191,916	14
Repower Coal with IGCC (RPIGC)	3.82	5.10	8	137,021	8
Advanced Pulverized Coal (APC)	4.84	6.58	13	178,862	13
Repower Coal with AFB (RPAFB)	4.29	5.86	12	156,086	12
Wind Turbine (WIND)	6.24	9.04	17	247,939	16
Press. Fluid. Bed Coal Comb. Cyc. (PFBCC)	5.53	8.07	15	218,845	15
Life Extend Conventional Coal (LECOAL)	3.39	5.56	10	144,070	9
T-8 Bulbs and Ballasts	1.13	1.86	2	49,437	2
Urban Wood Waste (WOOD)	5.96	10.41	18	280,652	18
Efficient Motors	0.81	1.74	1	45,341	1
Municipal Solid Waste (MSW)	6.82	26.85	20	728,742	20
Compact Fluorescent	1.60	3.50	5	91,516	5
Variable Speed Drives	2.33	5.82	11	154,102	11
Fiberglass Insulation	1.67	4.31	7	107,071	7

a/ Real levelized 1991 dollars.
b/ Number of jobs created (full and part time).

Comparison of New York State and the U.S.

Estimates of employment and earnings impacts for the select technology alternatives for the U.S. economy are presented in Table 4. The technologies continue to be listed according to the magnitude of their New York State employment impacts per million dollars of expenditure (as in Tables 1, 2 and 3) in order to highlight changes in ranking between the U.S. and New York State.

The results indicate that there is a considerable difference in ranking of technologies by their employment impacts in comparing the New York State with U.S.. For example, urban wood waste ranks fifteenth in New York State but ranks seventh in the U.S. in terms of employment impacts created per million dollars of expenditure.

Figure 5: New York State Employment Impacts per 10 Gwh of Output

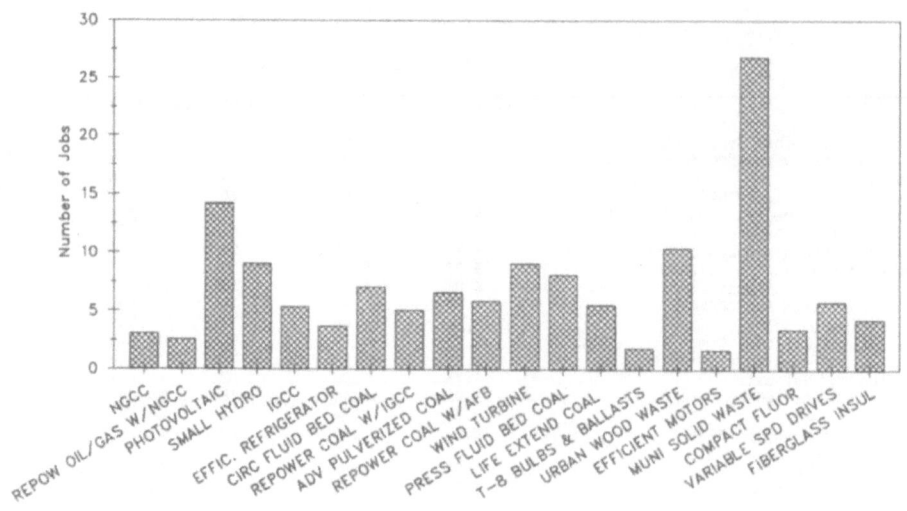

Figure 6: New York State Earning Impacts per 10 Gwh of Output

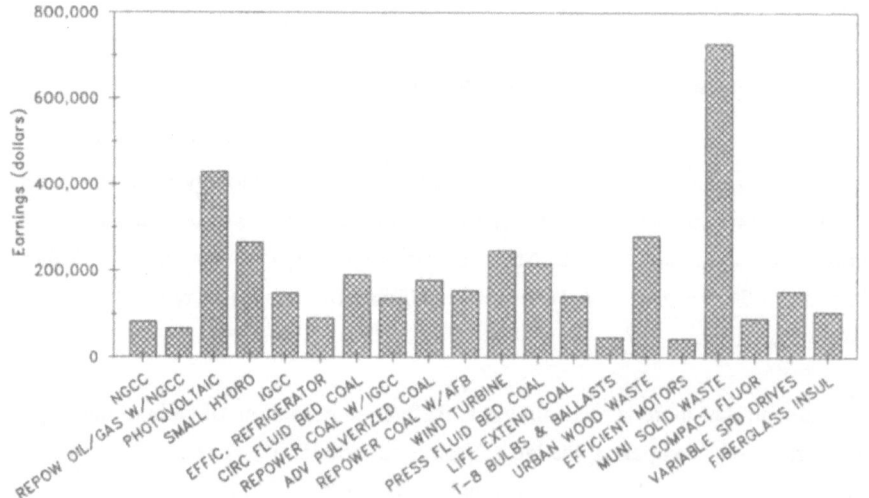

Table 4:   United States Employment and Earnings Impacts of Selected Technologies
            (per Million Dollars of Expenditure)

	Cost[a]	Employment Impacts[b]		Earnings Impacts	
	(c/Kwh)	(Jobs)	(Rank)	(dollars)	(Rank)
Natural Gas Combined Cycle (NGCC)	5.81	15.03	2	394,162	2
Repower Oil/Gas with NGCC (RPNGC)	4.69	15.35	3	399,875	3
Photovoltaic (PV)	17.29	14.88	1	381,588	1
Small Hydroelectric (HYDRO)	9.17	16.63	4	417,545	4
Integrated Coal Gasif. Comb. Cyc. (IGCC)	4.49	27.15	6	694,635	7
Effic. Household Refrigerator	2.90	39.25	17	958,400	17
Circulating Fluidized Bed Coal (CFB)	5.30	29.07	8	733,149	8
Repower Coal with IGCC (RPIGC)	3.82	29.90	10	757,560	11
Advanced Pulverized Coal (APC)	4.84	29.18	9	734,171	9
Repower Coal with AFB (RPAFB)	4.29	30.83	12	775,593	12
Wind Turbine (WIND)	6.24	23.91	5	579,316	5
Press. Fluid. Bed Coal Comb. Cyc. (PFBCC)	5.53	30.13	11	753,540	10
Life Extend Conventional Coal (LECOAL)	3.39	36.96	15	922,832	15
T-8 Bulbs and Ballasts	1.13	33.61	13	791,900	13
Urban Wood Waste (WOOD)	5.96	27.69	7	659,533	6
Efficient Motors	0.81	41.77	18	1,015,876	19
Municipal Solid Waste (MSW)	6.82	34.89	14	835,402	14
Compact Fluorescent	1.60	38.81	16	928,376	16
Variable Speed Drives	2.33	44.29	20	1,092,383	20
Fiberglass Insulation	1.67	42.83	19	1,009,244	18

a/ Real levelized 1991 dollars.
b/ Number of jobs created (full and part time).

Comparisons of U.S. and New York State employment and earnings impacts per million dollars of expenditure are illustrated in Figures 7 and 8. The results indicate that a given level of initial direct expenditure on a technology results in about two times greater employment and earnings benefits in the U.S. than in New York State.

The reason for this relationship is that the U.S. economy is better connected (i.e. it produces many more types of goods and services than those produced in the New York economy). As a result, even if all initial direct input requirements (expenditures) for a project could be met by the state economy, total economic impacts in the U.S. would still be larger than the impacts in the State. A large percentage of indirect and induced impacts associated with an initial direct expenditure are likely to be contained within the U.S. economy but may leak from a regional economy. For example, for an urban wood waste plant a large part of the initial input requirements are provided by the state economy, yet the U.S. impacts per million dollars of expenditure are more than fifty percent larger than the impacts which will be created within New York State.

Figure 7: Comparison of New York State and United States Employment Impacts (per Million Dollars of Expenditure)

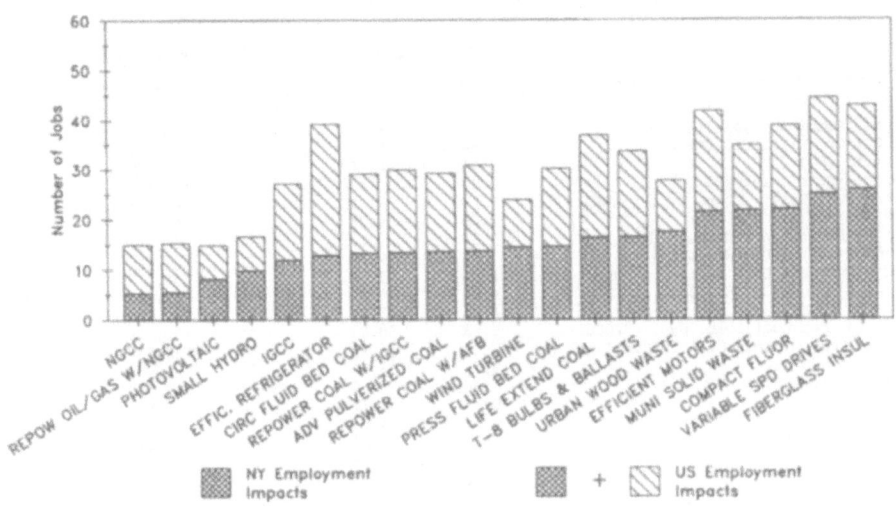

Figure 8: Comparison of New York State and United States Employment Impacts (per Million Dollars of Expenditure)

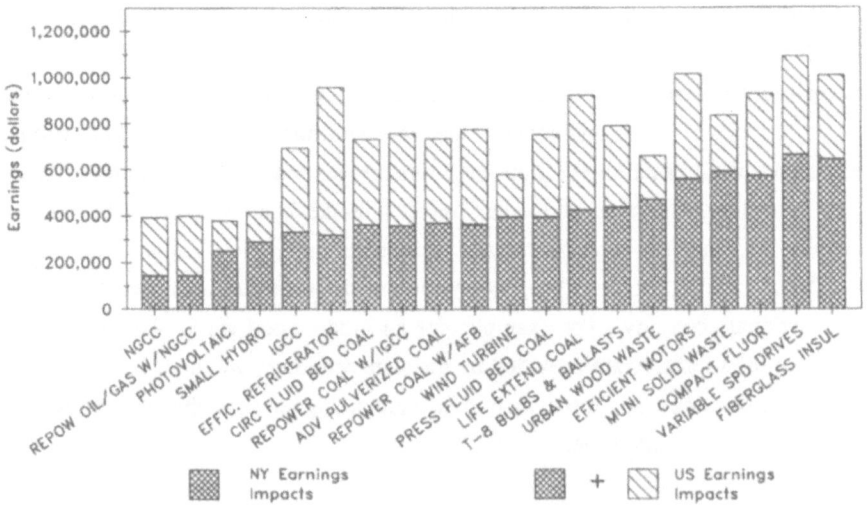

The divergence between U.S. and N.Y. impacts is also explained by the level of initial input requirements which is provided by the State economy. For example, a large portion of total expenditure on a natural gas combined cycle plant consists of natural gas purchases. Since natural gas is not produced within New York State, a large portion of initial expenditure on a natural gas plant constitutes economic leakage from the State. However, when the U.S. economy is considered as a whole, most of the natural gas is supplied by indigenous resources and economic leakage is averted. This factor further increases the divergence between U.S. and New York impacts. In the case of an NGCC plant the U.S. employment impacts are about three times as large as the employment impacts within New York State per million dollars of expenditure.

The above analysis shows that there could be substantial differences in regional economic impacts created by alternative electricity supply technologies. The preferred technology in one region (based on economic benefits) may not be the preferred technology in another region.

Discussion

The consideration of multiple societal goals in energy planning and utility resource procurement (bidding) programs is a laudable objective. However, there is no free lunch. As long as these goals can affect the ranking of alternatives, they have an implicit trade-off with price. Therefore, the weighting of these goals in the decision-making process must be done carefully to provide least cost electricity to ratepayers while maintaining an economically efficient use of resources.

The consideration of economic development attributes in decision making and resource procurement is not a recent phenomenon. The New York State Energy Office has been estimating economic development impacts for energy planning purposes since 1979, the most recent example being the HQ Economic Study (1992).[9] Also, in recently instituted utility bidding programs for procurement of new resources, three of New York State's utilities have included economic development attributes in their selection process.[10] However, the State's

---

9 New York State Energy Office, Hydro-Quebec Economic Study, Draft appendices, pp V102 - V111, May 1992.
10 The three utilities are New York State Electric and Gas Corp. (NYSEG), Niagara Mohawk Power Corp. (NMPC) and Central Hudson Gas and Electric Corp. (CHG&E). Of these, NYSEG and NMPC have given specific weights to economic development, although the weights proposed are considerably less than 1% of the maximum price factor score.

Public Service Commission has provided little guidance regarding how these factors should be considered.

As shown in the preceding sections, economic development indices, such as regional employment and earnings, vary considerably across electricity supply and DSM technologies, as do the costs (or price) of electricity from these technologies. Given this, a central issue confronting decision makers is how much additional value to place on employment and earnings attributes, in terms of a trade-off with price.

This issue becomes problematic because economic development attributes do not constitute true externalities associated with electricity production, nor is there a justifiable economic method of assigning stable monetary values to them. Valuation of economic development benefits derived from construction and operation of an energy facility is a subjective determination which will depend upon a region's economic condition and the political environment.[11] However, studies which attempt to value the creation of new jobs, for inclusion in the price of resources, have also appeared in the literature.[12][13]

The following presents a heuristic approach for including economic development considerations in utility competitive bidding programs, which is substantially based on the 1989 New York State Energy Plan. The approach is presented as an Economic Development Component (EDC) of the utility's bid scoring formula.[14]

The scoring method utilizes results of input/output analysis applied to generic project data to rank projects by the relative degree of economic impacts which would occur if each project were implemented. Further, the EDC scoring also considers special characteristics unique to each proposed project. Special characteristics represent extraordinary economic development benefits (beyond those related to the construction and operation of a project as perceived by the host utility). The criteria defining special characteristics are specified by each host utility in their solicitation for bids. An example could be a cogeneration facility which influences a large industrial customer to continue operations in the region. Since the continued operation

---

[11] Sanghi, Ajay K. "Should Economic Impacts Be Treated as Externalities in Utility Bidding Programs?" The Electricity Journal. Vol. 4, No. 2, March 1991.

[12] Hohmeyer, Olav. Social Cost of Energy Consumption: External Effects of Electricity Generation in the Federal Republic of Germany. Springer-Verlag, 1988.

[13] An ongoing study entitled "Social Cost of the Fuel Cycle" by Resources For the Future and Oak Ridge National Laboratory is attempting to evaluate the employment benefits of different energy supply technologies using an opportunity cost approach.

[14] New York State Energy Office. Draft New York State Energy Plan, Vol. VI, Electricity Issues: Issue 3B, Economic Development, May 1989, pp 5-6.

of the industrial facility could avoid a large loss in regional earnings and employment, it has greater economic value to the region than a generic cogeneration facility considered in its own right.

The criteria for scoring the EDC should assign the highest score to projects which possess special characteristics and a lower range of scores to projects which do not. As such, the maximum economic development score would only be awarded to projects when they can claim extraordinary economic value (proposals should not receive the highest ranking based on technology or fuel type alone). For example, a cogeneration facility would only qualify for the maximum score if the bidder can show that the facility is likely to result in saving jobs in a local industry. In other words, without the cogeneration facility the industry, which would use steam from the project, is likely to go out of business.

The scoring of proposals which do not possess special characteristics should be based on the ranking of generic technologies obtained using input/output modelling results. These results indicate that some non-fossil fuel fired plants (e.g., renewable resources such as municipal solid waste and wind) rank above fossil fuel fired plants because these technologies require no fuel to be purchased from outside New York State (creating minimal economic leakage). In addition, DSM and cogeneration projects (which do not qualify as having special characteristics) should be ranked ahead of fossil fuel plants. The reason being, DSM requires no fuel purchase and its expenditures create local earning and employment as energy-conserving equipment is purchased and installed on-site. Cogeneration is included because it may increase the competitiveness of the regional industry by reducing operating expenses, though not to an extent which could be construed as preventing a large firm from relocating.

An example of including an economic development component is presented in Table 5. This proposal assigns the EDC a weight equal to 4 percent of the maximum price factor score. For illustrative purposes, a maximum price factor score of 500 points is used, which equates to a maximum EDC score of 20 points. It is also assumed that the maximum point score assigned to each group is equal to 50 percent of the previous group's point score (the last group is assigned a score of zero).

As shown in Table 5, Wind, MSW, DSM and cogeneration technologies would receive an EDC score of 10 points (50 percent of the maximum score). Fossil fuel-fired technologies would receive an EDC score of 5 points (25 percent of the maximum score). There are no EDC points awarded to projects located outside New York, regardless of fuel type, as both the construction, and the operation and maintenance of these projects is likely to have minimal impact on earnings or employment in New York State.

Table 5:   Consideration of the Economic Development Component

Technology/ Fuel Type	Percentage of EDC Weight	Score
Special Circumstances (e.g. Certain Cogen- eration)	100%	20
Wind, MSW, and DSM	50%	10
Fossil Fuel Plants	25%	5
Out-of-State Projects	0%	0

Recommendations

- Economic development potential of energy technologies should be considered in the procurement of future energy resources.

- Utilities should assign a higher ranking for this factor to projects which, because of special circumstances, would bring about economic benefits which are significantly above and beyond the economic benefits related to the construction and operation of the project itself.

- In utility bidding systems, 1-4 percent of the weight of the price factor should be allocated to the economic development potential of alternative electricity resources.

References

Beemiller, Richard. "Hybrid Approach to Estimating Economic Impacts Using Regional Input-Output Modelling System (RIMS II)" Transportation Research Record 1274.

Bureau of Economic Analysis, U.S. Department of Commerce. Regional Multipliers: A User Handbook for the Regional Input-Output Modelling System (RIMS II), 2nd Edition, May 1992.

Hohmeyer, Olav. Social Cost of Energy Consumption: External Effects of Electricity Generation in the Federal Republic of Germany. Springer-Verlag, 1988.

New York State Energy Law, Section 6-106, Paragraph 5 [Senate Bill 4912-A].

New York State Energy Office, Hydro-Quebec Economic Study, Draft appendices, pp V102 - V111, May 1992.

New York State Energy Office. Draft New York State Energy Plan, Vol. VI, Electricity Issues: Issue 3B, Economic Development, May 1989, pp 5-6.

Sanghi, Ajay K. "Should Economic Impacts Be Treated as Externalities in Utility Bidding Programs?" The Electricity Journal. Vol. 4, No. 2, March 1991.

SUBJECT AREA 3:

**INSTRUMENTS AND APPROACHES FOR THE
INTERNALISATION OF SOCIAL COSTS**

15. Evaluation of Instruments for the Incorporation of Externalities

Rainer Friedrich

Institut für Energiewirtschaft und Rationelle Energieanwendung

Universität Stuttgart

Heßbrühlstraße 49a, D-7000 Stuttgart 80

1 Introduction

The neglect of environmental damage and other external effects when deciding about investments in the field of electricity supply may cause wrong allocations of resources. A first step to avoid such misallocations is the quantification and monetizing of the external effects. Once the external costs are known, optimal decisions resp. optimal energy supply systems can be identified.

To do that, we may start with a reference energy supply system (e.g. the actual energy system or an energy system without any measures for environmental protection), that statisfies a certain demand for energy services. Then, the posible measures to change the environmental impacts of the different elements (techniques) of the systems have to be identified.

These measures include not only the use of devices e.g. to clean the flue gas, but also substitution of fuels (e.g. from coal to gas), new technologies (e.g. fluidized bed combustion instead of grate firings) and measures to save energy.

An environmental protection strategy can then be defined as a bundle of measures for all users of environmental resources resp. for all elements of the energy supply system.

The optimal environmental protection strategy q is that strategy, where the sum of the monetized environmental damage and the sum of the costs for the measures are minimal:

$$\Delta D_q + C_q = min$$

where:

ΔD_q = damage, which is avoided by strategy q
(compared to the damage of the reference system s),

$$\Delta D_q = D_q - D_s$$

C_q = costs of strategy q.

If we assume convex and differentiable cost and damage functions, it is a necessary and sufficient condition for the optimal strategy, that the marginal damage and the marginal costs of environmental protection measures are equal.

Once the optimal strategy is known, the question arises, how, i. e. with what ecopolitical instrument, this strategy should be installed in reality.

The basic instruments differentiated standards, taxes or permits are especially suited to serve this purpose. In principle, all three instruments lead to the application of the optimal strategy. In practise, however, differences occur.

In a study financed by the government of Baden-Württemberg, a state in the south-west of Germany, the author has investigated the advantages and disadvantages of the three mentioned instruments.

In the literature these advantages often have been deduced from theoretical assumptions about the polluters and their costs for emission reduction, that do not reflect the real costs and the actual structure of the different pollution sources. So, in the study some of the features of the instruments have been examined by using the data of the existing emission sources (for SO_2 and NO_x) in Baden-Württemberg and the actual possible measures to reduce these emissions.

This means, that data have been processed for

- each of the firing plants in Baden-Württemberg with a thermal power of more than 1 MW, these are 2100 plants with 5000 furnaces,

- the 2.3 Mio smaller firings. E.g. for furnaces in private households 200 household types are considered, which are defined as a combination of building size, building age, type of furnace, fuel used and consumer habits,

- several millions of cars and trucks, that are operated on the 1400 road sections outside settlements and within the 1111 municipalities in Baden-Württemberg, where they cover about 70 billions km. Cars are divided into 24 car types, which differ according to their cylinder capacity, fuel type (diesel or gasoline) and specific emissions.

For each of these sources or source groups possible emission reduction measures have been identified, costs and emission reduction for these measures have been determined. The investigated measures include primary NO_x-measures, low-NO_x-burner, flue-gas-desulphurization and DENOX-plants, uncontrolled and controlled three-way-catalyst for cars, substitution of coal and heary fuel oil by light fuel oil and natural gas, insulation of buildings and many more.

To compare SO_2- and NO_x-reduction measures and to take into account measures or techniques, that reduce SO_2 and NO_x simultaneously, SO_2 and NO_x-emissions are made comparable by transforming them into 'pollutant equivalents' (PEQ). The relative importance or potential for damage of 1 kg SO_2 is thereby set equal to that of 1 kg PEQ and of 0,57 kg NO_x.

2   Evaluation of Instruments

For the evaluation of the instruments for the incorporation of externalities in essence four criteria are chosen:

- effectiveness
- capability to adjust to changing conditions
- incentive to achieve technical progress
- allocation of costs and competitiveness.

2.1   Effectiveness

The criterion 'effectiveness' measures the difference in welfare between the strategy r, that is realised by using the instrument j, and the optimal strategy q:

$$E_j = D_q - D_r + C_q - C_r.$$

The smaller the value of E, the better the criterion is fulfilled.

Theoretically, all three investigated instruments lead to the optimal strategy (E = 0). In practice, deviations from the optimal solution occur.

In the case of standards, the tresholds are normally not fixed individually, but are the same for a class of emission sources. This is a cause for non-optimal solutions, as individual circumstances are not fully regarded.

As an example for the 'effectiveness' of a policy with standards the effects of the actual environmental policy for air pollution control in Baden-Württemberg can be compared with the effects of the optimal strategy.

The measures for air pollution control lead - for the reference year 2000 - to the following emission reduction compared to the emissions, that would result without measures. SO_2-emissions are reduced from 240 000 t/a to 91 000 t/a, a 62 % reduction. NO_x-emissions are decreased from 368 000 t/a to 254 000 t/a (minus 31 %).

These emission reductions cause costs of 1030 million DM (ca. 670 million $) per year.

For comparing this strategy with the optimal strategy, a cost curve for efficiently reducing SO_2- and NO_x-emissions is calculated (Fig. 1). As the damage function is not known, the point of the optimal strategy is not known, but this point certainly is situated on the cost curve.

In Fig. 1, the point, that describes the actual environmental policy for reducing SO_2 and NO_x emission, is also marked. As can be seen, the actual policy point lies above the cost curve, that represents the possible optimal control strategies. From the result the following can be obtained:

- The actual control policy for SO_2 and NO_x reduction leads to annual losses (sum of additional costs and damage) of at least 230 million DM compared to the optimal policy, these are 22 % of the total costs for emission reduction.

- To get the same amount of emission reduction as with the current environmental policy in Germany, taxes have to be fixed at 3,50 DM/kg SO_2 and 6,20 DM/kg NO_x.

Of course, the use of taxes also leads to differences between the realised and the optimal strategy.

These differences are caused by

- lacking knowledge about measures for reducing environmental damage,
- problems with the application of these measures (e.g. no available space, problems with disposing waste),
- indolence,
- fear of loosing flexibility by tying up funds,
- lack of capital.

Figure 1: Curve of costs for reduction of SO$_2$ and NO$_x$ emissions in Baden-Württemberg

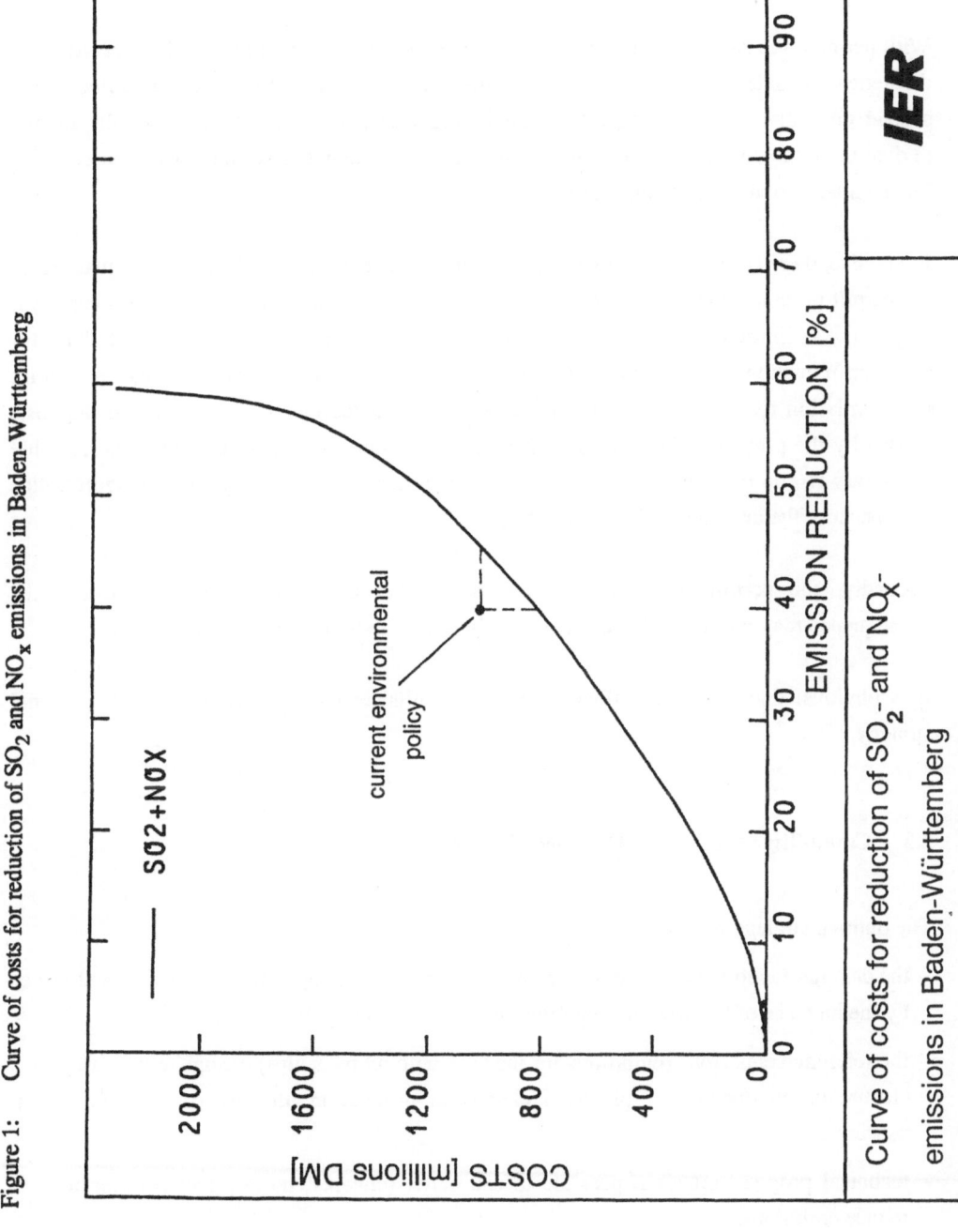

Curve of costs for reduction of SO$_2$- and NO$_x$-emissions in Baden-Württemberg

However, most of these impediments can be removed to a large extent with additional political measures (e.g. the supply of advice).

With permits, the same impediments as with taxes occur. The effectiveness is - according to theoretical considerations - better with permits than with taxes, if the absolute value of the second derivative of the damage function is larger than the absolute value of the second derivative of the cost function (i.e. if marginal damage changes faster than marginal costs) in the neighbourhood of the optimal strategy.

In practise, there are a number of additional problems with permits. Above all, fluctuations of the permit price have to be expected. One reason for that is, that the short-term procurement of permits is limited, as other polluters, that just have invested in a certain technique for emission reduction, are not able to change this technique immediately to another one (with more emission reduction and higher marginal costs); at least they have to regard the costs caused by the premature depreciation of the old technique. This may lead to a considerable temporary rise in the price of the permits, if a plant needs additional permits at short notice (e.g. because the demand for its goods rises).

Secondly, the permit price depends on the market behaviour of many actors - the consequence may be fluctuation of the price similiar to those at the stock exchange.

It is obvious, that such price fluctuations of permits lead to deviations from the optimal strategy.

2.2 Capability to Adjust to Changed Conditions

The optimal strategy changes, if

- the damage function is altered (e.g. caused by a change in population density or land use or by the increase of knowledge about the formation of damage);
- the reference emissions (emissions for the case without measures) change (e.g. caused by a closure or an opening of plants, a change in production, the use of new production processes);
- technical progress makes it possible to use better (cheaper, more effective) measures to reduce emissions.

The criterion 'capability to adjust to changed conditions' examines, wether resp. to what extent the new optimal strategy is realised without additional public action (e.g. the rise of the tax or the purchase of permits), after a change in the conditions has happened.

This criterion is based on the thesis, that an adjustment without the necessity of public action is better than an adjustment, that is steered by public authorities, because

- public action needs time,
- necessary public action may remain undone,
- considerations, that are irrelevant, may influence the decisions.

Table 1 shows, how the different instruments deal with changes in boundary conditions.

Table 1:   Effects of changed conditions

Changes	Effects with		
	standards	taxes	permits
raise of damage[2]	H	H	H
decrease of damage[2]	L	L	L
raise of reference emissions[3]	O/H[1]	O/H[1]	L
decrease of reference emissions[2]	O/L[1]	O/L[1]	H
technical progress I[4]	H	O/L[1]	H
technical progress II[5]	L	O/H[1]	L

O = the new optimal strategy is realised
H = emissions are higher than optimal
L = emissions are lower then optimal
[1]   with linear/convex damage function
[2]   despite constant emissions
[3]   reference emissions = emissions, that would appear without emission reduction measures
[4]   the new technique increases the reduction of emissions
[5]   the new technique implies lower costs, but higher emissions

The instruments generally do not react properly to changes of the damage function. Changes of the reference (base) emissions are considered of standards and taxes in an optimal way, if the damage function is linear. Adjustments to technical progress are made optimally in the case of a linear damage function, if taxes are used.

So taxes are the instrument, that fulfill the criterion best, if a linear damage function can be assumed.

Permits have advantages, if the marginal damage in the neighbourhood of the optimal strategy changes faster than the marginal costs of emission reduction.

However, the price fluctuations of permits already mentioned before worsen the assessment of permits.

If for example a plant closes or reduces its emissions, it will try to sell the surplus permits. Most possible customers for these permits are not able to change their emission reduction measures immediately. So they will only buy permits, if the price of the permits is lower than the running costs (excluding fixed costs and investment) of their emission control plant. This means, that the price for the permits will decrease below the value, that belongs to the optimal strategy (unless the government buys permits back to stabilize the price).

This can be illustrated by an example:

If we assume the realisation of an optimal strategy for SO_2 and NO_x control in Baden-Württemberg with permits, that allow emissions of 355 000 t SÄQ/a, the equilibrium price of the permits will be 4 DM/kg SÄQ · a. If we assume, that one refinery and one paper producing plant with emissions of 25 000 t SÄQ/a close (this has actually happened in the past), the new equilibrium price for the permits should be 3,40 DM/kg SÄQ · a.

However, short term the price might fall to 2,50 DM/kg SÄQ · a. With permits bought at this price non-optimal measures are installed.

2.3   Incentive to Achieve Technical Progress

The criterion 'incentive to achieve technical progress' measures, whether the instrument provides an incentive to improve the techniques for emission reduction.

Incentives for technical progress are most important, as technical progress provides the only possibility of decreasing pollution further (beyond the current optimal niveau) without decreasing welfare.

With a linear damage function, taxes give an optimal incentive for technical progress, as the use of new techniques, that lead to a new optimal strategy, automatically reduce the costs (sum of tax and costs for emission control measures) of the polluters. All operators of emission sources continously try to find ways to decrease the taxes they have to pay for their emissions.

With permits, an incentive to achieve technical progress also exists. However, the use of new techniques may reduce the amount of emissions and, so, the price of the permits, although marginal damage is not changed. This leads to a smaller incentive than for taxes (unless public authorities buy permits back or devaluate the permits).

With standards, there is only an incentive to realise the standard with less costs, the search for techniques, that reduce the emissions more than demanded by the standard, is not encouraged. However, this situation can be improved

- if the governement grants subsidies for the improvement of control techniques and

- if government policy is such, that institutions, that develop new techniques, can be confident, that the governement changes the standard, when new efficient products are available.

Nevertheless, an automatic encouragement of technical progress through market forces certainly is preferable to a process, that is directed by political decisions.

2.4 Allocation of Costs

Within this criterion, the distribution of the costs for the emission control measures is investigated. Furthermore it is discussed, whether these costs impair the ability of industrial plants or branchs to compete on the international market.

The expenditure of polluters is smaller, when standards are used or when permits are initially granted free of charge, because then nothing has to be paid for the remaining emissions. However, under current pricing procedures, this may lead to a price for the good (e.g. electricity), that is too small and so maybe to a consumption of the good, that is too high.

As an example, the allocation of the costs can be examined for the case of the SO_2 and NO_x control strategy in Baden-Württemberg.

As said before, the current environmental policy with regard to SO_2- and NO_x-reduction leads to costs of 1 billion DM/a. The industrial branchs are burdened by surplus costs, which amount to an average of 0.2 % of the net output value, the maximum is 1,5 % of the net output value. These figures do not only include the direct costs of the branchs for environmental protection, but also the indirect costs, that occur due to environmental measures of the suppliers of the branch. These costs are determined by means of a detailed input-output-analysis.

Introducing taxes, operators of emission sources have to pay not only the reduction measures, but also taxes for the remaining emissions.

When taxes are fixed at a level of 3,50 DM/kg SO_2 and 6,20 DM/kg NO_x, operators of emission sources not only pay 0.8 billions DM per year for emission reduction measures, but also taxes of 1.9 billions DM. If these additional taxes are not compensated by abatements of other taxes or expenses, this means surplus costs of 0.5 % of the net output value on an average and up to 3 % of the net output value for some branchs of the basic industry (glass, cellulose and pulp, cement, iron).

So, using taxes leads to expenditures for branchs of the basic industry, that may affect the capability to be competitive.

However, if the additional revenue of the governement is spent for a general release of expenses and taxes for trade and industry, the input-output-analysis shows, that all industrial branchs (except energy) have to bear less costs than before despite the tax on SO_2 and NO_x.

Surplus costs are then totally borne by the private households (which get a better environment in return) and the energy sector.

3 Summary

- If a linear damage function (constant marginal damage) is assumed, taxes are the instrument best suited for internalizing external costs.

- If the damage function is not known, the assumption of constant marginal damage is a useful hypothesis in many cases. For those cases taxes should be given preference to permits.

- In many cases, a linear damage function can be postulated only for a certain range of values. If for instance the emissions and - subsequently - ambient concentration of pollutants reach values, where human health is considerably affected, marginal damage might increase rapidly. In that case, a combination of a tax with a limit for the maximum ambient concentration of pollutants is recommended.

- Permits are the best suited instrument, if the optimal strategy is located near a discontinuity (step) of the damage function or of the first derivative of the damage function.

  However, even if a discontinous damage function is assumed, taxes might be preferable to permits, if no information about the location of the discontinuity (step) is available.

- In theory permits are preferable to constant taxes, if marginal damage changes faster than marginal control costs near the optimum. However, the ups and downs of the price for permits may qualify this statement.

- Standards possess the following disadvantages:
  -- poor efficiency,
  -- limited incentive to gain technical progress.

  This disadvantages could be partly reduced by
  -- establishing individual standards,
  -- supporting the development of new technologies,
  -- rapid adjustment of the standard, when new techniques are available.

- The additional revenue for the public authorities should be compensated by general tax abatements.

4   Conclusion

From the results described above, the following elements could be the base for a policy to internalize the external effects of air pollution:

- A tax per kg of pollutant is charged to all emission sources.

  (Taking the current aims of environmental policy in Germany and the current costs of measures for air polution control as a base, the level of the tax for SO_2 and NO_x could be approx. 3,50 DM/kg SO_2 and 6,20 DM/kg NO_x).

- Information about techniques for pollution control should be provided by public authorities.

- In addition to the tax, limits for the ambient concentration of pollutants are fixed to protect human health.

- The additional revenues for the public authorities caused by the eco-taxes should be compensated by general tax abatements for industry and trade, or by reduction of the wage-cost factor.

References

Friedrich, R.: Darstellung und Bewertung verschiedener Maßnahmen zur Internalisierung externer Kosten; ed.: PROGNOS AG, Basel, 1992

Boysen, B., Mattis, M., Friedrich, R., Voß, A.: Kosten-Effektivitäts-Analyse von Maßnahmen zur Minderung von SO_2- und NO_x-Emissionen in Baden-Württemberg für alle Emittentengruppen. Projekt Europäisches Forschungszentrum für Maßnahmen zur Luftreinhaltung (PEF), Bericht KFK-PEF 54, Karlsruhe, 1989

Endres, A.: Umwelt- und Ressourcenökonomie, Darmstadt, 1985

Friedrich, R., Voß, A.: Emission Inventories and Cost-Effectiveness Analysis, the Base for Sound Environmental Control Policies. In: J. Fenhann (ed.): Environmental Models; Emissions and Consequences; Amsterdam 1990a

Friedrich, R.: Umweltpolitische Instrumente zur Luftreinhaltung - Analyse und Bewertung. In: Projekt Europ. Forschungszentrum für Maßnahmen zur Luftreinhaltung (PEF): 6. Statuskolloquium. PEF-Bericht 61, Band 2; Karlsruhe, 1990b

Friedrich, R.: Taxes as a Political Instrument for Air Pollution Control and their Impacts on the Energy Systems. In: IAEE(Hrsg.): Integrated Energy Markets and Energy Systems; Conference Proceedings, Copenhagen, 1990c

Friedrich, R.: Umweltpolitische Instrumente zur Luftreinhaltung - Analyse und Bewertung; Institut für Energiewirtschaft und Rationelle Energieanwendung (IER). Bericht, Stuttgart, 1992, in preparation

Heister, J., Michaelis, P., Mohr, E.: Praktische Einsatzmögichkeiten für Zertifikate im Rahmen der marktwirtschaftlichen Umweltpolitik in der Bundesrepublik Deutschland und der EG, - Kohlendioxid - Manuskript, Institut für Weltwirtschaft, Kiel, 1990

Nichols, Albert, L.: Targeting Economic Incentives for Environmental Protection, Cambridge, London, 1984

Pigou, A.C.: The Economics of Welfare, 4th. ed., London, 1932

Siebert, M.: Ökonomische Theorie der Umwelt, Tübingen, 1978

Wicke, L.: Umweltökonomie, 2. Auflage München 1989

16.   Pollution Taxes - The Preferred Means of Incorporation of Environmental Externalities

Richard L. Ottinger
Pace University Center for Environmental Legal Studies
78 North Broadway, White Plains, NY 10603

Introduction[1]

Internalizing the environmental costs imposed on society by polluters is the wave of the future in addressing environmental degradation. By signalling to industry the true societal costs of their operations, inclusion of environmental costs in the price of goods produced gives an economic incentive to industry to reduce pollution. This can be an important supplement to regulation.

Government regulation of pollution has proved to be generally inadequate to address the severe threats to the planet posed by global warming, stratospheric ozone depletion, acid rain, urban smog and toxic contamination of our air, water and food supplies. While some improvements have resulted from regulation, particularly where contaminants have been totally prohibited as with most uses of asbestos, lead in gasoline and DDT, generally pollution increases stay well ahead of mandated regulatory controls. Inevitably economic growth, in both developing and industrialized countries, receives higher government priority than environmental protection.

Governments are just starting to consider supplementing pollution regulation with pollution taxes or fees that will introduce into the marketplace prices that reflect the damages to society inflicted by polluting resources. The OECD published in 1988 a review of pollution levies indicating a total of 85 pollution taxing regimens in six of its principal countries.[2] The U.S. House of Representatives' Ways & Means Committee held hearings on pollution taxes (March 6, 7 & 14, 1990).

The U.S. environmental organizations, which historically have resisted pricing environmental impacts on grounds that it would constitute a license to pollute, have now embraced the idea. Daniel Dudek of the Environmental Defense Fund has been a leading advocate of marketplace treatment of pollution; he helped draft the Administration's Clean Air Act proposal to create emissions trading rights, which was adopted as a part of the 1990 U.S. Clean Air Act Amendments.

There has been much recent international discussion of carbon taxes to address the threat of global warming to which CO_2 is the principal contributor. In 1992 the Commission of the European Communities proposed a combined carbon/energy tax equivalent to $10 per barrel of oil to be instituted by the Community subject to its adoption by other OECD countries. Individual European countries have adopted or proposed a wide variety of emissions taxes and fees. The United States recently adopted a tax on chlorinated fluorocarbons to facilitate the phase-out of these stratospheric ozone-depleting chemicals. Thus, for the first time, pollution taxes and fees are achieving international attention.

As mentioned above, currently most governments are combating environmental pollution by regulating industries that pollute. These regulations establish pollution levels based on technology standards or emission levels. The problem with sole reliance on regulations is that they take too long to promulgate and set standards that fail to eliminate pollution damages to a level acceptable to society.

Furthermore, regulations only encourage industries to lower their pollution to a set level without offering incentives to surpass that level. Thus, abatement of pollution only proceeds as quickly as the government can establish regulations that reflect both currently available pollution reduction technologies and politically feasible enforceable levels. Even if regulations were adequate, the penalties for failure to meet the standards are likely to be so low that it is cost effective to pay a fine rather than purchase pollution control technology.

With the high current public concern about environmental issues, governments are realizing that regulations are not attaining pollution reductions that satisfy the public or adequately preserve the environment. Thus, governments are starting to examine and institute pollution taxes, also referred to as environmental taxes, as a supplement to regulations.

Pollution taxes are also politically attractive. Governments requiring additional revenues may find that it is much more acceptable to tax emissions, making polluters pay for the damages they impose on society, than to further tax gasoline or other energy sources independent from their environmental consequences, which would directly affect the taxpayer, even though the pollution tax will be indirectly reflected in higher energy and product prices. It also is more politically attractive to tax a negative such as pollution than taxing a positive like one's personal income.

To address the impacts which higher pollution taxes would create for the economy and low-income individuals, governments can offset pollution taxes with a reduction of other taxes. They can even make pollution taxes revenue neutral if this adds to their political acceptability.

Long advocated by economists, pollution taxes internalize the cost of the damages caused to human health and the environment by pollution that is emitted or associated with industrial goods and processes.

Rationale of Pollution Taxes

Much literature on the uses of fiscal mechanism to protect the environment lumps fees and taxes together, treating them as a common entity. Fees, such as administrative fees or user fees, are imposed to cover administrative expenses of environmental programs or the cost of treatment or disposal of a pollutant in the environment. This is in line with the "Polluter Pays Principle", adopted by the OECD in 1975,[3] which holds that a polluter should be charged by the government for any control or prevention measures for which it is responsible.

A pollution tax, on the other hand, is a revenue-raising impost, imposed to capture the damages caused to society by pollutants. It thus usually requires the polluter to pay more than the costs of legislated controls.

While damages are the more relevant costs to capture, both effluent charges and pollution taxes approach the same objective of internalizing the costs pollution imposes on society, so both are here considered.

The substance of the pollution tax theory was set forth by N.K. Lee:

The skeleton of the analysis is plain. From economic theory we adopt three axioms: (i) individual self-interest underlies human behavior; (ii) prices act as signals guiding this behavior; (iii) trade-offs - - more elaborately, decisions derived from cost-benefit analyses - - can be steered by suitable management of prices. This set of assumptions permits, in principle, the implementation of an optimal allocation of resources, within the famed - -if ever unobserved - - free market. From our recent environmental past, we add three conditions: (i) goods, formerly free, including air, water, and some land, have become scarce; (ii) but the market has not properly assigned them prices, because they are in a crucial respect, common property - - namely, the costs of maintaining environmental quality are difficult to apportion, even where the benefits to be derived from despoiling the environment are quite divisible; thus (iii) intervention by government is required.[4]

Hence, the ultimate objective of a pollution tax, in the economic sense, is to correct a failing of the marketplace in that the environment is a common good the value of which is not reflected in the purchase price of the polluting good or process. The pollutant is taxed so that the price of the good or process which produces it reflects the true cost. To save money the polluter will look for ways to avoid paying the tax. Profits and fiscal self-interest will act as a motivating force to abate pollution.

One way to set the level of a pollution tax is to determine the value of the damages to society caused by the relevant pollutant emitted. A great deal of work has been done in the United States on valuing the damages caused by electric power plant pollution. Pace University did a comprehensive review of literature valuing such damages[1] and a number of utilities and utility regulatory commissions have adopted values used in integrated resource planning and selection of utility resources. These values are set forth in Tables 1 and 2.

Table 1:   Summary of Externality Values Expressed in cents/kWh generated

	Existing Power Plants		New Power Plants			
Valuation Method	Existing unscrubbed coal plant	Existing scrubbed coal plant	CTU #2 oil .3% S	CC firm gas	AFBC .5% S	IGCC .45% S
1. Ontario Hydro		0.41				
2. Pace University	10.3	4.0	2.6	0.77	2.6	2.1
3. Massachusetts DPU	7.7	5.2	4.0	1.4	3.6	3.0
4. CEC In-state	30.3	10.9	5.3	0.6	5.9	3.8
5. CEC Out-of-state	3.8	2.2	1.5	0.4	1.4	1.0
6. New York PSC	2.5	1.3	0.8	0.3	0.7	0.5
7. Nevada PSC	7.9	5.3	3.9	1.4	3.7	3.0

Where the damages caused by a pollutant, such as carbon dioxide, cannot be adequately determined, control costs or an arbitrary tax figure will have to be adopted, high enough to provide an incentive to avoid pollution, yet not so high as to be seriously damaging to the economy or sensitive individuals. This socially optimal tax on pollution may not be the ideal economic one, but is it politically feasible and workable. For instance, the United States has imposed a tax on CFCs even though damage done to the ozone layer can not be measured monetarily. A few European countries have also set taxes on carbon dioxide emissions without calculating damages. Though the taxes have been imposed only recently, early indications are that the taxes are changing consumption and production patterns.

Table 2: Comparison of Monetized Values of Externalities in the U. S.

EXTERNALITY	PACE EXTERNALITY VALUES	CALIFORNIA ENERGY COMMISSION OUT-OF-STATE VALUES	MASSACHUSETTS DPU VALUES	NEW YORK PSC VALUES	NEVADA PSC VALUES
SO2	2.03	0.54	0.75	0.41	0.78
NOx	0.82	1.46	3.25	0.89	3.40
VOC's	NE	0.16	2.65	NE	0.59
CO	NE	NE	0.43	NE	0.46
Particulates	1.19	0.43	2.00	0.26	2.09
CO2	0.006	0.0035	0.011	0.001	0.011
CH4	NE	NE	0.11	NE	0.11
N2O	NE	NE	1.98	NE	2.07
Water Use (cents/kWh)		NE	NE	0.10	site specific
Land Use (cents/kWh)		NE	NE	0.40	site specific

All units in 1989 $/lb. except Nevada, which is in 1990 $/lb.

Current and Proposed Pollution / Emissions Taxes

Europe is the leader in advocating and imposing pollution taxes and emissions fees. European countries have imposed a wide variety of pollution charges including taxes or fees on carbon, carbon dioxide, air and water emissions, wastes, virgin products and other environmentally damaging products like plastic packaging and pesticides. Japan has imposed taxes only on air emissions, while in the United States the only pollution tax is on CFCs; hearings have been held in Congress about a carbon tax, virgin materials tax, pesticides tax and a sulphur tax. The discussion of the various taxes and fees will be by pollutant source.

U.S. CFC Taxes

The U.S. CFC tax was passed to accelerate compliance with the Montreal Protocol of 1987 which sets limits on the uses and emissions of CFCs. The rates are designed to encourage the polluter to switch to an alternative source. The impact of the tax is indicated in Table 3.

Although opponents of any pollution tax argue that it will hinder competitiveness, Congress effectively countered the argument in designing its CFC tax. The taxing statute imposes a tax on the production or import of 20 listed ozone depleting chemicals and on the import of any product which contains at least one of these chemicals. If the content is not known, the Treasury is allowed to levy a tax based on the estimated CFC content.[5]

Industries argued that plans to limit, recycle, or find alternative sources were underway before Congress introduced a CFC tax. Regardless, with the implementation of the tax, CFC use has declined and this tax can serve as a model for other pollution taxes.

Besides the U.S., both Denmark and Finland introduced taxes on CFCs and halons. The rates are DKr 30 (US $ 5) and FIM 30 (US $ 6) per kg of CFCs or halons.

Carbon & CO_2 Taxes

Concern about global warming and what can be done to limit human contribution of greenhouse gas emissions has caused many European countries to call for a reduction in carbon dioxide emissions. Several of these countries have committed to a stabilization or a reduction policy as described in Table 4. CO_2 is the largest contributor to the greenhouse effect, and this share will increase when CFC use is eliminated pursuant to the Montreal Protocol and subsequent London amendments applying stricter CFC phase-out standards.

234

Table 3:   United States Tax Rates for Ozone-Depleting Chemicals

Ozone-depleting chemicals	Ozone-depletion factor	Year	Base Tax Amount (US $ per pound)	Tax (US $ per pound)
Methyl chloroform	0.1	1990, 1991	1.37	0.14
		1992	1.67	0.17
		1993, 1994	2.65	0.26
		1995	3.10	0.31
		1996	3.55	0.35
		...	...	...
CFC-115	0.6	1990, 1991	1.37	0.82
		1992	1.67	1.00
		1993, 1994	2.65	2.19
		...	...	...
CFC-113	0.8	1990, 1991	1.37	1.10
		1992	1.67	1.34
		1993, 1994	2.65	2.12
		...	...	...
CFC-11, CFC-12, CFC-114, CFC-13, CFC-111, CFC-112, CFC-211, CFC-212, CFC-213, CFC-214, CFC-215, CFC-216, CFC-217	1.0	1990, 1991	1.37	1.37
		1992	1.67	1.67
		1993, 1994	2.65	2.65
		...	...	...
Carbon tetrachloride	1.1	1990, 1991	1.37	1.51
		1992	1.67	1.84
		1993, 1994	2.65	2.91
		...	...	...
Halon-1211	3.0	1990, 1991	1.37	4.11
		1992	1.67	5.01
		1993, 1993	2.65	7.95
		...	...	...
Halon-2402	6.0	1990, 1991	1.37	8.22
		1992	1.67	10.02
		1993, 1994	2.65	15.90
		...	...	...
Halon-1301	10.0	1990, 1991	1.37	13.70
		1992	1.67	16.70
		1993, 1994	2.65	26.50
		...	...	...

Source: 26 U.S.C.A. Sec. 4681 and Sec. 4682 (1992).

Table 4:   Unilateral Commitments to Reduce Carbon Dioxide

Country	Commitment
Australia	agreed to stabilize greenhouse gases not controlled by the Montreal Protocol at 1988 levels by 2000
Austria	agreed to 20% reduction in CO2 from 1990 levels by 2005
Belgium	agreed to a 5% reduction in CO2 by 2000
Canada	stabilization of CO2 and other greenhouse gases at 1990 levels by 2000
Denmark	20% cut in CO2 from 1990 levels by 2000
European Community	community-wide stabilization of CO2 at 1990 levels by 2000
Germany	25-30% cut in CO2 from 1987 levels by 2005
Italy	20% cut in CO2 by 2005 (committed to by Environment Minister)
Japan	Stabilization of CO2 on a per capita basis at 1990 level by 2000
Netherlands	Stabilize CO2 at 1988-89 levels by 1994-95
New Zealand	20% cut in 1990 CO2 emissions by 2000
United Kingdom	called for stabilization of CO2 at 1990 level by 2005

Source: BNA International Environmental Reporter, Vol. 14, No.4, pg. 120.

The combustion of fossil fuels is the principal source of human contribution to emissions. Though there is some dispute in the scientific community about the validity of the global warming theory (on which the U.S. government has unfortunately seized to argue against immediate imposition of carbon taxes), such taxes have been placed on the top of many countries' agendas, and in the United States, the National Science Foundation has recommended immediate action.

In June of 1992, the United Nations Conference on Environment and Development (known as the "Earth Summit"), was held in Rio de Janeiro, Brazil. One of the primary objectives of the Conference was to limit the dangers posed by global warming. A report released by the Third Preparatory Conference in Geneva proposed that the Conference consider a tax on pollution from all finite energy sources, including a carbon tax. The preparatory International Conference on Global Warming and Substantial Development advocated imposition of carbon taxes and taxes on transport of oil and airline passengers. Recommendations for carbon taxes were also seriously considered by preparatory conferences in Nairobi, Kenya, a conference of Arab environmental ministers in Cairo, Egypt and a meeting of non-governmental organizations in Zimbabwe.

Currently technology to control carbon emissions is uneconomic. Thus, society must reduce its use of carbon intensive fuels if the dangers of global warming are to be avoided. A carbon tax that is high enough to encourage either a reduction in fuel consumption or a transfer to cleaner or renewable energy sources has been advocated. Though such a tax preferably would be multilateral in scope, Norway, Sweden, Finland, Denmark and the Netherlands have imposed carbon taxes unilaterally. Other European countries and the European Community are seriously considering a tax. The possibility of a carbon tax is increasing as governments look for revenue and try to manage and resolve this highly important environmental issues.

The following is a brief description of the actions taken by various countries on a carbon or tax:

Norway. Norway imposed a carbon dioxide tax which took effect in January, 1991. As of July 1992 some of the rates were increased and the tax was extended to coal. The rate is now NOK .30 (U.S. $ .05) per liter for oil; NOK .80 (US $ .13) per liter of gasoline; NOK .80 (US $ .13) a cubic meter of natural gas; and NOK .30 (US $ .05) per kg coal. The tax on natural gas applies only to gas burned offshore on the Norwegian shelf oil and gas platform. The tax was expected to raise $222 million in its first year with one-sixth of the revenue to be channeled into a global climate fund to help developing countries clean up the environment. Companies have investigated ways to reduce their emissions. One suggestion was to bring in electricity generated by hydroelectric plants to the gas platform instead of burning gas on the platform to generate electricity. However, it was deemed to be uneconomic because of the substantial investment needed to distribute the power from onshore. Now the industry is evaluating more efficient gas use on the offshore platforms. Even with a low tax rate industry examined options to reduce their tax burden.

Denmark. Despite strong objections from Danish industry the Danish legislature imposed a carbon dioxide tax on electricity consumption in the amount of DKr 100 ($ 17.70) per metric ton, which went into effect on May 15, 1992. The tax is determined as if all electricity is produced solely by coal and does not differentiate between types of fuel. It applies currently only to households, but the industrial sector will become subject to the tax as of

1993, although at a rate reduced by 50%. The tax is expected to yield about DKr 3,600m yearly in gross revenue, but the legislation calls for refunding Dkr 1,000m and a fund of DKr 600m for energy conservation, district heating, etc.

In addition, the Danish legislature introduced a carbon dioxide tax on heating fuels. The tax rate is Dkr .27 (US $ .05) per liter of heating oil; Dkr .29 (US $ .05) per kg of natural gas; and Dkr 242 (US $ 43) per metric ton of coal. Expected state revenue is Dkr 3.5 - 4 billion, with the heavier burden falling on industry. One third of the revenue raised will be used to give rebates to energy intensive industries that reduce their energy consumption. Danish based industries have threatened to move because a unilaterally imposed tax leaves them at a competitive disadvantage. However, as the European Community should adopt a carbon tax applicable to the entire European community this disadvantage will disappear.

Sweden, Finland & Netherlands.   Sweden, Finland, and the Netherlands have also imposed a tax without waiting for multilateral action. See Table 5. The Swedish tax which took effect in 1991, establishes a tax rate of SEK 250 (US $45) per metric ton of carbon dioxide emitted which is equivalent to a tax of SEK 620 (US $111) per ton of coal, offset by some decreases in other taxes. As a result of the tax Sweden expects a reduction in CO_2 emissions of 5-10 million tons, 15,000 tons of sulfur and 25,000 tons of nitrogen dioxide by the year 2000.

Table 5:   Carbon Tax Rates, Finland, the Netherlands and Sweden ($US), 1990

Fossil Fuel	Finland	Netherlands	Sweden
Coal	$3.89/ton	$1.49/ton	$97.69/ton
Gasoline and Diesel Fuel	11.51/bbl.	2.73/bbl.	16.03
LPG	?	1.94/ton	118/ton
Natural Gas	.08/10^3 cu.ft.	.029/10^3 cu.ft.	2.63/10^3

Source: Poterba, James, <u>Tax Policy to Combat Global Warming: On Designing a Carbon Tax</u>, Table 2.

The tax imposed by Finland and the Netherlands is $4 and $2 per ton of coal, respectively. There is no indication yet of the effectiveness of this rate.

European Communities.   The European Communities (EC), which agreed to a community wide stabilization of carbon dioxide at 1990 levels by 2000, stirred the discussion about carbon dioxide taxes with a proposal by the EC Commission for a combined carbon/energy tax.[6] The EC emits thirteen percent of the world's carbon emission. The tax under discussion would be equivalent to an extra $10 per barrel of oil raising more than $42 billion a year in revenue. The tax would be collected by each government to be used as it sees fit. The tax would also be phased in year by year.

According to the proposal 50% of the tax would be imposed on carbon emissions from fossil fuels, while the other half would be levied on all energy sources (except renewables) depending on their calorific value. The energy tax would apply to nuclear power which would take an advantage away from France which gets much of its energy from nuclear sources; environmentalists do not wish to encourage the use of nuclear power by making it cheaper than other energy sources.

There also has been much debate in the EC because some countries feel that if Japan and the United States will not impose a carbon tax then EC countries will be at a competitive disadvantage. British industries state that the tax could cost British firms $7 billion a year. Addressing these concerns the tax proposal contains a conditionality clause setting forth that "the Community will be unable to apply the tax until such time as its main competitors within the OECD have introduced a similar tax or measures having an equivalent financial impact." In other words: the tax should not be imposed unless the U.S. and Japan take similar steps.

A study by Data Resource, Inc. of Paris estimated that this tax, phased in from 1993-2000, would lower the G.N.P. of the eight leading European countries on average by 0.07% a year (varying from .12% in Portugal to .03% in Spain) and inflation would rise by 0.2% a year during 1995 to 2005.[7] The study stated that the EC would benefit from the tax through better air quality, transportation efficiency and an improved trade balance.

United States.   The United States Congress held hearings in March of 1990 on pollution taxes and examined a carbon tax.[8] The U.S. Congressional Budget Office made a study of the effects of a carbon tax on the U.S. economy.[9] The report examined a ten-year phased in tax, starting at $10 per ton and ending in the Year 2000 with a tax of $100 per ton ($160 per ton with inflation). This tax would be expected to stabilize U.S. carbon dioxide emissions in the Year 2000 at 1988 levels, 12% below projected levels.

The study assumed that the tax would be levied at the point where fossil fuels enter the economy, at coal mines, wellheads or docks. Table 6 shows that in the Year 2000, the tax would amount to $60.50/ton of coal, $12.99/barrel of oil and $1.63/mcf of natural gas. This would triple the price of coal, increase the price of a barrel of oil by 50% and add an extra

$.30 per gallon of gasoline, and increase the price of natural gas by 53%. When phased in by 2000, the tax would raise revenue of $110-120 billion per year.

Table 6:   Impact of Proposed United States Carbon Tax of $ 100/ton in 2000

	Coal	Oil	Natural Gas
Unit of measurement	ton	barrel	10^3 cu.ft.
Tons of carbon/ Unit of fuel	.605	.130	.016
Average price, 1989	$23.02	$17.70	$1.78
Tax	60.50	12.99	1.63
Carbon Tax/as a Percentage of price	263%	73%	92%

Source: Poterba, James, <u>Tax Policy to Combat Global Warming: On Designing a Carbon Tax</u>, National Bureau of Economic Research, Inc., Table 3.

Table 7 shows the price increase and use decrease from projected levels of fossil fuels by the Year 2000. Overall energy consumption would decrease by 7% from projected levels without the tax. Demand for renewable energy resources would increase. Coal consumption would drop by 2 quadrillion Btu, while natural gas would supplant 17 tons of coal use, and distillate or residual fuel would replace 15 tons of coal.

Table 7:   Effect of $ 100 Carbon Tax on Fossil Fuel Use and Price in the United States in 2000

	Oil	Natural Gas	Coal	All Energy
Prices	21	16	161	NA
Use	-6	-4	-13	-7

Source: United States Congressional Budget Office, August 1990.

Reduction of carbon dioxide is expected to occur as a result of changes in consumption patters, changes in investment behavior, reducing the energy content of production activities, switching production fuels and development of new technologies.

The report found that if the tax were imposed without being phased in, the economic consequences would be disastrous. A phased in tax would allow the economy, consumers and producers time to adjust. The phased in tax would reduce GNP by 2% and increase inflation by only about .33% in 2000.

The uses of the revenue were found to have an impact on how the tax effects the economy. If the revenue were used to offset the deficit, the taxpayer would have to assume the extra tax burden and consequent economic effects. If the carbon tax were to be offset by reduction in other taxes, of course, the overall burden on taxpayers would be neutral, but there would still be a considerable impact on energy consumers.

Air Pollution

In some countries in Europe air pollution is subject to a pollution tax. Since 1985, France has been levying a tax of $23/ ton of sulphur dioxide emission from large industrial firms. Annual revenue of $15-16 million a year is collected. The purpose of the tax is to finance subsidies for pollution control equipment for these plants. France also taxes the industrial emissions of other sulfides, nitrous oxides and hydrochloric acid. Sweden imposes a sulphur tax of 5.4 cents a kilogram. The tax is expected to reduce Swedish emissions by 15,000 tons a year. A tax of $6.75 per kilogram of nitrogen oxide emitted which is likely to lower emissions by 5,000-7,000 tons was also introduced. West Germany considered imposing an air pollution tax on sulphur dioxide and nitrogen oxides, but opted for strict emission standards because of difficulty in establishing a tax rate and imposing it. Japan levies a variable pollution tax on sulphur dioxide emissions from industrial plants to compensate victims of illness caused by air pollution. Table 8 describes various European air pollution taxing regimens.

Water Pollution

Water effluent charges in Europe serve two purposes. One is to raise revenue for investment in improved waste water treatment while the second acts as an incentive for industries to lower their pollution discharges. France and the Netherlands impose a charge based on the former to households and industries. France raises $300 million a year while at a tax of $15-41/ unit of discharge the Netherlands raises $550-600 million a year. Italy and Germany have a water pollution charge that encourages industries to comply with a regulated level. This charge encourages investment in pollution abatement equipment so that the polluters can reduce their pecuniary burden. Table 9 describes the European water pollution charges and fees.

Table 8: Air Pollution Emission Taxes in Operation

Country	Purpose	Start year	Coverage	Tax base	Rate	Revenue raised
France	Raise revenue to subsidise investment in pollution control equipment	1985	Industrial plants over 50 MWe capacity or emitting over 2,500 t/y of sulphur or nitrogen oxides	Actual sulphur dioxide emissions	23 $/t	15–16 $M/y
Japan	Raise revenue to compensate sufferers from air pollution-related illness	?	Industrial plants over 1.5 MWe capacity	Actual sulphur dioxide emissions	Variable (fixed annually to meet revenue requirements)	Variable (depends on demand for compensation)

Source: Vernon, Jan L., _Market Mechanisms for PollutionControl_, IEA Coal Research, 1990, Table 3.

Table 9: Water Pollution Emission Taxes in Operation

Country	Purpose	Start year	Coverage	Tax base	Rate	Revenue raised
France	Raise revenue to subsidise investment in water pollution control equipment	1969	Households, local authorities, industry	(Flat rate)) actual discharges) of suspended solids, oxidisable matter, soluble salts, toxic matter, nitrogen, phosphorus	Variable	Approximately 300 $M/y
FRG	Incentive to reduce water pollution discharges below levels required by regulations	1981	Households, industry	Actual emissions of suspended solids, oxidising substances, mercury cadmium, substances toxic to fish	23 $/per unit of discharge (reduced if emissions are below regulation levels)	Approximately 150–200 $M/y
Italy	Incentive to encourage rapid compliance with regulations	1976	Industry	Estimated volume of discharge, estimated quality (based on type of plant discharging) and average costs of treatment	Variable	Not known
Netherlands	Raise revenue to subsidise investment in water pollution control equipment	1972	Industry, water boards, households	Bio-degradeable matter, suspended solids, toxic substances, heavy metals	15–41 $/per unit of discharge	550–600 $M/y

Source: Vernon, Jan L., _Market Mechanism for Pollution Control_, IEA Coal Research, 1990, Table 4.

Waste Emission

Waste emission charges are another successful pollution abatement mechanism in Europe. The charge is levied to encourage treatment or recycling. Denmark levies a charge of 130 kroner/ per metric ton of waste which is dumped or incinerated. The charge is refunded when the waste is recycled. Starting in 1981, Belgium imposed a charge on industry treating or disposing of waste at the rate of $0.1-2.6/ ton or cubic meter of waste. The Netherlands levies a minimal charge of $0.1-0.3/ ton of manure disposed of to raise revenue for research. Table 10 describes the various European waste emission taxes and fees.

Table 10:   Waste Emission Taxes in Operation

Country	Purpose	Start year	Coverage	Tax basis	Rate	Revenue raised
Belgium	Incentive to encourage treatment of waste prior to disposal	1981	Industry treating or disposing of waste	Amount of waste composted, incinerated or landfilled (recycled materials are exempt)	0.1-2.6 $/t or cubic metre	Not known
Denmark	Incentive to encourage recycling of waste	1987	Households, industry	Amount of waste generated (wastes classified as harmless, for example straw, are exempt)	6.5 $/t	Not known
Netherlands (surplus manure charge)	Raise revenue for research and pilot projects (some incentive effect expected)	1987 (not yet fully operated)	Farms	Phosphate content of manure produced beyond that permitted to be dumped on land	0.1-0.3 $/t	Not known (not yet fully operated)
USA (federal and state)	Raise revenue for restoration of hazardous waste disposal sites after closure	1987	Hazardous waste disposal site operators	Amount of chemical waste disposed of	2.3 $/t (federal) up to 72 $/t (state)	Not known

Source: Vernon, Jan L., <u>Market Mechanism for Pollution Control</u>, 1990, Table 5.

Noise & Other Emissions

Germany, Switzerland, Japan, the Netherlands, and the United Kingdom all impose a minimal noise emission tax on airlines. Switzerland raises $5 million a year from the tax which is used to finance noise abatement structures for the effected communities. To boost the demand for unleaded gasoline in the United Kingdom, an extra tax of 32 cents was added to the price of leaded gasoline since 1988. Denmark imposes a tax of 5 kroner per metric ton of sand or gravel, imported or domestic, to encourage the building industry to recycle. In fact Denmark has been very successful in collecting pollution taxes. The Danish Environmental

Protection Agency collects an estimated $1.350 billion per year from environmental taxes, charges and fees to cover 95 percent of public environmental protection and administrative costs.[10] Table 11 describes the various noise emission taxes and fees.

Table 11:   Noise Emission Taxes in Operation

Country	Purpose	Start year	Coverage	Tax basis	Rate	Revenue raised
FRG	Raise revenue for insulation programmes around airports	1976	Airline companies	Surcharge on landing fees, higher if aircraft is non-noise certificated	Not known	Not known
Japan	Raise revenue for noise abatement measures around airports	1975	Airline companies	Surcharge on landing fees based on weight of aircraft and sound level on landing and take-off	Not known	Not known
Netherlands	Raise revenue for insulation programmes	1983	Airline companies	Surcharge on landing fees based on aircraft weight and noise characteristics	Not known	Around 5 $M/y
Switzerland	Raise money to finance insulation programmes	1980	Airline companies	Noise level of aircraft (charge doubles for each 10dB increase)	10% of landing fees	Around 5 $M/y
United Kingdom	Incentive to encourage quieter aircraft	1987	Airline companies (at some small airports only)	Differential landing fees for 'noisy' and 'quieter' aircraft	'small'	'minor'

Source: Vernon, Jan L., Market Mechanism for Pollution Control, IEA Coal Research, 1990, Table 6.

Pesticides & Other Products

Sweden taxes pesticides at a rate of $.75/Kg of active ingredients and $5.00 per hectare of land that it is applied to. The purpose of this tax is to raise revenue for agriculture and forestry programs. Norway also places a surcharge on pesticides to fund programs in sustainable agriculture. However, the tax rates are too low to reduce use and create a change to less polluting substances.

The U.S. Congressional Budget Office studied a tax on the sale of agricultural pesticides and fertilizers. These chemicals cause harm to the environment through farmer exposure, surface and groundwater pollution and pesticide residues. The effect of the tax would be a reduction in use by the farmer. A ten percent tax on the sale of agricultural chemicals in the United States would raise about four billion dollars in 1991-1995. The effect of his tax on farmer income is expected to be slight and a relatively small increase for the consumer.

Table 12 shows the taxes imposed by various countries on pesticides and other environmentally deleterious products such as plastics. In Italy there is a U.S. $.01 cent tax on retailers who use plastic bags. This is to discourage use and any bag that is biodegradable is exempt from this tax. As the table shows the main purpose of these taxes are to enhance recycling programs and fund environmental programs. A higher tax needs to be imposed to obtain substantial change in consumer patterns and reductions in pollution.

Table 12:   Product Taxes in Operation

Country	Coverage	Start year	Purpose	Tax basis	Rate	Revenue raised
Finland	Non-returnable beverage containers	1976	Incentive to support the deposit – refund system for bottles	– beer containers – soft drinks in glass, metal containers – soft drinks in other containers	$0.09 $0.07 $0.25	Approximately 2 $M/y
	Lubricant oils	1987	To finance improved waste oil collection and processing	Standard tax per tonne of oil	35 $/t	3.6 $M/y
	Fuels: coal, peat natural gas, heavy and light fuel oil, petrol, diesel	1990	Raise revenue	Standard tax per MWh or litre of fuel	Coal 0.6 $/t Peat 0.5 $/t Gas 0.3 $/t Heavy fuel oil 0.5 $/t	Not known
France	Lubricant oil	1981	Finance for waste oil collection and treatment	Standard tax per tonne of oil	7 $/t	Approximately 5 $M/y
FRG	Lubricant oil	1969	Finance for waste oil collection and treatment	Standard tax per tonne of oil	115 $/t	Approximately 75 $M/y
Italy	Lubricant oil	1985	Finance for waste oil collection and treatment	Standard tax per tonne of oil	4 $/t	Approximately 25 $M/y
	Plastic bags used as retail containers (manufacture and import)	1989	Incentive to discourage plastic bag use	Standard tax per bag (bags over 90% biodegradeable are exempt)	0.08 $/per bag	Not yet known
Netherlands	Fuels: petrol, fuel oil, gas oil, coal, LPG, blast furnace gas, natural gas, petroleum coke	1988	Finance to cover environmental policy expenditure of the Ministry of Environment	Standard tax per litre, tonne, cubic metre or gigajoule of fuel (rebate available on coal or heavy fuel oil charge where user has flue gas desulphurisation)	Petrol 0.1/2.0 $/litre, heavy fuel oil/coal $3.0 per tonne	Approximately 155 $M/y
Sweden	Oil products	1984	Finance for environmental programmes, including timing of lakes	Standard tax per cubic metre of oil	About 2.5 $/cubic metre	Not known
	Fertilizers	1984	Revenue raising for environmental programmes	Tax per kilogramme of Nitrogen (N) and Phosphorus (P) contained	N 0.1 $/kg, P 0.05 $/kg	Not known
	Pesticides	1984, 1987	Revenue raising for agriculture and forestry programmes	Tax per kilogramme active ingredient (1984) and per hectare of land applied (1987)	0.75 $/kg, 5 $/ha	Approximately 1–2 $M/y
	Batteries containing mercury or cadmium	1987	Incentive to reduce use of cadmium in mercury in batteries	Tax per kilogramme of batteries sold	3.9 $/kg	Approximately 13.2 $M/y (added t general revenue)
	Beverage containers	1984	Incentive to encourage use of deposit/refund systems and prevent littering	Tax depends on container size	0.2–0.05 $/container	Approximately 13.2 $M/y (added t general revenue)
USA	Chemical feedstocks (petroleum, base chemicals, derivatives)	1981 (revised 1986)	Revenue for 'superfund' to clean up abandoned hazardous waste sites	Tax per barrel of petroleum, per tonne of base chemicals and derivatives	0.08–0.12 $/barrel, $0.03–0.11 per tonne	Average 950 $M/y

Conclusion

The European countries and Japan have experimented with a variety of pollution and effluent taxes, generally with favorable results on pollution reduction and without adverse effects on their economies. The European Community is actively considering a carbon tax. The U.S. has imposed a tax on CFCs and is considering a variety of other pollution taxes. U.S. utilities and their regulatory commissions have lead the way in establishing and incorporating environmental externality values. Finally, the "Earth Summit" in June of 1992, is actively considering recommendations for a carbon tax and other pollution taxes.

Thus, there is a lot of momentum behind the pollution tax concept. If the taxes are high enough to affect industry behavior, pollution taxes are likely to have a significant effect on pollution reduction and technological innovation to reduce pollution.

Pollution taxes are best imposed internationally to prevent competition between countries on the basis of the leniency of their pollution tax laws. If they are imposed unilaterally, care must be taken to avoid damage to domestic economies by leavening their effect with reduction in other taxes. Care must also be exercised to make sure that the net effect of pollution taxes will not place a burden on those who can least afford to pay them.

References and Endnotes

[1] That portion of this paper which deals with incorporation by U.S. utilities of environmental externalities is based on a study performed by the Pace University Center for Environmental Legal Studies (Center) for the New York State Energy Research and Development Authority (NYSERDA) and the U.S. Department of Energy (DOE), *Environmental Costs of Electricity*, Oceana Publication, Inc., Dobbs Ferry, N.Y. (1990).

[2] *Money from Greenery*, The Economist, London, U.K., October 21, 1988, at p. 16.

[3] OECD, *The Polluter Pays Principle*, Paris (1975).

[4] Lee, N.K., "Options for Environmental Policy," *Science*, Vol. 182, pp. 911,912 (November 30, 1973).

[5] 26 U.S.C.A. Sec. 4681 and Sec. 4682 (1992).

6   Commission for the European Communities, *Proposal for a Council Directive Introducing a Tax on Carbon Dioxide Emissions and Energy*, (COM [92] 226 final), Brussels (1992).

7   Data Resource, Inc., *The Economic Impact of a Package of EC Measures to Control CO_2 Emissions*, Final Report (November 1991); and: Data Resource, Inc., *Impact of a Package of EC Measures to Control CO_2 Emissions on European Industry*, Final Report (January 1992); both prepared for the Commission of the European Communities DG XI, Brussels.

8   Hearings held before House of Representatives Committee on Ways and Means, *Long Term Strategies on the Environment*, March 6, 7, and 14, 1990.

9   U.S. Congressional Budget Office, *Carbon Charges as a Response to Global Warming*; *The Effect of Taxing Fossil Fuels*, U.S. Government Printing Office, Washington, D.C. (1990).

10 BNA International Environmental Reporter, Vol. 14, No. 1, pg. 23 (1991).

17. A Prudent CO_2 Reduction Policy: Melding Top-Down and Bottom-Up Approaches

Ajay K. Sanghi; Anthony L. Joseph
New York State Energy Office
Albany, New York 12223

Introduction

The recent amendments to the Clean Air Act will drastically reduce emissions of criteria pollutants regulated under National Ambient Air Quality Standards (SO_x, NO_x, CO, Ozone, PM & Lead). As of yet, emissions of CO_2, considered to be the major cause of global warming problem, are not covered under the Clean Air Act, thus are not required to be controlled. However, given that international pressure is mounting to limit emissions of CO_2, their control has become the focus of attention at both national and state levels.

A prudent CO_2 reduction policy is necessary otherwise it could seriously dislocate energy markets. This paper discusses limitations in applying traditional command and control and market-based methods for reducing CO_2 emissions. Further, it shows how technology specific CO_2 control strategies (the "bottom-up" approach) and carbon taxes (the "top-down" approach) can be melded to formulate a prudent CO_2 reduction policy.

Difficulties With Command-and-Control and Market-Based CO_2 Emission Control Policies

Traditional Command-and-Control Approach

Currently, there are no commercially available technologies for reducing CO_2 emissions either by combustion modifications or post combustion flue gas scrubbing technologies.[1]

---

[1] Different types of fossil fuels (e.g., coal, oil and natural gas) vary in terms of the carbon content per unit of energy. During combustion, the carbon in these fuels converts to carbon dioxide (CO_2). An advanced CO_2 scrubbing technology has been proposed by M. Steinberg and H.C. Chen. "Advanced Technologies For Reducing CO_2 Emissions." Brookhaven National Laboratory. December 1987. However, the estimated cost for such a technology is extremely high, and a feasible solution to the disposal of the huge amount of recovered CO_2 has not been found. For this reason CO_2 scrubbers are not discussed in literature as viable options.

Therefore, traditional command and control approaches of legislating technology specific control requirements and emission rate requirements are not suitable for controlling CO_2 emissions at this time.

The Bottom-Up Approach

A wide range of technology specific measures can be used to reduce CO_2 emissions. These include such measures as switching to relatively lower carbon content fuels, reforestation, CAFE standards, appliance efficiency standards, energy conservation, methane recovery, cogeneration, CFC reductions, alternative fuel vehicles, replacement and repowering of utility sources and use of renewable energy resources.

Policy makers have assessed the use of these control measures in terms of their ability to achieve a desired CO_2 reduction objective. Determining the CO_2 control measures required to achieve a specific CO_2 reduction objective is termed a "bottom-up" approach.

The bottom-up approach provides a useful tool for policy makers, in that it gives rise to the least-cost combination of control measures which can be used to achieve a specific CO_2 reduction objective. However in its own right, the bottom-up approach does not possess an administrative or regulatory mechanism to bring about implementation of the desired control measures. Given this, policy makers have been discussing contemporary market-based approaches like marketable permits, offsets and pollution taxes in order to bring about the implementation of the bottom-up control measures.

CO_2 Marketable Permit Program

Since passage of the marketable permit program in the Clean Air Act Amendments of 1990 for controlling SO_2 emissions, such programs have been discussed among policy makers for controlling CO_2 emissions.[2] However, in applying such a program to CO_2 emissions, the following question must be addressed. How successfully can a marketable permit or allowance trading program of the type designed for reducing SO_2 emissions in Title IV of the Clean Air Act Amendments be used to reduce CO_2 emissions?

---

[2] Decision Focus Incorporated. "CO_2 Trading Issues". May 1992.

Assessing the salient characteristics of the Title IV SO_2 marketable permit program provides insight in answering this question. The SO_2 program is only targeted at the electricity sector. This is because over 60 percent of nationwide SO_2 emissions come from the electricity sector, therefore a meaningful reduction in SO_2 emissions can be achieved through further controls in this sector. In addition, the electricity sector is heavily regulated and generally characterized by large stationary sources, for which good data on SO_2 emissions is available. Given these characteristics (good data and a manageable number of sources), application of a SO_2 marketable permit program is feasible and promises significant total reductions in emissions at least cost by harvesting market forces.

In contrast, CO_2 emissions are relatively evenly dispersed across the electricity, buildings and transportation sectors.[3] Therefore, a meaningful overall reduction in CO_2 would require participation across these sectors, unlike the SO_2 marketable permit program which is largely directed at the electricity sector. However, the buildings and transportation sectors are generally characterized by numerous and diverse groups of small sources. In addition, CO_2 is an unregulated pollutant, in which case CO_2 emission inventory data are limited or unavailable. When combined, these factors will likely hinder establishment of baseline CO_2 emissions for each of the millions of small stationary and mobile sources, as required in order to allocate CO_2 allowances, and also will likely hinder the monitoring of allowance trades.

If only the electric sector is targeted for achieving a meaningful reduction in total CO_2 emissions through a marketable permit program (as in the SO_2 allowance trading program), there will likely be an unacceptable increase in statewide electricity rates. For example, it is estimated that a CO_2 marketable permit program which requires a 50 percent reduction in electric utility CO_2 emissions from 1988 levels by the year 2000, will increase electricity rates in New York by about 20 to 30 percent, and only reduce statewide CO_2 emissions by about 15 percent. In comparison, a SO_2 marketable permit program which requires a 50 percent reduction in electric utility SO_2 emissions from 1988 levels by the year 2000 (similar to the Title IV SO_2 reduction requirements) will only increase electricity rates in New York by about 1 to 2 percent, and reduce statewide SO_2 emissions by about 35 percent.

CO_2 Offset Program

A partial variant of the CO_2 marketable permit program is a requirement for new plants to offset their CO_2 emissions. This type of approached has been introduced by Representative

---

[3]   In New York State each of these sectors contributes about one third of total statewide CO_2 emissions.

Jim Cooper (D-TN) and Senator Liberman (D-ME) as an amendment to the Clean Air Act entitled "CO_2 Offsets Policy Efficiency Act of 1991".[4] Under their proposal, new major sources which emit 100,000 tons or more of CO_2 annually (greater than a 15 Mw coal-fired power plant) will be required to possess (at the end of each year) sufficient CO_2 emission reduction credits to offset their annual CO_2 emissions.[5]

If passed in the congress, this legislation would greatly impact the nation's CO_2 emissions and attendant energy use profile in the long run. However, because the bill only affects new stationary sources, mostly electricity generating facilities, it is not expected to reduce CO_2 emissions significantly in the foreseeable future. For example, New York State CO_2 emissions in the year 2008 are projected to increase by about 88 million tons from 1988 levels of about 232 million tons.[6] In comparison, new electricity generating capacity (estimated at about 6,000 Mw by the year 2008) only accounts for approximately 21 million tons of the increase in statewide CO_2 emissions. Thus, the expected reduction from offsetting emissions from new sources will likely fall short of achieving even the modest goal of stabilizing CO_2 emissions in New York State at 1988 levels by the year 2008.

A Carbon Tax (The Top-Down Approach)

Another approach for reducing CO_2 emissions which has been receiving attention in many environmental forums is a carbon tax (top-down approach). A desirable feature of a carbon tax policy is that the cost burden of the resultant reduction in CO_2 emissions will likely be shared relatively uniformly across the different sectors of the economy. However, a carbon tax policy too has inherent hurdles to overcome if it is to be a prudent policy.

The major hurdle in the top-down carbon tax approach is that it generally has the highest per unit cost of control for a given reduction objective, compared with the other methods of

---

[4]   The bill proposes to create a national CO_2 offset bank for the purpose of ensuring adequate supplies of CO_2 offsets. A procedure will be established to identify the owners of CO_2 offsets, their addresses, the tonnage available, the source of the CO_2 credit and the minimum asking price.

[5] CO_2 emission reduction credits can be obtained by similar measures discussed under the bottom-up approach. For example low carbon content fuel switching, reforestation, increase of CAFE standards, increase of appliance efficiency, energy conservation, methane recovery, cogeneration, CFC reductions, alternative fuel vehicles, replacement and repowering of utility sources and use of renewable energy resources.

[6]   New York State. Analysis of Carbon Reduction in New York State. Prepared by the NYS Energy Office in consultation with the NYS Department of Environmental Conservation and the NYS Public Service Commission. June 1991.

control. This is because the effect of a carbon tax is to influence producer and consumer behavior through market response to the tax. Thus, the tax rate must be set equal to the highest marginal cost of abatement measure required to achieve the desired level of reduction. As shown in Figure 1, a required tax of tax of T would be necessary to achieve the desired pollution reduction OA. This level of abatement occurs because it is less expensive for sources to abate the pollutant over this range of reductions than to pay the tax. If the pollution tax is set below T, market signals will not be sufficient to induce the desired level of pollution reduction.

Figure 1:   Reducing Emission through a "Top-Down" Tax

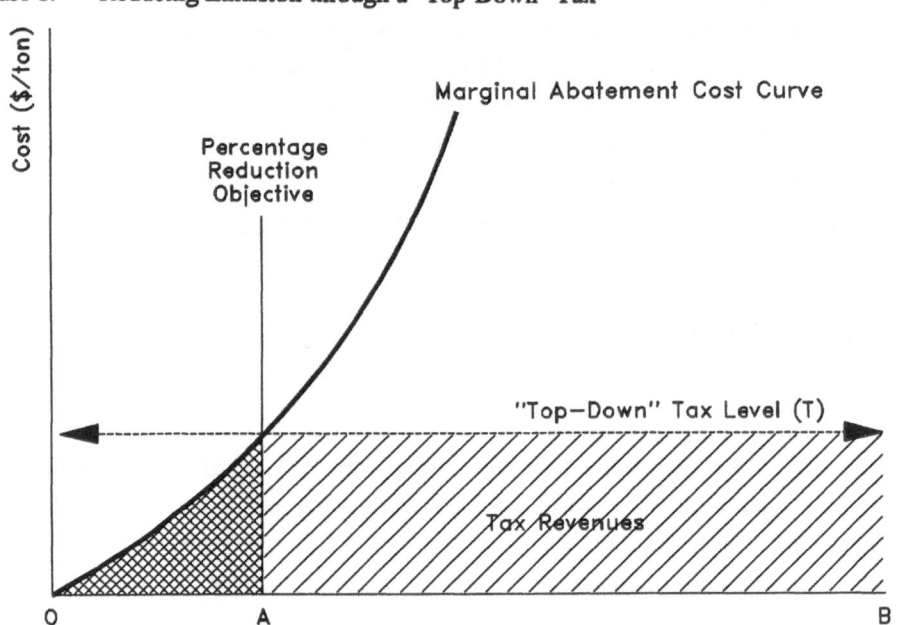

The cost to energy consumers under the top-down tax policy will be equal to the cross hatched area plus the shaded area. The cross hatched area represents increased costs to energy consumers resulting from the tax induced implementation of pollution abatement measures by polluting sources (the least cost combination of measures defined by the bottom-up approach). The slashed area represents the revenues collected from the tax, being that it is less expensive for sources to pay the tax on this level of emissions rather than to abate.

Since most CO_2 reduction targets are set below 20 percent, tax revenues collected by the government (the slashed area) will likely be much larger than the cost of the tax induced implementation of abatement measure (the cross hatched area). This is because the residual emissions (i.e., 80 percent of the total) generate tax revenues. Thus, the appropriation of carbon tax revenues raises an important policy issue.

While it is conceivable (theoretically) to refund the revenues generated under the top-down tax approach back to energy consumers in order to mitigate the associated cost impacts, the political feasibility and even "do-ability" of such approaches are questionable. For example, it is estimated that a 5 percent reduction in carbon emissions in New York State from 1990 levels by the year 2010 would require a top-down carbon tax rate of about \$217 per ton.[7] Such a tax will increase the total energy bill of New York State by an estimated \$15 billion, of which only about \$0.6 billion represents expenditures on carbon reduction measures. The remaining \$14.4 billion is revenues paid to government treasuries which must be returned to the taxpayers.[8]

Some advocates of carbon taxes have addressed the recycling of tax revenues issue by reforming federal income tax structure.[9] However, such a strategy is likely to create serious income distribution problems.[10]

Melding the Top-Down and Bottom-Up Approaches

While the top-down carbon tax appears to be the most applicable policy for controlling CO_2 emissions, it is problematic in that placing tax revenues in government treasuries gives rise to a very high per unit CO_2 control cost relative to that under the bottom-up approach. This limitation can be mitigated to the extent the carbon tax is used to generate tax revenues only in the amount sufficient to finance the bottom-up combination of least-cost abatement

---

[7] Draft New York State Energy Plan 1991 Biennial Update. Volume III. Issue 9: Energy and Environmental Taxes. July 1991.

[8] For this reason the "top-down" tax is referred to as a General Revenue Tax in the Draft New York State Energy Plan 1991 Biennial Update, Volume III, Issue 9: Energy and Environmental Taxes, July 1991.

[9] A discussion of carbon taxes and tax revenue allocation issues is presented by Roger Dower and Mary Beth Zimmerman in "The Right Climate for Carbon Taxes: Creating Economic Incentives to Protect the Atmosphere." World Resources Institute. August 1992.

[10] A further discussion of this point is presented in Ajay K. Sanghi and Anthony L. Joseph. "Taxing Pollution Instead of Labor: Is It a Good CO_2 Reduction Policy." The Electricity Journal (Forthcoming December 1992).

measures.[11] This approach is termed a Trust Fund (TF) carbon tax, since the revenues are only collected in the amount sufficient to fund the bottom-up abatement measures required to achieve the pollution reduction objective and are specifically earmarked for this purpose.

As shown in Figure 2, the TF tax rate is derived by distributing the total cost of the abatement measures required to achieve the desired pollution reduction (the cross hatched) over the tons of pollution on which the tax will be paid, or the tax base (AB tons). Since the TF tax is only designed to generate revenues in the amount of the total abatement cost of the pollution reduction objective, the total cost to energy consumers is equal to the cross hatched area (total abatement cost) or the slashed area (tax revenues), since the two areas are equal (by definition). For this same reason, the required tax rate will be lower under the TF tax approach than under the top-down carbon tax approach, as shown in Figure 2.

Figure 2:   Reducing Emission through a Trust Fund Tax

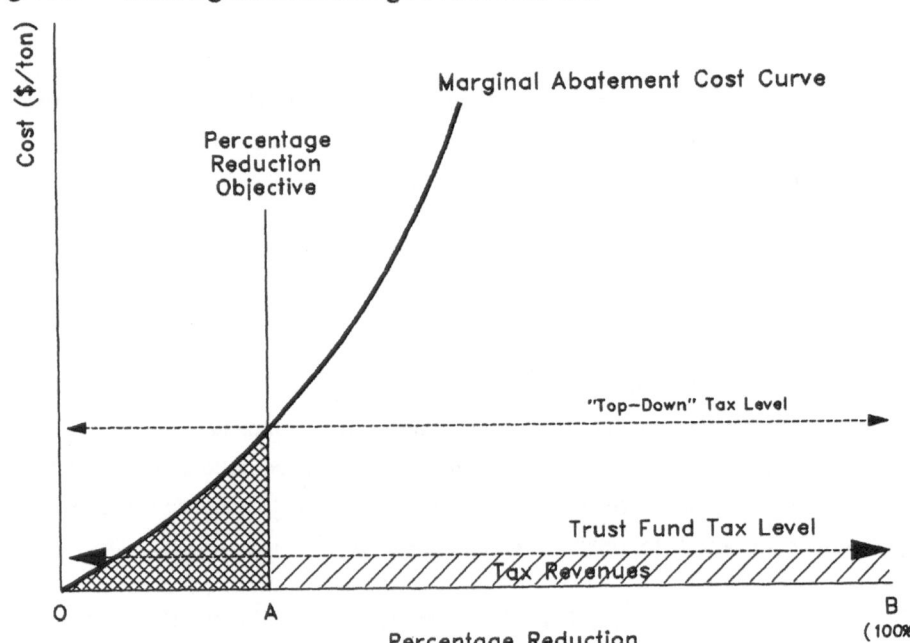

---

11 A good exposition on "top-down" and "bottom-up" approaches is presented by Florentine Krause, Jonathan Koomey and David Oliver. "Energy Policy in the Greenhouse." Volume 2, Least Cost Insurance Against Climate Risks, The Cost of Cutting Carbon Emissions. International Project for Sustainable Energy Paths. September 1992.

Conclusions

The "bottom-up" approach gives rise to numerous measures which can be used to achieve a desired reduction in CO_2 emissions. Of the more contemporary approaches being considered for harnessing market forces to encourage implementation of these measures in a least cost manner, a federal carbon tax seems to hold the most promise. Alternatively, a CO_2 marketable permit program appears to be the least practical. And, an offset program which requires new stationery sources to offset their CO_2 emission, while being a workable market-based CO_2 reduction policy option, may not even achieve a CO_2 stabilization objective.

This paper has shown that although the carbon tax holds promise as a CO_2 reduction policy, it has an inherent hurdle to overcome before it can be considered a prudent policy. That is, unless the resultant tax revenues are dedicated for CO_2 remediation purposes, the carbon tax rate will have to set at levels which are likely to be considered politically unacceptable. For example, it is estimated that a top-down CO_2 tax of about $27 per ton of CO_2 ($100 per ton of carbon) will be required to stabilize national CO_2 emissions at 1990 levels by the year 2000.[12] This tax level is likely to draw significant politically opposition, as it could be seriously disruptive to U.S. energy markets through its impact on relative energy prices. In comparison, it is estimated that the CO_2 stabilization target could be achieved through a tax of about $0.50 per ton of CO_2 ($2 per ton of carbon) by utilizing a trust fund mechanism.[13] Thus, the trust fund tax is likely to be more politically acceptable than the top-down tax.

In absence of the trust fund mechanism, the nationwide reduction in CO_2 emissions resulting from the $0.50 per ton CO_2 tax will fall far short of stabilizing CO_2 emissions at 1990 levels. Therefore, a small carbon tax will only serve as a "token" CO_2 reduction policy (i.e., it will do very little in the way of encouraging CO_2 reductions) unless revenues are dedicated to pay for CO_2 abatement measures, as called for under the Trust Fund tax approach.

---

[12] Stabilization of CO_2 emissions is being considered as a national policy objective as a result of the United Nation's Earth-Summit meetings in Rio DeJaneiro in June of 1992.

[13] It is estimated that the $0.50 per ton trust fund tax will generate on the order of $3.5 billion in revenues, which is sufficient to pay for the CO_2 mitigation measures required to stabilized CO_2 emissions at 1990 levels in the year 2000.

References

M. Steinberg and H. C. Chen. "Advanced Technologies For Reducing CO_2 Emissions." Brookhaven National Laboratory. December 1987.

Decision Focus Incorporated. "CO_2 Trading Issues". May 1992.

New York State. Analysis of Carbon Reduction in New York State. Prepared by the NYS Energy Office in consultation with the NYS Department of Environmental Conservation and the NYS Public Service Commission. June 1991.

New York State. Energy Plan 1991. Draft. Biennial Update. Volume III. Issue 9: Energy and Environmental Taxes. July 1991.

Roger Dower and Mary Beth Zimmerman. "The Right Climate for Carbon Taxes: Creating Economic Incentives to Protect the Atmosphere." World Resources Institute. August 1992.

Ajay K. Sanghi and Anthony L. Joseph. "Taxing Pollution Instead of Labor: Is It a Good CO_2 Reduction Policy." The Electricity Journal (Forthcoming December 1992).

Florentine Krause, Jonathan Koomey and David Oliver. "Energy Policy in the Greenhouse." Volume 2, Least Cost Insurance Against Climate Risks, The Cost of Cutting Carbon Emissions. International Project for Sustainable Energy Paths. September 1992.

18. Consideration of Environmental Externality Costs in Utility Buy Back (PURCHASE) Rates

Sury N. Putta
New York State Department of Public Service
Albany, New York 12223

Background

The business of electric power generation which was once the nearly exclusive domain of investor owned and public utilities is now open to a more diverse group of participants on a competitive basis. The passage of the Public Utility Regulatory Policies Act (PURPA) in 1978 was instrumental in cracking the utility monopoly and opening opportunities for diverse forms of electric generation by independent power producers (IPPs). One of the purposes of PURPA is to reduce dependence on fossil fuels through the use of renewable energy resources and more efficient use of non renewable resources which were not a significant part of the utility resource mix. To achieve its objective, the PURPA required the utilities to purchase power from certain IPPs at a price equal to the utility's avoided cost--the levelized life-time cost in cents/kwh that the utility would incur if it had to produce the energy itself. The price offered by the utilities for power purchases from the IPPs and its own customers is labeled as the "buy back rate" in utility jargon.

Additional momentum followed by the state regulatory policies which also served to encourage non-utility generation (NUG) by mandating the utilities to procure future electric resources through competitive bidding. These federal and state actions brought increasing pressure on utilities to develop accurate methods for determining the avoided costs and for evaluating various factors in awarding purchase power contracts to the IPPs. This paper examines the need and methods for considering environmental characteristics of the NUG facilities in determining the rates and award purchase contracts to the IPPs which may encourage them to develop and use environmentally clean resources and cost effective pollution control methods.

Avoided Cost-derivation

The long run avoided cost (LRAC) estimate is a measure of the stream of future resource costs avoided by a utility when it purchases a new resource or implements demand side management (DSM) measures. The utilities use the LRAC estimates in determining the buy back rates for the power purchases from the IPPs. Currently, only the avoided economic costs are used in estimating the LRACs. No consideration is given to the societal costs of avoided environmental impacts. The LRAC estimates consist of avoided energy and capacity components which are derived periodically for a block of future years. Prior to the date when new capacity is needed, only avoided energy costs are estimated. Production costing programs are used to develop estimates of total system production costs based on input assumptions on many factors such as peak load and energy forecasts, fuel prices, off-system sales and purchases, generating unit characteristics, maintenance schedules, inflation rate, energy production from IPPs already contracted and expected level of DSM measures. The LRAC estimates generally decrease when increasing IPP capacity is added to the utility system because the last unit of IPP energy displaces lower cost utility power than the first unit of IPP energy. Since the addition of IPPs and incremental DSM decrease marginal cost of generation and with it future levels avoided energy costs which are used to determine the utility buy back rates, it is important for the utilities to consider the effect of adding IPP energy in small blocks on the future avoided costs. Figure 1 illustrates the effect of future resource additions either by the utilities or IPPs on the LRAC estimates in New York State. Figure 2 illustrates the LRAC estimate in the year 1993 for each block of IPP energy added to the utility system in 1000 MW intervals. The utilities are considering the use of block LRAC estimates to make payments to IPPs. Under this proposal the first block of offer by the IPPs will receive payment at the rate equal to the highest LRAC estimate and other IPPs will receive successively lower payments determined by the production cost model which takes into account the energy contracted to earlier IPPs.

Environmental impacts - a factor in choosing and pricing IPPs energy

If a generation resource imposes external costs or benefits, it will do so whether it is owned by a utility or by an IPP selling power to the utility. This raises several questions such as (1) Whether the utility should account for the value of environmental externalities in comparing the IPP proposals, (2) Whether the rate offered for the purchased power account for the value of externalities, and, if so, what effect it will have on utility rates.

Figure 1: New York Statewide Average Avoided Energy Costs

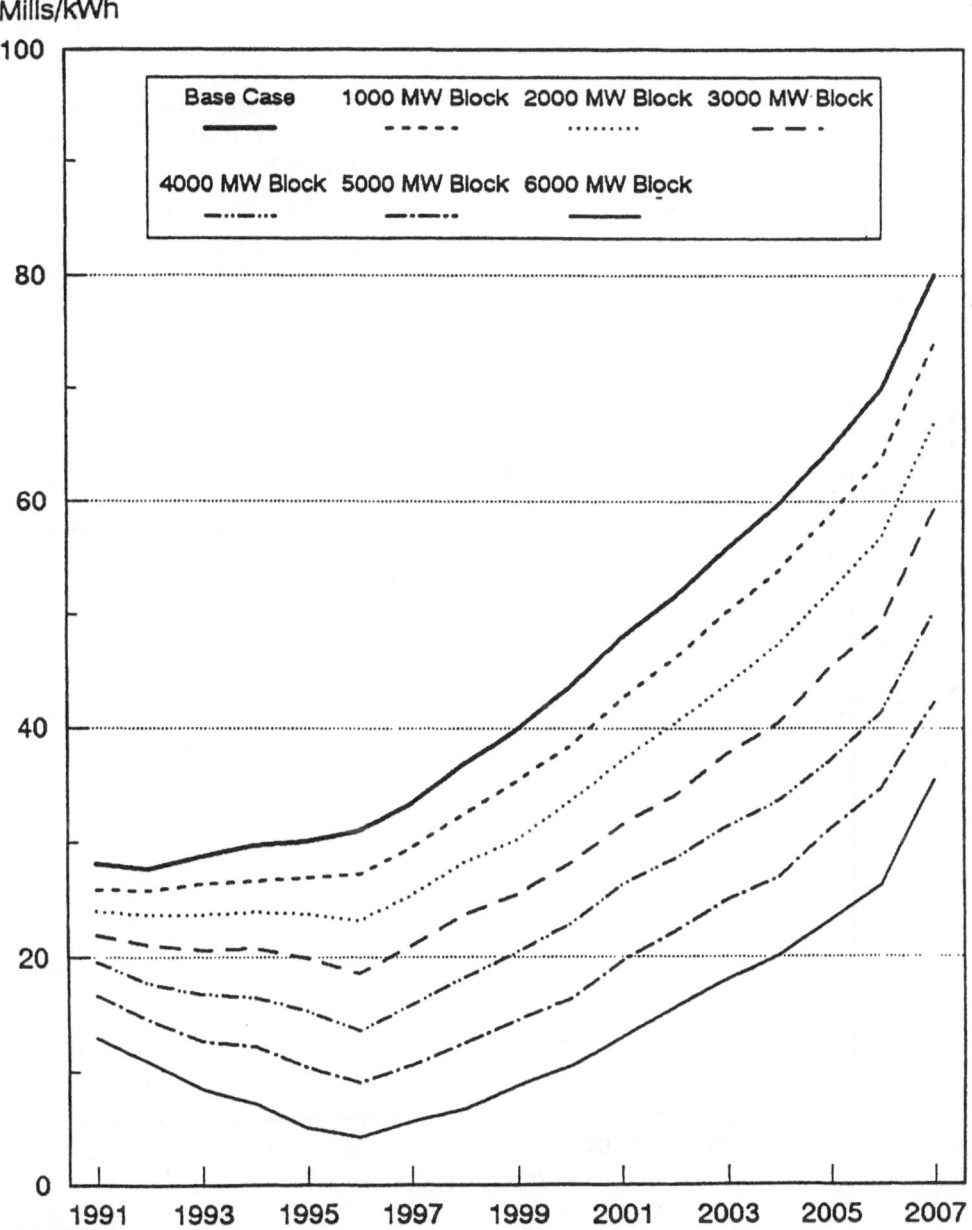

*
NYS Public Service Commission Proceeding 91-E-0237
Long-Run Avoided Cost Estimates by NY Power Pool

Figure 2:   Projected Statewide Average Avoided Energy and Environmental Costs - 1993

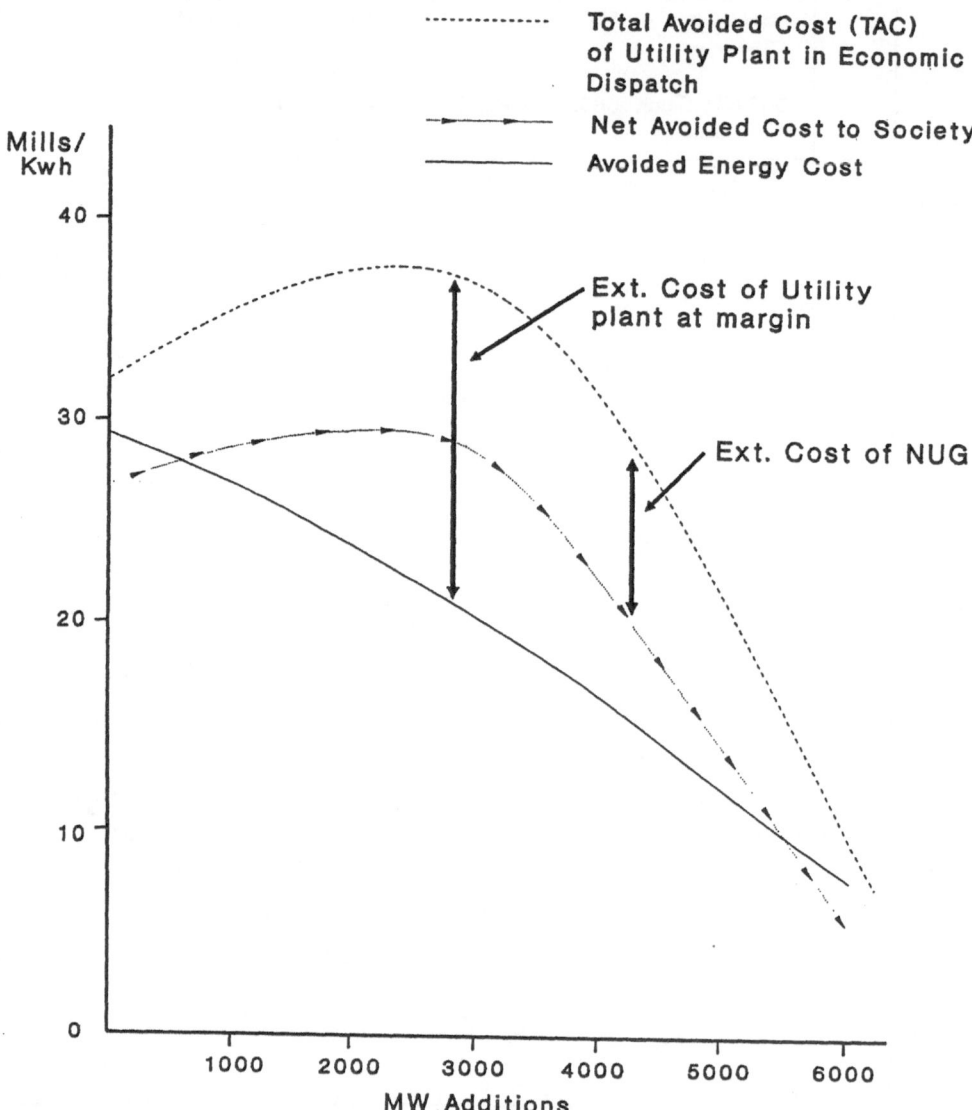

Currently the utilities do not consider the value of environmental externalities in awarding contracts to IPPs and purchase power at a single rate from all resources regardless of their environmental characteristics; i.e., coal, oil and gas or renewable energy resources are paid at the same rate. If this practice continues, the IPPs will choose the least expensive electric resources which meet minimum environmental standards and will have no incentive to search for environmentally cleaner resources or cost effective control methods as these choices increase costs and reduce their profits. On the other hand if the buy back rates reflect the value of avoided environmental impacts, it will result in increased payments to the IPPs whose residual impacts are less than those of the existing utility generating facilities and those increased costs will be passed on to consumers in the form of higher electric rates. Whether the increased rates are worth the environmental benefits and, if so, how society can get the most benefit for the increased costs incurred is a matter that rests in the hands of the utility regulators who are responsible for protecting the welfare of the ratepayers. Some utility regulators are currently considering to adopt policies aimed at encouraging the utilities and the IPPs to search for environmentally cleaner resources and develop pollution control methods which offer most benefit at a least cost to the society.

Methods for Considering Avoided Environmental Impacts in Evaluating IPP Proposals

Several methods are examined for considering the avoided environmental impacts in evaluating the IPP proposals and award purchase contracts that benefit the consumers. Application of those methods will affect the ratepayers and the IPPs as those methods may alter payments to IPPs which are otherwise determined traditionally by the avoided dollar costs to the utility. Three methods are discussed in this section for accounting the environmental impacts of the IPPs and those impacts avoided at the utility plants in awarding purchase contracts to the IPPs.

Method 1 - Environmental Ranking Method (ERM)

In this method the purchasing utility designs an environmental scoring system to rank the IPP proposals received in a three to six month period in a descending order based on their scores. The utility then uses the declining avoided cost curve similar to that illustrated in Figure 2 to award the highest avoided cost to the highest ranked proposal and offer successively lower costs along the curve to the remaining proposals based on their environmental rank. The rates offered by this method will encourage the IPPs to consider environmentally superior facilities and compete on environmental score as the environmental rank determines the payments for their power. Since the utilities will pay the IPPs no more than the actual avoided cost to the utility, the ratepayers will have the opportunity to select environmentally cleaner electric service at no additional cost. One disadvantage of this method is that the buy back rates offered by the utility may not be sufficient to attract IPP proposals as those rates determined by the avoided production costs of the existing utility plants which are subject to less stringent environmental standards may be lower than the production cost of the IPP facilities which must meet more stringent new standards. Under this method the utilities will initially attract a limited number of IPP proposals until the utility buy back rate is higher than the IPP production cost. Later, when the avoided costs decline, offers from IPPs will fade and electric generation will be left to the utility plants which may be environmentally inferior to the IPP proposals.

Method 2 - Total Avoided Cost (TAC) Method Based on Economic Dispatch

The TAC is the sum of the production cost and the value of incremental air, water and solid waste pollution impacts which will be avoided by reducing generation at the marginal utility plant. In this method the utility offers to buy power from IPPs at a rate equal to the TAC less the value of environmental impact of the IPP generation. If new capacity is needed to meet the forecasted demand for power, then the capacity cost is included in the TAC. To estimate the TACs, the utility first establishes a method for estimating the value of the environmental impacts of power generation. Then the TACs are determined in two ways. One way would be to simply add the value of the avoided environmental impact of the marginal utility plant to the LRAC estimate as illustrated in Figure 2. Since the LRAC estimates are based on the economic dispatch method, the first utility plant avoided will have the highest production cost and it is more likely that it will avoid much smaller environmental impact than that of the next facility in the system. The TACs and the net societal costs avoided as a result of adding an incremental capacity of the IPPs will initially be smaller which will increase as additional blocks of IPP capacity are added. Hence, the utilities will pay less for the first block of IPP power than to the next block until the the avoided environmental cost of the marginal utility plant reaches maximum as illustrated in Figure 2. This method of rewarding the laggard IPPs with higher payments may frustrate utilities seeking energy contracts from IPPs as they wait on the sidelines to get the highest rate from the utilities for their energy.

Figure 3:   Comparison of Avoided Costs in Total Cost Dispatch with Economic Dispatach

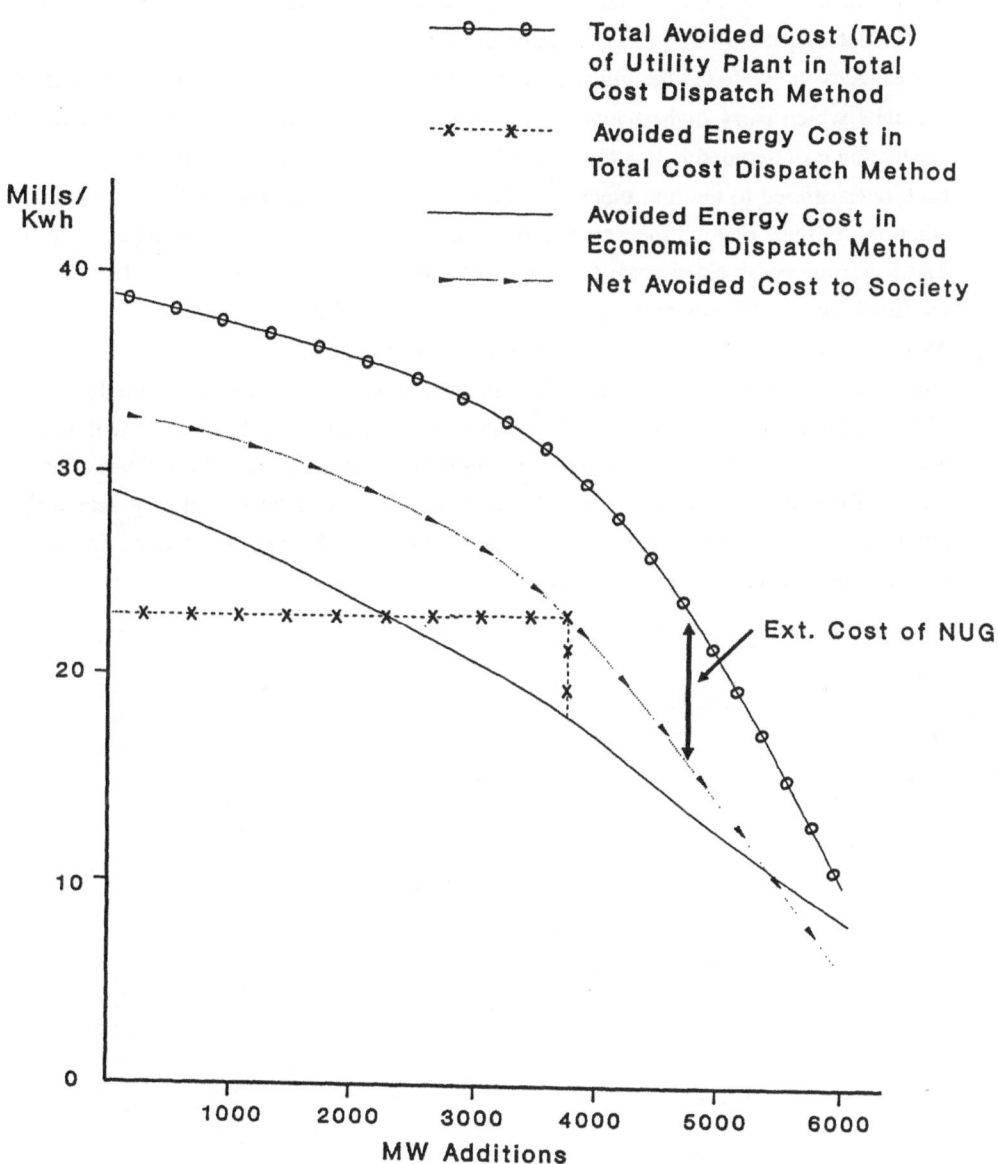

Method 3 - Total Avoided Cost (TAC) Method Based on Total Cost Dispatch (TCD)

In this method the utilities simulate the operation of the statewide utility system based on Total Cost Dispatch (TCD) and determine the TACs as illustrated in Figure 3 when blocks of IPP capacity are added to the utility system. In the TCD method the utilities will curtail first the units which carry highest total cost (production cost + environmental cost) even if the production cost avoided is less than that of other units in the system. In this method the buy back rates offered to the first block of IPPs will likely be much higher than the rates offered when the utilities operate their system based on conventional economic dispatch. Both the TACs and the net avoided costs decrease consistently with increasing added IPP energy and the first block of IPPs which displace generation at the utility plants will receive a higher rate compared to the next block of IPPs which displace less dirtier utility units. While the higher buy back rates determined by the TCD method would attract environmentally clean IPP offers first to displace the dirtiest utility plants, the ratepayers will be charged with the higher costs paid to the IPPs. Whether the utility regulators will allow the utilities' use of the TCD method for making payments to the IPP energy will depend on how they perceive the the ratepayer responsibility to reduce environmental impacts of energy production and the value of environmental benefits to the society.

19. The Indispensability of Externality Valuation in Least-Cost Planning

Paul Chernick, President
Sabrina Birner, Research Analyst
Resource Insight
18 Tremont Street, Boston MA, 02108 USA

0 Introduction

Utility regulators have been expanding the range of costs included in evaluating utility decisions such as the selection of new supply sources or the analysis of conservation programs. In addition to known direct costs to the utilities, regulators are including various allowances for risks and for environmental and economic externalities. Considerable controversy has arisen regarding the inclusion of environmental externalities and the specific methods used in setting externality values.

This paper begins with a brief introduction to environmental externalities and their monetization. While most of the arguments could be extended to social and economic externalities, only environmental effects are considered here. The bulk of the paper then reviews three groups of proposed alternatives to monetizing externalities: reducing or eliminating emissions through environmental regulations; non-monetary inclusion of externalities in resource planning; and internalizing externalities without monetizing them. None of these approaches is able to entirely replace monetization.

Certain policies may be helpful in achieving a socially optimal level of pollution or in minimizing the societal costs of controlling emissions. In particular, emission caps can be useful in guiding planning efforts, including the selection of monetized externality values. Policies that harness market forces to achieve emission caps (such as offsets and tradeable allowances) may reduce the cost and increase the equity of meeting emissions limits, while providing monetized values for the externalities in the planning process. Monetized externality values remain an indispensable tool for making least-cost resource allocation decisions.[1]

---

[1] Much of the material in this paper is drawn from Chernick and Caverhill, 1992.

1   The Emerging Roles of Monetized Externalities

Utility planners and regulators are generally responsibility for acting in the public interest, maximizing public welfare, and meet customer needs at the minimum cost to society. These decision-makers are engaged in an activity that has been dubbed "least-cost planning." In order to minimize total social costs, decision analyses must account for externalities. Externalities may be generally defined as costs of planning decisions that are not reflected in the direct price customers pay for the service being rendered.

One large and important group of externalities created by virtually all energy-related decisions are environmental effects, including production of solid wastes, heating of rivers and streams, consumption of water, and risks of accidents. The environmental externalities that have attracted the greatest attention are air emissions of ozone precursors, particulates, acid gases, toxic materials, and greenhouse gases. These environmental costs are not internalized by traditional environmental regulation. Utilities are typically required to install control equipment, which reduces emissions and increases internal costs. There is generally no incentive, however, to choose a cleaner technology or better control equipment than that required. Nor is there usually any incentive to reduce the intensity of usage of plants that are already installed.

The potential external effects of a power plant fall into three categories. First, some external effects are eliminated by the imposition of controls and are replaced by internalized costs that are usually less costly than the eliminated externalities. Second, some of the cost of the residual emissions may be internalized in the utility's direct costs for emission taxes, fees, or allowance purchases. Third, the remainder of the residual externalities, which are neither eliminated nor internalized, must be reflected in planning decisions through monetized values if the utility is to serve its customers at the lowest feasible social cost.[2]

Hence, utility regulators in several states (California, Massachusetts, New York, Nevada, and to a lesser extent New Jersey and Wisconsin) have required that externalities be explicitly taken into account in utilities' resource allocation decisions. Utilities in these states assign a monetized value (in $/MMBtu, or $/ton of emissions) to the externalities associated with each resource they are considering. Under such a system, a resource with high direct costs but low external costs (e.g., a gas turbine) might be deemed the least-cost resource for meeting the

---

[2]   Projecting the costs of required controls and of internalized environmental costs can be difficult and contentious; utilities may have very different expectations than do their regulators or other parties.

utility's needs, whereas traditional analysis might have favored a resource with low direct costs but high external costs (e.g., a coal plant).

The monetizing of externalities expresses the environmental costs of any resource in the same units as its direct costs. This offers planners a simple and flexible method for achieving a societally least-cost provision of energy services. Because externalities are treated as any other cost in a benefit/cost analysis, they can be incorporated into diverse resource allocation analyses, from selecting a supply source to assessing the cost-effectiveness of increased ceiling insulation.

Rather than attempt to make an independent assessment of the benefits of reducing environmental effects, utility regulators have generally inferred externality values from the rules set by legislatures and environmental regulators, using what is referred to as the marginal-cost-of-control method. This method relies on legislatures and environmental regulators, rather than utilities and their rate-regulating agencies, to assign dollar values to the external effects of utility operations.

While the cost-of-control methodology has some limitations, and must be used with due care, it is generally well suited to the needs and abilities of utility regulators. The analyses are flexible enough to recognize differences in externality valuation between regions, to differentiate between pollutants, and to recognize anticipated future changes, while being straightforward enough to be conducted and reviewed by non-specialized regulatory staff and intervenors.

2 Proposed Alternatives to Monetization of Externalities

2.1 Reducing or Eliminating Environmental Effects

Some observers have argued that the monetization of environmental externalities in utility planning is duplicative of environmental regulation and that, if additional clean-up is desired, the environmental agencies should require additional emission reductions. In many states, the utility regulators can directly require additional control equipment for new utility plants, in connection with issuance of a certificate of convenience and necessity. If all cost-effective controls can be required for each emission source, the argument goes, it is not necessary to add another level of environmental regulation in the guise of monetized externalities.

This argument has three serious flaws. First, without some form of monetization, it would not be possible to select the highest cost-effective level of control. Environmental regulators usually set dollar limits on the costs of controls they are willing to impose to avoid a ton of emissions; utility regulators should also use those values in their analyses.

Second, many pollution sources are difficult to regulate, for legal, administrative, or equity reasons. Existing plants are often exempt from environmental regulations imposed on new units, and small sources (such as residential gas furnaces) may be prohibitively expensive to regulate. Utility regulators can achieve reductions from some sources more easily than can environmental regulators.

Third, even if all sources were subject to regulation, reducing emissions from specific facilities would not capture all cost-effective opportunities for reducing pollution. Environmental regulators issue permits for specific effects, such as the release of pollutants to the air, and establish generic requirements, such as statewide limits on fuel sulfur content. Utility regulators have additional opportunities for cost-effective pollution reduction. For example, the utility regulator can order the utility to pursue demand-side management (DSM) more aggressively before approving additional supply resources, reducing the environmental burden of supplying power to the additional demand. The same is true for decisions to switch fuels at existing plants, select a cleaner fuel for a new plant, or increase boiler efficiency.

To the extent that utility decisions lessen the burden on environmental regulators (and society) to make more expensive pollution reduction decisions in the future, the utility regulators have provided cost-effective pollution control. In order to determine whether a resource option is cost-effective, utility regulators will need monetized values for the changes in external effects.

2.2 Non-Monetary Methods for Including Externalities in Utility Planning

A number of non-monetary methods have been proposed to incorporate externalities without explicitly estimating the value of environmental benefits. These range from the entirely qualitative recognition of externalities, to quantitative ranking and weighting approaches.

2.2.1 Qualitative Methods

Utilities often claim that environmental attributes are considered in planning as tie-breaking resource attributes, so that the environmentally superior option will be selected if the direct costs of the options are very close. Taken literally, where the environmental effect is continuously variable, as through the efficiency or pollution removal level, this approach would always result in selection of levels slightly higher than those that produce the lowest direct cost. In practice, utilities rarely apply qualitative considerations quite so continuously; if they did, they would have to decide **how much** improvement could be justified by "tie-breaking."

The most important problem with any truly qualitative approach is that it is completely subjective, and hence usually under the control of the utility. As there is no uniform measure against which to compare projects or check the reasonableness of the weight assigned to environmental factors, no independent evaluation of the utilities' resource decisions can be undertaken.

Traditionally, utilities have been suppliers of energy, and while they routinely comply with environmental laws, they are generally not expert in environmental tradeoffs. Furthermore, the utility does not necessarily represent the public in its preferences for trade-offs between, for example, mercury accumulation in food supplies versus higher utility bills. Regulators (and the public they represent) have no way of knowing what factors the utility considered in its decision. The participation of other parties is necessary to determining which environmental benefits are worth pursuing and are socially cost-effective.

A second problem with qualitative methods is the lack of a way to ensure that externalities are treated consistently from one decision to the next. For instance, the utility may choose a supply option that is desirable for its SO_2 benefits, but adopt DSM only up to its direct avoided costs, ignoring the SO_2 benefits of DSM, which may be larger than those of the supply option. If these benefits were explicitly estimated, at least a measured decision considering the uncertainties could be made. Even for similar resources, there is no way to ensure that environmental factors are considered equally under a purely qualitative approach.

2.2.2 Weighting and Ranking

Another non-monetized approach to recognizing externalities involves the rating of the relative importance of each externality and the scoring of each resource option on that scale.

For instance, acid rain might be assigned a weight of 10 points, ozone formation 15 points, and global warming 20 points. A resource option would then be rated on a numerical scale (say 1-5 points) for each externality, depending on the perceived extent of the resource's contribution to the externality. The rating of each externality is multiplied by its weight; the scores are then summed over externalities to create a score for the resource option (Destribats, et al., 1990).

The subjective nature of the relative values of reducing particular externalities is probably the greatest, but certainly not the only, weakness in this method. The externalities score is compared to price in utility resource decisions, but as there is no unique method of weighing the score against price, results are not often reproducible. Further, weighting and ranking relies on utilities' subjective judgement about environmental matters in virtually every element: in the relative weights, the total weight of the externality category in relation to price, the rating scale, and the individual ratings assigned to particular projects.

This method of estimating the value of reducing externalities is so crude as to be virtually useless. As the Massachusetts DPU said in rejecting the proposal of the New England Electric System (NEES) to use ranking and weighting:

> in order to design an effective weighting and ranking approach, environmental impacts would have to be estimated so that appropriate weights could be determined. . . . If weighting and ranking systems require quantification of externality values in order to determine the appropriateness of the weights, forming weights becomes unnecessary because the quantified externality values could be monetized and added directly to project costs to assist in the determination of the mix of resources that minimizes cost and environmental impact simultaneously. . . .

Accordingly, the comments in this proceeding convince the Department that externalities should be monetized to the full extent possible. (MDPU 89-239)

By 1992, NEES appeared to have abandoned rating and weighting in favor of explicit monetization. Other utilities have also tried this method, but no utility commission appears to have has expressly sanctioned its use.

2.3   Proposed Policies for Internalizing Externalities

2.3.1 Taxes as Price Signals

Some critics of externality valuation have argued that taxes would provide better signals than regulatory valuation (Rowe, 1992; Leone, 1992; Joskow, 1992). The basic arguments for this preference are:

- the tax would apply to all sources, not just electric utilities (or utilities in general),

- the tax would apply throughout the state, region, or country, depending on the externality involved,

- polluters could determine the least-cost set of measures for reducing emissions, implementing only those that cost less than the value of the externality tax.[3]

These arguments are correct, as far as they go. However, externality taxes have not been widely applied, for a number of practical reasons. At this writing, the European Community is moving toward conditional approval of a combined carbon and energy tax that would be equivalent to $10/bbl of oil by the year 2000, but the fate of that initiative is not clear.

Neither the limitation of externalities to regulated utilities nor the limitation to particular states is a particularly important drawback of externality valuation. Indeed, the universality of taxes may not be a significant advantages, for two reasons. First, other than externalities that are determined by the composition of the fuel (e.g., carbon and sulfur), determining emissions and assessing taxes will be impractical for smaller sources. No tax on actual NO_x emissions is likely to be applicable to the majority of sources. Second, externality values depend on local conditions and local attitudes, and thus vary regionally. It is not clear that the same pollution taxes should be applied throughout the country, except perhaps for CO_2. Third, since so many externality sources are regulated largely on the state level (e.g., utilities, building codes), national regulation of emissions may not be practical.

The most serious bar to the extensive use of externality taxes, however, are their inefficiency and high cost. Studies that have estimated the cost of using taxes to reduce CO_2 emissions generally have found total costs that are greater than those found by studies that estimate the

---

[3]   John Rowe of NEES has also argued that "it is much easier to price the impact of taxes, than it is to determine the price of an externality rule" (Leone 1992). It is not clear why this should be the case. The cost of a tax may be either lower or higher than the tax itself, depending on what other taxes it replaces and how the economic effects propagate through the economy.

costs of requiring specific cost-effective actions (Chernick and Schoenberg, 1991). This finding is consistent with the experience of utilities with consumer behavior, where market barriers impede low-cost adaptation to price changes. The desired goals may be achieved at less social cost with intelligent and directed regulation than with the blunt instrument of taxes.

The high costs of achieving environmental improvements through tax policy limits the size of the tax that is practical from the perspective of equity and efficiency. The European carbon tax proposals have excluded energy-intensive industries, since applying the tax to those industries would make them non-competitive, and since the industries are not expected to be able to reduce their carbon efficiency much anyway.

The limits of high-cost externality taxes are also implicit in US utility proposals for carbon taxes. John Rowe, president of New England Electric System, has proposed that the Massachusetts DPU carbon dioxide valuation of $22/T be replaced by a tax of no more than $2/T (Leone, 1992; Rowe, 1992).[4] The resulting effect on the cost-effectiveness of investments in efficiency, fuel switching, renewable energy, and other low-carbon would only be one eleventh of the effect of the DPU externality value.

Despite this huge difference in effect, the cost to NEES would be about the same under either approach. The tax would charge $2/T for every ton of CO_2 emitted, which NEES estimates would cost it $30 million annually, implying emissions of 15 million tons of CO_2 (Leone 1992).[5] The carbon emissions of a new gas combined-cycle power plant would cost only about 0.07¢/kWh, which would increase avoided costs by only about 1%, barely affecting the cost-effectiveness of DSM and renewables. The externality valuation would result in NEES paying between nothing and $22/T for the CO_2 reductions actually achieved. If carbon valuation reduced emissions by 20% (in addition to actions that would have been cost-effective without valuation), or 3 million T/year, and the average cost of the savings were $10/T, the cost to NEES ratepayers would also be $30 million/year.[6] Thus, for the same burden on ratepayers, the externality valuation reduces carbon emissions by 20%, while the carbon tax does virtually nothing.

---

[4] It is not clear whether the proposed tax would be $2/T of carbon, or $2/T of CO_2. This analysis assumes the latter; $2/T of carbon is only about $0.73/T CO_2.

[5] These emission values appear to be too small for NEES, given its sales and fuel mix. This example will use the 15 million T/year value for purposes of comparison.

[6] Supply curves for DSM tend to curve upward, suggesting that the cost of the average measure would be less than half the cost of the most expensive measure.

2.3.2 Emissions Caps: the Centralized Planning Approach

An emissions cap is a limit on total aggregate emissions within a prescribed region. The most familiar application of an emissions cap is the national SO2 emissions cap required under the U.S. Clean Air Act Amendments of 1990. Another example is Ontario Hydro's cap on SO_2 and total acid gas emissions. Caps can also be set globally, such as for CO_2 emissions, or regionally, such as for ozone precursors. Some advocates of caps have suggested that they are preferable to the monetization of externalities (e.g., Krause, et al., 1992).

In principle, a cap could be set at a technically determined emission level, without regard to cost. This approach has been suggested for CO_2 (IPCC, 1991), and is nominally the approach taken in the US Clean Air Act for requiring compliance with the ambient ozone limits. However, it is unusual for any important environmental restrictions to be imposed without some analysis of the cost and feasibility of compliance. More commonly, legislators and regulators examine the cost and achieving various levels of emission control, and select a level that balances costs and benefits. This second approach is used by Krause et al. (1992). They derive a partial supply curve for reducing carbon emissions in New England, and then suggest that emissions be capped at a level that achieves significant reductions at acceptable costs.

An emissions cap can be implemented in one of two ways. Under the centralized planning approach, regulators or legislators can issue command-and-control decisions as to how the cap will be met. Alternatively, through the use of offsets or tradeable permits, they can allow the market to determine the price and the means for meeting the cap. This section will discuss the selection of the cap, and implementation through centralized planning. Market implementation is considered in the following section.

In the command-and-control model, the analysis of options used in setting the cap may produce the list of actions necessary to meet the cap. If the reduction level is set exogenously, regulators would then need to determine the cheapest way in which to reach the cap.[7]

---

[7] An additional complication to setting caps is the allocation of reductions to emissions-contributing sources. For example, a 20% reduction target for Canadian CO_2 emissions may not translate into the same target for each province: if Ontario is richer than other provinces, or has more low-cost options, it may be appropriate for Ontario to reduce emissions more than 20%. Ontario would then have to allocate its reductions among utilities, transportation, and other sources withinä the province. Utilities might be required to reduce their emissions more than the provincial average because of difficulty in achieving additional efficiency improvements in the transportation sector or other factors. Similarly, the individual utility targets may be more or less than those for utilities on average, reflecting relative costs of marginal reductions.

Emission caps are very useful tools in ideal centralized planning. If regulators had perfect information, and if their modeling included all possible choices, they would be able to issue command-and-control decisions to ensure that the caps set emissions at a level that maximized societal net benefit, and that the caps were met at least cost.

Actual implementation costs and constraints, however, may differ from those assumed in the regulators' analyses of the cheapest way in which to meet a cap. For example, a regional planning analysis might conclude that 500 MW of wind farms was part of the least-cost plan with a CO_2 cap. But wind could prove to be more (or less) expensive than the analysis had predicted, or it could meet with siting difficulties. If a resource's cost or availability change, the planners' set of resource selections for meeting the cap may no longer be least-cost.

Central planning does not deal well with such changes. Under extreme inflexibility (asä in the Soviet system), exactly 500 MW of windmills would be built because they were in the plan; the cap would not affect the design of the windmills (e.g., capacity factor or reliability). Under extreme planning flexibility, the entire least-cost plan would have to be reanalyzed every time an input changed or a new option emerged; the optimization model would have to reflect current data every DSM measure (including options, such as R-30 versus R-22 walls, and for every application), every power plant option (e.g., fuel switching, boiler rehabilitation, turbine modernization, retirement) for every existing unit, and every new supply resource. Since even highly simplified optimization analyses can take many months or more, the latter approach is just not feasible on the level of detail on which real energy decisions are made.

Emission caps on a particular pollutant thus do not remove the need for a monetized externality value for that pollutant in screening of specific DSM, new supply, enhancements to existing supply, and emissions controls. The benefit/cost analysis of each option must includeä the option's contribution to reducing the cost of meeting the cap. The analyses of the cost of meeting emissions level, which are generally produced in setting the caps, can usually provideä the cost of meeting each emission constraint.[8] Far from being a substitute for monetizing an externality, an emission cap can be useful in deriving the value of the externality. The cap operates at the centralized planning level, while the externality value guides detailed implementation decisions.

---

[8] In optimization jargon, the cost of meeting a constraint is the "shadow price" of the constraint, and are expressed in the units of the objective, which in utility planning would be dollars.

2.3.3 Tradable Allowances and Offsets

The alternative to implementing an emissions cap through centralized planning decisionsä is to maintain the cap through the exchange of allowances, permits, or offsets. For every ton of emissions from a source (either for any source or for new sources, depending on the control scheme) in the prescribed region, emissions must be reduced an equal amount from existing sources elsewhere in the same region so that total emissions do not exceed the cap. Emissions offsets can be required even in the absence of a formal cap. For instance, all new large emissions sources in areas that are out of attainment with the ozone provisions of the U.S. Clean Air Act Amendments (CAAA) of 1990 must obtain offsets.

These offsets and tradeable permits do have some theoretical advantages over the traditional command-and-control policies. An offset system would encourage a wide range of parties to find less expensive means for emissions reductions. Equity constraints can be largely eliminated, as vulnerable parties can be excluded from compliance requirements and can then be paid by regulated polluters to reduce emissions and produce offsets. Administrative costs go down in most cases, but they may also go up. For example, offset systems are subject to self-selection and verification problems. EPA is having difficulty working out the opt-in provisionsä in Title IV of the Clean Air Act Amendments, due to problems in determining whether production and emissions are being shifted to unregulated units, and whether reductions (especially from shutdowns) are really incremental.

The marginal control costs with an offset trading system will tend to be less than costs without trading. However, establishing a workable emission trading scheme can be very difficult, as demonstrated by the experience of SCAQMD (which probably has more data and greater control over pollution than any other environmental agency in the country) in establishing a trading system for NO_x and VOC.

In a functioning competitive market, each unit of a fungible commodity, such as an air pollution offset, should have essentially the same price. For example, under the post-2000 sulfur-trading scheme of the CAAA, each utility will have an incentive to find the lowest-cost ways to reduce sulfur usage. A market-clearing price (which now looks likely to be about \$600/T SO_2, 1991\$) will emerge. Any extra allowances can be sold for that price, and any shortfall in allowances will have to be purchased at that price. Hence, every pound of sulfur emission will have the imputed price, which is the marginal cost of meeting required reductions.[9] In determining their bids, QFs will have to include \$600/T for SO_2; in evaluating

---

[9] This is true if the emission reduction requirement reflects all the costs of the emissions. In some regions, the value of sulfur reductions is greater than implied by the national cap, as indicated additional limits on oil sulfur content.

DSM, life extensions, and repowerings, utilities should similarly include $600/T.[10] All sulfur emissions will be effectively priced at $600/T (or wherever the true marginal cost ends up).

Though the CAAA sulfur market is not yet functioning, we can be fairly certain that it will be functioning (at least for futures transactions) within a few years. Most potential offset markets are not functioning, however. They will not be fully competitive for a long time, and cannot be made to function by utility regulators, and perhaps not even by environmental regulators. For example, there is no international requirement that all fossil fuel users offset CO2 emissions from new sources, or that they use offsets or tradeable permits to reduce emissions by (say) 1% annually.[11] Hence, there is no current market price of CO_2 offsets.

The declaration of a cap does not immediately internalize the costs of pollution, nor does it ensure that the proper values will be used in planning. The costs of allowances or offsets will reflect their marginal value to society only in the presence of a competitive market. While offsets may be obtained internally by the utility, or purchased at a very low cost prior to the establishment of a market, the marginal value of the offsets is their replacement cost, which will also be their price once the market is functioning. While all participants should be encouragedä to find lower-cost options for reducing emissions, utilities must value those reductions at their marginal value to society, rather than at their acquisition cost.

As with other environmental regulation, the existence of an offset system, even if it meets social goals for environmental clean-up, does not eliminate the need for externality valuations in planning. Under most offset systems, many sources are too small or too hard to verify (e.g., residential furnaces, lawn mowers), and are not covered in the trading system. Hence, regulators must ensure that environmental values are included in all decisions, either as an estimate of the costs to be internalized or as explicit externality values. The most expensive offset required to clear the market would establish the market price for internalized offsets and could also be used to define the marginal cost of control, which can then be applied as an environmental to all decisions.

---

[10] Not many have yet done so.

[11] Obviously, directly imposing these requirements on many small CO_2 sources (home heating, automobiles) would be administratively difficult, other than through fuel taxes.

3   Conclusion

Many of the proposed alternatives to monetized externalities have very little value for utility planning decisions. Stricter environmental controls do not eliminate externalities, and would not always be cost-effective. Qualitative treatments of externalities, including ranking and weighting, are too subjective to be useful tools for least-cost planning.

Certain environmental policies, such as emissions caps implemented through tradeable permits or offsets, can play an effective role in decreasing emissions levels at the lowest societal cost. These policies do not obviate the need for monetized externality values in utility resource allocation decisions, and are often useful in setting externality values. Monetized values are still essential inputs to benefit/cost analyses that determine the level of DSM to pursue (e.g., identifying the optimal insulation thickness or equipment efficiency level), or that select between competing supply options. Monetized externality values are flexible tools that guides regulators, utilities, and other parties in selecting the least-cost resource mix for meeting a region's energy requirements.

References

Chernick, P. and Caverhill, E., "Monetizing Externalities in Utility Regulations: The Role of Control Costs," *Proceedings from the NARUC National Conference on Environmental Externalities;* October 1990.

Chernick, P. and Caverhill, E., "Externalities," in *From Here to Efficiency*, Pennsylvania Energy Office, 1992.

Chernick, P. and Schoenberg, J., "Determining the Marginal Value of Greenhouse Gas Emissions," *Energy Developments in the 1990s: Challenges Facing Global/Pacific Markets*, International Association for Energy Economics, July 1991.

Destribats, A.F., Hutchinson, M.A., Stout, T.M., and White, D.S., "Environmental Costs and Resource Plannng Consequences: New England Electric's Rating and Weighting Approach," NARUC Externalities Conference, October 1, 1990.

Intergovernmental Policy Coordinating Committee (IPCC), "Climate Change," *The IPCC Response Strategies*, Island PRess, 1991.

Joskow, Paul L., "Dealing with Environmental Externalities: Let's do it Right!," Edison Electric Institute Strategic Planning Department #61, 1992.

Krause, F., et al., "Incorporating Global Warming Risks in Power Sector Planning - A Case Study of the New England Region," Lawrence Berkeley Laboratory, Draft March 1992.

Leone, M. "The Problem with Externalities," *Electrical World,* May 1992, p. 26.

Massachusetts Department of Public Utilities, Decision in Docket No. 89-239, August 31, 1990.

Rowe, J.W., Abstract of Testimony in U.S. House of Representatives Hearing on Domestic and International Climate Change Policy, New England Electric Systems, March 3, 1992.

20. State Externalities Policy and Carbon Dioxide Emissions: Who Bears the Risks of Future Regulation?

Ralph Cavanagh, Ashok Gupta, Dan Lashof, and Marika Tatsutani
Natural Resources Defense Council, NY 10011

ITEM: In January 1991, representatives of 38 state consumer advocacy offices and 17 environmental organizations warned utilities that failures to anticipate future carbon-dioxide-emission cost increases "will open those responsible to prudency challenges."

ITEM: In October 1991, the Pacific Northwest's Bonneville Power Administration announced that it would seek from prospective sponsors of fossil-fueled generators assurances that they could absorb future regulatory costs of greenhouse gas emissions.

ITEM: In December 1991, the European Community's joint council of environmental and energy ministers directed its executive commission to draw up a formal proposal for carbon taxes, with the proceeds (up to $21 per ton of carbon dioxide) to be used to reduce other taxes.

ITEM: In April of 1992, California's PUC ordered the state's utilities to condition long-term fossil generation purchases on the suppliers' assumption of all financial risks associated with future carbon dioxide taxes or controls.

ITEM: In June 1992, President Bush and 153 other world leaders sign the United Nations Framework Convention on Climate Change at the Earth Summit. The treaty establishes an ultimate objective of stabilizing "greenhouse gas concentrations in the atmosphere at a level that would prevent dangerous anthropogenic interference with the climate system."

0   Introduction and Overview

Greenhouse gas emissions, which cost nothing today, are a source of increasing financial risk. Nowhere is this clearer than in the utility industry, whose deliveries of electricity and natural gas account for more than half of U.S. releases of carbon dioxide, the most significant greenhouse gas. This article's principal focus is strategies for anticipating and avoiding potentially crippling costs that no prudent utility manager can ignore any longer.

Our thesis is independent of any technical nuance of climate science. We do not propose here to summarize the literature on whether and to what extent the planet is exposed to potentially catastrophic climate change. It is enough to observe that greenhouse gas concentrations are increasing at rates that alarm most of the world's technical and political leadership; within just the past thirty years, for example, we have seen carbon dioxide and methane concentrations grow by 12% and 35%, respectively. The nations of the earth collectively are conducting a gigantic and unauthorized climate experiment, and efforts to suspend it will continue. While some dispute the urgency of remedial action, none of these skeptics has offered to indemnify emitters against future taxes or emissions limits on greenhouse gases in general or carbon dioxide in particular.

The risk is most palpable for coal combustion, which accounts for more than half of current U.S. electricity production and figures heavily in many U.S. utilities' plans for future generating capacity.[1] But all fossil fuels face surcharges that are likely to grow over time. Fortunately, mechanisms for anticipating and minimizing those costs are ready at hand.

Our analysis has four parts. We begin by explaining why many states and utilities now disregard potential costs associated with carbon dioxide emissions. We then offer an overview of the regulatory and financial risks associated with carbon dioxide emissions, including several highly plausible scenarios under which substantial costs are imposed and utilities are prevented from passing them through to customers. We also provide case studies in the emergence of a risk shifting approach from California and the Pacific Northwest. We close with specific recommendations for prudent utility managers and regulators.

1   The Case for Apathy

As of this writing, the United States exacts no tax or limitation on the emission of any greenhouse gas. Many powerful political elements stand ready to oppose such initiatives. Skeptics can be found who will question whether global warming has begun and whether it is likely to assume dangerous proportions.

No single state or industry accounts for an appreciable fraction of greenhouse gas emissions, and the damage associated with particular tonnages of individual gases is speculative in quantity and cost. No state or company has anything obvious to gain by restricting its own emissions in advance of a national or international framework for achieving broad reductions. Under the circumstances, why isn't inaction the best policy, at least for individual states and industries, at least for now? Or at most, should not such entities focus solely on "no regrets" policies that are fully justified by non-greenhouse considerations?

2   Apathy Reconsidered

In January 1991, representatives of 38 state consumer advocacy offices and 17 environmental organizations jointly rejected the case for apathy on greenhouse emissions. Specifically, they warned utilities that failures to anticipate future carbon-dioxide-emission cost increases "will open those responsible to prudency challenges." The authors called on utilities "contemplating substantial investments in long-lived fossil fuel technologies" to "begin explicitly to take . . . into account" anticipated requirements to reduce fossil fuel use:

> Ratepayers' income, utility shareholder investments, and environmental quality will all be at risk, if the utility industry fails to take into account future costs of greenhouse gas emissions in its resource planning. Conversely, all of our constituents stand to gain when utilities cost-effectively substitute what amount to climate defense technologies for additional greenhouse gas emissions. We jointly pledge our best efforts in helping regulators to gauge utilities' performance and to respond appropriately.[2]

A complementary initiative by the Interfaith Center on Corporate Responsibility will target electric utilities starting in 1993. ICCR members hold nearly $30 billion in investments, and they will be asking electric utilities to act in the interest of their shareholders by developing strategies to reduce carbon dioxide emissions.

Utilities across the United States already have begun to respond. In the vanguard are Los Angeles Department of Water and Power, New England Electric, Pacific Power & Light, Pacific Gas & Electric, and Southern California Edison; all are acting to cut emissions substantially through some combination of energy efficiency, renewable energy, fuel substitution and reforestation investment.[3]

Buttressing the case for anticipatory action are at least three plausible scenarios under which greenhouse gas emissions could soon become increasingly costly. None is certain, but neither can any be dismissed with anything approaching certitude. And for utilities at least, the very plausibility of these outcomes is grounds for taking them seriously now, lest lack of foresight be found imprudent -- and its costs disallowed -- in some future proceeding.

2.1   Scenario 1: International Treaty Commitments

The United Nations Framework Convention on Climate Change was signed by 154 countries at the Earth Summit in Rio this June, and the United States Senate can be expected to ratify the treaty promptly. The Convention includes an implicit benchmark of limiting carbon dioxide emissions to 1990 levels by 2000. Perhaps more important, the treaty establishes an ultimate objective of stabilizing "greenhouse gas concentrations in the atmosphere at a level that would prevent dangerous anthropogenic interference with the climate system." The commitments in the treaty are to be reviewed on a regular basis in light of this objective, until stabilization is achieved. Because stabilizing greenhouse gas **concentrations** requires **emission** reductions of more than 50%, the treaty's objective will generate sustained pressure to strengthen emission reduction requirements over the 40 year lifetime of power plants built today.

Current federal policy relies on a series of voluntary emission reduction efforts, such as EPA's Green Lights and Coal Seam Methane Capture programs, coupled with state-level utility initiatives, to limit emission increases. But if existing efforts appear to be falling short of international goals, the pressure for more stringent market interventions will mount under any Administration. Even if a comprehensive tax or permit trading scheme is not enacted, the utility sector is likely to be among the first targeted; it is the largest single contributor to U.S. carbon dioxide emissions.

2.2   Scenario 2: Deficit Reduction

The federal deficit for the fiscal year ending September 1993 is expected to approach $400 billion; cumulative deficits since 1981 are on the order of $3 trillion. To test the limits of understatement, this is not a fiscal record that necessarily admits of indefinite extension. As a broad based means of budget balancing, carbon dioxide taxes already are attracting substantial attention.[4] With current U.S. emissions running at about five billion tons, a $20 per ton tax would raise some $100 billion while adding only about 18 cents to the price of a gallon of gasoline. The surcharge on a typical electricity bill would be less than 20 percent. By contrast, raising $100 billion in annual revenues from a gasoline tax alone would raise the price at the pump by a dollar per gallon. Collecting the same amount by increasing income taxes would extract an extra $1000 per year from the average household's paychecks.

2.3   Scenario 3: Tax Reform

Even if net increases in the overall tax burden are not feasible, a strong political consensus could form around shifts in that burden and could result in support for substantial carbon taxes. "Tax pollution, not jobs" is the rallying cry of an incipient tax reform movement. For example, one recent study pointed out that a $25 per ton carbon dioxide tax would allow the Congress to cut social security and unemployment compensation taxes by more than 50 percent.[5] In December 1991, the European Community's joint council of environmental and energy ministers directed its executive commission to draw up a formal proposal for just such a tax reform initiative; proceeds from the new taxes (up to $21 per ton of carbon dioxide) would be used to reduce other levies.

Any of these scenarios would dramatically shift the life-cycle costs of a fossil-fueled generating plant. Coal would be the biggest loser; a 250 MW facility operating at a 75% capacity factor would face an annual tax liability of some $38 million if carbon dioxide were taxed at $20/ton;[6] that amounts to a surcharge of almost 2.5 cents per kilowatt-hour. For fluidized bed facilities, an equivalent tax on nitrous oxide emissions could increase that total by another 50%.[7] Gas-fired generation would pay substantially less, but the toll for such units under a $20/ton tax would still exceed 1.3 cents per kWh; very possibly enough to tip the scales to renewable energy generation or some of the more costly energy-efficiency options.

3   Case Studies in Risk Shifting

Regulators and utilities that ignore these risks do not eliminate them; they merely convert electricity customers and stockholders into involuntary insurers for the fossil fuel industry. In the past six months, two major institutions have declared that result intolerable. More will almost certainly follow.

3.1   The Bonneville Power Administration

As the Bonneville Power Administration prepared to solicit bids for 300 MW of new generating resources in the Spring of 1991, the agency decided to incorporate environmental costs explicitly in its evaluation of potential new power supply sources. Its staff devised a detailed accounting system for environmental costs, which included a specific dollar value for emissions of carbon dioxide, nitrogen oxides, sulfur oxides, and total suspended particulates.

On March 28, 1991, a letter demanding withdrawal of the CO_2 adjustment was mailed to DOE Secretary Watkins by four Senators acting on behalf of the coal industry, which correctly feared competitive disadvantage.[8]

The Natural Resources Defense Council worked with BPA and DOE to devise a mutually acceptable alternative way of addressing the potential costs of carbon dioxide. We reached agreement on a BPA pledge to "seek assurances that [resource] sponsors are prepared to absorb costs should their plants be subject to future regulations on greenhouse gases."[9] As a result, BPA's subsequent agreement to purchase a gas-fired generator included requirements to fund carbon-dioxide absorption strategies and to hold the agency harmless against costs of future carbon-dioxide emissions limits.[10] Agreement on how to treat carbon-dioxide taxes, however, was deferred to negotiations that will be convened promptly after imposition of such levies;[11] attempts by the plant sponsor to procure some form of insurance or indemnity against the tax risk had proved unavailing. That is itself sobering news for those who contemplate long-term commitments to resources with significant CO_2 emissions.

3.2   The California Public Utilities Commission

In April of 1992, California's Public Utilities Commission went one step further than BPA. Its comprehensive regulations on acquisition of new power supply resources included a key provision on greenhouse-gas risks:

> [Utilities] should undertake a long-term purchase [of fossil generation] only if the supplier provides assurance that it alone will bear the cost of meeting any future costs resulting from a carbon tax, acquisition of tradeable emission permits, retrofits, or any other carbon emission control strategy or regulation applicable to the supplier's plant(s).[12]

In short, California regulators were unwilling to defer the tax liability issue; they resolved it unambiguously against those whose emissions would constitute the source of that liability. In the contracts that emerge under the new policy will come the first test of whether resources can be insured against future carbon taxes -- and at what price.

4 Conclusions and Recommendations

Debates rage across the nation over whether and how to incorporate environmental costs explicitly into utilities' resource investment decisions. Greenhouse gas emissions are by far the most important pollutants from the perspective of externalities policy, given the extent to which costs associated with other emissions already have been internalized by state and federal regulations. Precisely because nothing yet has been done to regulate greenhouse gases, they potentially represent the dominant terms in evaluations that assign dollar costs to environmental damage.

This paper has explored an approach that should have appeal across the spectrum of externalities analysis: shifting the risk of future greenhouse costs to those who create the risk itself, and letting the marketplace determine the resulting premium associated with purchases of fossil-based resources. Northwest and California precedents point the way toward what ought to become a universal policy.

As that policy emerges, creators and enforcers should insist on more than simply a recitation that the fossil resource's sponsor will assume the risk of carbon dioxide taxes; some tangible evidence of reasonable insurance should be required. Otherwise sponsors might be tempted to make an empty pledge and gamble that they could renegotiate a shift of new costs back to utility customers if and when the costs emerged, given the likely unpalatability of an interruption in power supply based on a sudden and technically unrecoverable tax increase. Explicit requirements of this kind will, in turn, spur the development of the very insurance mechanisms whose absence partially frustrated BPA's pioneering risk-shifting efforts.

References and Endnotes

[1] See, e.g., R. Johnson & C. Solomon, Coal Quietly Regains a Dominant Chunk of Generating Market, Wall Street Journal, August 20, 1992 (contending that "[t]he 54 largest electric plants currently on utilities' drawing boards are all designed to burn coal").

[2] Open Letter to the Managers of the U.S. Utility Industry from National Association of State Utility Consumer Advocates, Natural Resources Defense Council, National Audubon Society, Sierra Club et al., January 31, 1991 (available from NASUCA, 202-727-3908).

[3] See Testimony of John Rowe before the U.S. House of Representatives Committee on Energy and Commerce, Hearing on Domestic and International Climate Change Policy

(March 3, 1992), at 3 (New England Electric has set a goal for the 1990s of reducing carbon dioxide emissions by 20%, or 3 million tons/yr); Joint Statement of Southern California Edison and Los Angeles Department of Water and Power, May 20, 1991 (20% carbon dioxide emissions reduction goal for 2010); C. Petit, PG&E Chief Urges Bush to Set Emissions Goal, San Francisco Chronicle, May 8, 1992 (PG&E seeks to avoid 20 million tons of CO2 emissions over the next decade); Pacific Power & Light, Strategy in Focus, at p. 6 (pledging to "investigate and test strategies for offsetting carbon dioxide emitted by burning fossil fuels in plants to generate electricity).

[4] For an exhaustive recent survey, see R. Dower & M. Zimmerman, The Right Climate for Carbon Taxes: Creating Economic Incentives to Protect the Atmosphere (World Resources Institute: August 1992).

[5] See Natural Resources Defense Council et al., America's Energy Choices, Executive Summary at 20-21 (1991).

[6] A tax of $20/ton is a conservative estimate. Already, some states have adopted a tax $22 per ton.

[7] The estimate is from Professor Bob Williams of Princeton (March 23, 1991 memorandum); it is based on test results suggesting that up to thirty percent of fuel-bound nitrogen can be converted to nitrous oxide in fluidized bed boilers, whose relatively low temperatures encourage such formation.

[8] The signatories were Quentin Burdick, Kent Conrad, Alan Simpson and Conrad Burns.

[9] Bonneville Power Administration, Department of Energy Information, December 10, 1991, at p. 2.

[10] Power Purchase Agreement executed by Bonneville Power Administration and Tenaska Washington Partners II, L.P. (July 16, 1992), at 7 ("The purchase price shall not be adjusted for any costs of additional equipment or modifications made to the Project required to comply with any future . . . regulation of carbon dioxide emissions"); a Letter of Intent executed by the parties on July 16, 1992 further provides (p. 5) that Tenaska will set aside $1 million to sequester carbon dioxide emissions from the project.

[11] The contract (note 9 above) provides that in the event such taxes are imposed "the Parties will conduct good faith negotiations to determine an adjustment, if any, to the purchase price" (p. 7).

[12] D. 92-04-045, at 28-29 (April 22, 1992).

21. Incorporating Global Warming Externalities through Environmental Least Cost Planning: A Case Study of Western Europe

Florentin Krause, Ph.D.
IPSEP, El Cerrito, California
Jonathan Koomey, Ph.D.
Lawrence Berkeley Laboratory's Energy and Environment Division,
Berkeley, California
David Olivier
Energy Advisory Associates, Herefordshire, UK

Introduction

With growing acceptance of the need for preventative action, the international debate on greenhouse policies has shifted from scientific issues to economic questions. Broadly speaking, proposals for incorporating global warming externalities move along the same lines as the debate over other environmental externalities - valuation-based price signals (monetized surcharges or carbon taxes) versus quantity constraints (emission caps or reduction targets). However, the global warming issue has produced specific forms of these alternative approaches. Most important for the current discussion are the so-called insurance buying approach (Krause et al. 1989) and the "no (economic) regrets" approach.

Insurance Buying Versus "No Regrets"

These two alternatives have become central to the international politics surrounding the negotiation of an international climate treaty. The "no regrets" approach underlies the official policy of the current U.S. administration, while the insurance buying approach is implicit in proposals by various Western European countries and other nations.

In the insurance buying approach, some insurance premium is considered acceptable as the price of lower climatic risks. In the "no regrets" approach, preventative action would be limited to those measures that are entirely justified by economic benefits other than reduced risks of climate change.

The insurance buying proposal is a quantity constraints policy. In the analogy of buying insurance, the degree of emission reductions (the emission reduction target) is equivalent to an upper warming limit, and thus to the amount of insurance coverage bought. The cost of insurance against climate risks is the cost of changing current investment plans and business-as-usual emission trends.

The "no regrets" policy is a special version of the monetized valuation approach to environmental externalities. The monetized damage cost orientation of this policy is somewhat obscured by the assumption peculiar to that policy, i.e., that damage costs are zero. Though the limited body of valuation research on global warming suggests non-zero damage costs (Cline 1992), the "no regrets" approach has special policy significance: it defines worthwhile reductions of greenhouse risks as limited to those that can be obtained as a side-effect of policy measures that are not or not mainly aimed at preventing climate change.[1]

The Limits of Valuation Methods in Dealing with Global Warming

Much of the recent literature on the economics of global warming centers around carbon taxes (U.S. DOE 1991, CEC 1991b). However, contrary to what the monetary nature of such taxes might suggest, most proposals for carbon taxes are not based on damage cost estimates or on pollution control technology costs. Rather, they are proposed and analyzed as a specific implementation measure for a quantity constraints policy: carbon taxes are seen as one of two economically efficient means of implementing quantity constraints, the other option being emission rights trading.

This predominant focus on quantity constraints is due to several factors: First, current damage cost estimates are orders of magnitudes more uncertain than those for classical air pollutants, and thus largely irrelevant for guiding policy. Second, there is no viable exhaust stack control technology for non-stationary applications. Third, there is no viable means of long-term storage for carbon sequestered from stationary sources. Fourth, the use of trees for

---

[1] The term "no regrets" is potentially misleading, notably if the potential for emission reductions from zero-cost options should be small. In that case, such a policy would not achieve the goal of reducing the risk of catastrophic warming. While avoiding mitigation costs in the near-term, it could lead to large economic damages and costs in the OECD countries and globally in the longer-term. Short of detailed analyses that specify the range of emission reductions such a policy could achieve, the term "no regrets" is thus deceptive. A more neutrally descriptive term for this policy would be to call it a"zero-net-cost" policy rather than using the potentially misleading term "no regrets" policy.

sequestering carbon from the atmosphere is limited in contribution and fraught with land-use, permanence and verification issues. All these special factors make it difficult or impossible to use the analytic proxies of the externality valuation approach.

Environmental Least Cost Planning

The difficulties environmental policy makers face in dealing with the global warming issue are having an important fertilizing effect on the entire development of externality policies. By foiling both the traditional command and control approaches based on best available control technology (BACT) at the exhaust stack, and the valuation approach based on estimating monetized externality costs, the global warming threat leads us to a new kind of air pollution policy: its recasting as a specific version of energy policy - or of energy policy as a specific version of air pollution policy.

Specifically, the idea that certain normatively defined goals of air pollution control should mainly and principally be pursued through adjustments in the mix of energy resources and in the efficiency of their utilization has given birth to an alternative to the externality monetization movement. This alternative employs the knowledge and concepts of modern, least-cost oriented energy resource planning. We have therefore dubbed this alternative Environmental Least-Cost Planning (ELCP).

ELCP is based on the normative establishment of quantity constraints. In this respect it does not differ from conventional policy principles. But rather than focusing on a single best available control technology (BACT), control strategies are developed on the basis of a comprehensive least-cost approach that combines the gamut of energy technology options with applicable pollution control technology options into one integrated approach.

From an environmental policy perspective, the attraction of ELCP is that it retains the classic principle of normative environmental quality goals and environmental risk limits while overcoming the economic inefficiencies of traditional BACT regulation. Furthermore, unlike in a damage valuation approach, the uncertainties associated with the policy are on the side of the economy, and not on the side of the environment.

ELCP has a direct market-based equivalent: emission rights trading under a normatively established emission cap. As experience is gained with emission rights trading regimes for air pollutants in various sectors and end-uses, ELCP may be able to progressively move from a planning-oriented approach to a mixed planning and market approach. However, the need for ELCP-type analyses will probably always remain, if only to verify and assure that emission rights markets, once created, will continue to fulfill the societal functions they were designed to serve.

The IPSEP Study

This paper summarizes and clarifies the ELCP approach on the basis of a case study: an estimate of the cost of reducing fossil carbon emissions in Western Europe, as a function of the degree of emission reduction desired (Krause et al. 1992a, 1992b, 1992c). Our paper covers the second part of a two-phase research project that began in 1987. In its first phase, current scientific knowledge about climate stabilization requirements was translated into risk-minimizing emission reduction targets and milestones for fossil carbon and other greenhouse gases, and then allocated among industrialized and developing countries on the basis of global equity, ability to pay, and developmental considerations (Krause et al. 1989).

Findings on Reduction Targets

To stop the accumulation of carbon dioxide in the atmosphere, ultimate global carbon emissions reductions in the neighborhood of 60% relative to current levels have been identified by international scientific bodies (IPCC 1990). The schedule for these reductions depends on the degree of warming risk that is considered acceptable. Our earlier analysis of carbon emission trajectories for developing and industrialized countries under a risk-averse 2° C warming limit (Krause et al. 1989) shows that industrialized countries would have to reduce carbon emissions by 20 percent below recent levels around the year 2005 to 2015, and by 50 percent in the period between 2020 and 2030 to allow for Third World development and achieve global climate stabilization goals.

The international Toronto conference of 1988 called for a reduction in carbon emissions of 20% by 2005. A number of industrialized countries have made it their policy to freeze carbon emissions by the turn of this century, or to reduce them up to 25% over the next twenty years. At the UNCED conference in Brazil, signatories adopted the stabilization of carbon emissions in industrial countries as a desirable initial goal, without, however, agreeing on a specific date. The European Community, meanwhile, has announced a plan to stabilize its emissions at 1990 levels by the year 2000.

Estimating the Costs of Alternative Energy Futures: Top-Down Versus Bottom-Up

The above findings on climate stabilization requirements suggest that it is important to learn the cost of cutting emissions by large percentages within the next two to three decades. By contrast, the focus in the current policy debate is on only the most modest steps, such as a freezing of emissions at current levels by the year 2000. This hesitancy reflects, in part, a widespread presumption about the likely cost of carbon reductions to society, i.e., that any significant reductions would be expensive, and large reductions prohibitively so.

This presumption appears to be confirmed in some studies on the cost of low carbon energy futures, while being strongly challenged by others. These contradictory findings are rooted in a major analytic schism among energy analysts. When it comes to calculating the costs of energy policies and determining least-cost energy strategies, they split into a predominantly econometric, top-down school and a predominantly engineering-economic bottom-up school. Because of the importance of this schism for the implementation of ELCP in general and for greenhouse policy formulation in particular, our study encompasses a detailed review and critique of these competing paradigms.[2]

Notably in the U.S., the debate on greenhouse policies has been dominated by findings from macroeconomic studies that predict losses of one to five percent in GNP if modest carbon constraints were implemented by means of a carbon tax (Cline 1992). More recently, new macroeconomic studies have shown that these losses could be avoided or that GNP output could even be enhanced if carbon tax revenues were recycled into lower employer payroll taxes or investment taxes (Brinner et al. 1992), or into direct incentives for efficiency improvements and other low-carbon resource options (CEC 1991a, Sanghi et al. 1991).

The Top-Down Approach

Top-down studies are based on an econometric modeling approach in which it is assumed that energy service markets work well enough to approximate them by a perfect market model (U.S. DOE 1991). In that model, energy, labor, and capital are already being employed in a welfare-maximizing mix, so that the introduction of carbon constraints will necessarily create losses in economic output. Given the perfect market assumption, the finding that losses occur

---

[2] See Krause et al. 1992a, Chapters 1-4.

(unless the tax system is restructured) is tautological. The only modeling question of interest is how large these losses might be under various targets and schedules.

Critique

The top-down approach is a good starting point for developing business-as-usual scenarios, but it is inherently unsuited for identifying least-cost energy futures or levels of energy use. The reasons are that explicit engineering-economic supply-curves for demand-side technology options are not developed in such analyses, and/or because the role of market failures in consumer decision-making is ignored. Furthermore, because the method relies on aggregate, fixed elasticity coefficients that are derived from historic data, top-down models also fail to capture saturation trends and structural changes that will shape the future demand for energy services. In combination, both factors lead to inflated energy demand projections, and to inefficient capital allocation from a societal least cost perspective.

The Bottom-Up Approach

The bottom-up approach is perhaps best exemplified by the new resource planning practices that have emerged in the U.S. utility sector. In most of the U.S. utility industry, the perfect market assumption for the competition between consumer efficiency investments and energy supplies has been abandoned for some time. Instead, the bottom-up approach (variously known as least cost utility planning (LCUP) or integrated resource planning (IRP)) has become a standard method for developing optimal resource plans (Krause and Eto 1988).

The need to apply bottom-up analysis not only in the utility sector but also on a national, all-sector scale is being increasingly recognized as governments and citizen groups seek economically efficient ways of addressing environmental problems. Several recent studies in both the U.S. and Europe (NAS 1991, OTA 1991, UCS et al. 1991, CEC 1991b, Enquete Kommission 1990, Mills et al. 1991, Russ et al. 1991) illustrate this trend.

The Payback Gap: Market and Regulatory Failures

In the bottom-up approach, it is recognized that consumer efficiency investments are shaped by major market failures, price distortions, and inadvertent regulatory disincentives. These prevent the delivery of energy services at least economic cost.

While utilities and other energy companies invest in powerplants and other energy production facilities with a long-term perspective that involves payback periods of ten to fifteen years, industrial, commercial, and residential consumers buy or rent energy-using equipment with only limited attention to its efficiency. The key factors are

- The limited importance of energy expenditures in personal budgets and business production costs;

- lack of credible and easily accessible information;

- lack of financing and an emphasis on first costs rather than life cycle costs;

- risks in trying new technologies;

- plans to move to a different home or facility before efficiency investments would pay back;

- diverging interests among landlords and tenants;

- competing demands for managerial and consumer attention, and

- Generally high costs in time and bother (high transaction costs, or "hassle factor").

When actual purchasing choices are compared with choices that are least cost on a life cycle basis, it is found that industries and consumers will buy cost-effective energy-saving equipment only when the extra first cost pays back within a few months to two to three years.

While this consumer behavior is quite rational under status quo conditions, the resulting payback gap between demand-side and supply-side energy investments hurts the economy. Large amounts of capital are invested inefficiently when utilities and governments could help industries and consumers with cost-effective programs that take the hassle out of being energy efficient. Efficiency standards, rebates, free installations, and other financial incentives could ensure that energy consumers make least-cost choices from the same long-term perspective that is used by energy producers.

Economically wasteful outcomes do not only result from a blind reliance on laissez-faire markets; they can also be caused by deficient government regulations and interventions. Key examples are regulatory policies that impede access to the electricity grid by industrial cogenerators and other independent power producers. Another example is the set of utility rate-making regulations found in all EC countries that couple utility profits to energy sales rather than to the provision of energy services like heating or lighting. Heavy subsidies for road transport, and subsidies for fossil fuels and other energy forms in general, also undermine energy and economic efficiency. Finally, energy efficiency norms themselves can have this effect, as illustrated by the woefully outdated energy efficiency standards for buildings found in many EC countries.

On account of these empirical conditions, significant cost-effective opportunities to improve end-use efficiency exist in the various energy service markets.[3]

Critique

A shortcoming of the bottom-up approach is that it does not capture important economic feedback effects. One of these is the impact of reduced fossil fuel consumption on fossil fuel prices, and from there, on further substitution investments, on the demand for energy services, and on industrial migration should carbon reduction policies be applied only to one region. These feedbacks will tend to make actual reductions in carbon emissions smaller than predicted unless specific policy measures are undertaken to compensate for them. Top-down analyses have made an important contribution in identifying and modeling these effects under various carbon tax regimes.

In state-of-the-art bottom-up analyses, several of these impacts are addressed through extensive sensitivity analyses in which fuel prices, energy service demand, and other key inputs are varied. These sensitivity analyses substitute for the calculation of secondary feedbacks normally done with econometric models, which are inherently less transparent and fraught with great uncertainties of their own. Meanwhile, top-down studies are moving toward greater use of engineering-economic details to improve their forward-looking capabilities. However, the integration of top-down and bottom-up approaches into one coherent modeling approach has remained elusive.

What Approach Should Be Used in Policy Development?

In the absence of such integration, it is important to recognize that the sign of macroeconomic impacts (i.e., whether the growth rate of GNP is higher or lower relative to the business-as-usual case) is always determined by the sign of changes in society's energy service bill.[4]

---

[3] Annual spending on demand-side management programs in the U.S. utility industry is approaching US\$ 2 billion, the majority of which is for energy efficiency. Experience with utility programs and government programs in the U.S. and various European countries has demonstrated that demand-side resources can be flexibly mobilized through incentive programs for customers, dealers, or manufacturers, through efficiency standards, and through regulatory reforms that provide utilities with profit opportunities on the demand-side.

[4] See Zimmerman (1990). Assuming, of course, a constant structure of taxes on capital, and labor.

That sign, which indicates whether society's energy service bill needs to rise or could be lowered as carbon reductions are sought, cannot be determined in an analytically sound manner without the use of bottom-up analyses. For this reason, bottom-up analyses should be the primary basis of greenhouse policy development.

Once bottom-up analyses have established the sign of the economic impact of various reduction targets from an integrated, least-cost perspective, top-down models can offer important refinements. They reveal how fiscal reforms involving carbon taxes could be used to increase economic efficiency, while also highlighting the limits to a purely carbon tax based approach, and the need for legally binding emission constraints.

The IPSEP Study: Key Procedures and Assumptions

Our case study of Western Europe is based on the bottom-up methodology, combined with systematic sensitivity analyses. We also examine how international trade and competitiveness issues and other indirect effects could limit unilateral reduction goals in the European Community to less than what bottom-up cost-effectiveness calculations alone suggest.

Our basic approach is straightforward. We begin by defining a business-as-usual scenario and calculate the annual cost of providing energy services (heating, lighting, driving, manufacturing, etc.) for the final year. We then modify the resource mix. Here, we investigate two kinds of scenarios: economically oriented energy strategies that aim at minimizing costs (least cost scenarios); and environmentally oriented strategies that aim at minimizing carbon emissions and other major environmental risks, with costs only a subordinate consideration (minimum risk scenarios).

In the least cost case, we meet the demand for energy services (normalized to that of the business-as-usual case) by combining unconventional resource and low-carbon resource options, such as energy efficiency improvements, cogeneration, renewables, and fuel switching, with conventional resource options, such as nuclear and coal-fired powerplants and fossil fuels for direct combustion, on a strictly economic basis.

In the minimum risk case, low-carbon and non-carbon resources are combined to minimize carbon emissions, subject only to resource limitations and system reliability and dispatchability constraints. Costs considerations enter only in the selection of options when the combined low-carbon resource potentials exceed requirements.

One concern of particular interest in Europe is the role of nuclear power in a climate-stabilizing energy strategy. A number of countries in Europe have implemented de facto or declared nuclear moratoria or phase-outs, while France continues to rely on reactors as the mainstay of its electricity system. We address the poles in the European debate by examining a least-cost scenario in which nuclear power competes on a strictly economic basis with all other resource options, and a minimum environmental risk scenario in which carbon reductions are pursued while at the same time phasing out nuclear power.

The CEC Business-As-Usual Scenario

The basis for our business-as-usual case is the "Energy for the Next Century" report (CEC1990) prepared by the European Commission's Directorate-General for Energy (DGXVII). The business-as-usual scenario (S1) from that report is extrapolated by ten years to the year 2020 to capture the longer-term dynamics of capital stock turnover. Gross national product in 2020 is 2.6 times higher than in the base year, the value of industrial output increases two-fold, and energy service demand (weighted by base year energy intensities) rises by 48 percent.

Critique

The Commission's study develops several scenarios including ones in which carbon emissions are roughly stabilized and reduced by up to 18 percent, but does not quantify their economic costs. Using the data and procedures described below, we find that under business-as-usual energy policies, market and regulatory failures make Western Europe's bill for energy services considerably higher than it needs to be:

• Under business-as-usual trends, Western Europe would likely spend at least 15 percent and as much as 35 percent more for energy services in the year 2020 than it would under a least cost strategy.

• In absolute terms, these excess costs amount to 45 to 70 billion ECU in real (1990) ECU per year by 2020, or about 400-650 ECU per household per year.

When environmental damages and military costs for securing oil supplies are considered, these excess payments rise further. All these drains on Western Europe's economies could be avoided by least-cost oriented energy policies.

Input Assumptions for IPSEP Analysis

Range of Resource Options

In addition to conventional resources (coal, oil, gas, nuclear, hydro), we assessed these technologies:

- demand-side efficiency improvements
- switching from electricity to gas
- industrial cogeneration
- combined heat and power production in district heating
- small hydro generation
- wind power
- biomass-fired thermal generation
- solar-electric generation

We omit tree planting from our portfolio, though this option can provide inexpensive emission offsets. Excluding this option represents a conservatism in our work, and also avoids the various verification and accounting issues it presents.

Uncertainty Analysis

To capture real-world complexities, we explore several dimensions of uncertainty.[5] First, we define high and low capital costs for the different resource options. Second, we vary the amount of low carbon resources that are included in our resource portfolios by 25% increments, to capture implementation uncertainties. Third, we use fuel price forecasts with large differences in natural gas and oil prices. Finally, we examine the effect of alternative assumptions about future energy service demand in the electricity sector, in.

Cost Inputs

In our study, we go to considerable length in developing resource potential estimates and documenting plausible ranges for high and low cost assumptions for alternative resource options.

---

[5] The sensitivity analyses for the electricity sector are cnsiderably more comprehensive in the European study, where both new coal and nuclear plants play a major role in the reference case.

Figure 1 shows our fuel price assumptions relative to those of the European Commission and to prices in the base year.[6] For oil and coal, our high and low fuel price forecasts broadly parallel the projections used in the studies of the European Commission, but we widen the uncertainty range and assume somewhat higher prices for natural gas.[7]

Figure 1:   Fuel price assumptions, IPSEP scenarios (1989 Pf/kWh)

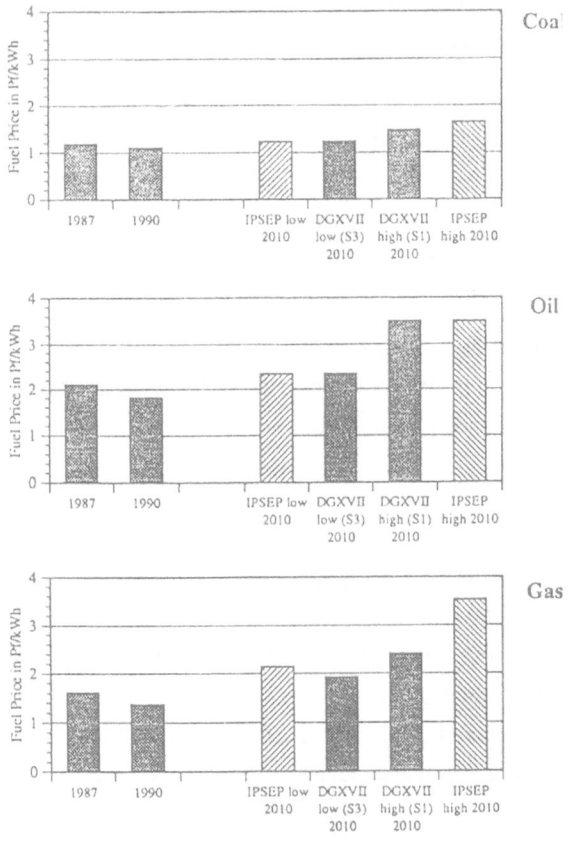

---

[6] In purchasing parity terms, one Pf is equivalent to about 0.5 cents. DGXVII refers to the Directorate-General for Energy of the European Commission (CEC 1990).

[7] In effect, we assume that the EC will prevent international price drops for coal and oil that might arise in response to carbon reductions in the EC from being passed on to energy consumers directly. Instead, these benefits would be captured by the treasury through carbon taxes and passed on through increased public investments or through tax relief for capital and labor. In the case of gas, carbon reduction strategies could lead to higher prices than in the reference case. We assume that these would be directly passed on to energy consumers.

In our technical appendices, we address a number of important analytic issues, notably the treatment of system aspects such as T&D credits and remoteness costs for modular power supply options, the maximum feasible fraction of non-dispatchable resources in the grid, back-up costs for wind farms and photovoltaics, methods for estimating cogeneration potentials, and heating system credits of very high efficiency buildings. Our demand-side resource costs are primarily based on the European prices of currently commercial products and technologies, and include estimates for program administration costs and T&D and other system credits.

Our cost assumptions for the nuclear option are developed in a detailed technical report (Krause et al. 1992c) that examines nuclear cost experience in France, Germany, and the United Kingdom. Low nuclear power costs are based on a review of French estimates (Ministere de l'Industrie 1990),, and high costs on the recent British attempt to privatize the nuclear power industry (British Energy Committee 1990). Data for other central stations and for cogeneration technologies were taken from utility industry sources (Friedrich et al. 1989), equipment supplier data, and various government studies.

Availability of Resource Potentials

Implementing the full potential of efficiency improvements and other unconventional supply options will require strong and persistent policy action, even when these resources are cost-effective. Given the political economy surrounding energy issues, it is sensible to allow for the possibility that only a fraction of unconventional resource potentials can be mobilized. We therefore calculate percentage carbon reductions (and costs) under the assumption that 25%, 50%, 75% or 100% of low-carbon resource potentials can be made available.

Principal Findings

Western Europe's Low-Carbon Resource Options

We find that the European Community's proposal to stabilize emissions at 1990 levels by 2000 would not nearly exhaust the region's low-carbon resource potentials from efficiency improvements, fuel switching, cogeneration, and renewables.

The largest low-carbon resource is found in currently unexploited opportunities for increasing the energy efficiency of the EC-5's buildings, vehicles, appliances, and industrial plants.

- If fully implemented in all sectors, demand-side efficiency resources are equivalent to 50 percent of year 2020 business-as-usual demand for final energy.

- Hydro, wind, biomass, and solar electric resource potentials are equivalent to 20 percent of all-sector demand.

- The technical potential for gas-fired cogeneration is equivalent to another 15 percent of all-sector demand.

- Overall, these resource potentials are equivalent to about 85 percent of year 2020 demand for final energy.

These aggregate numbers do not tell the whole story. A disproportionate amount (about 60 percent) of Western Europe's low carbon resources are electrical resources, while electricity demand represents only about 25 percent of overall demand in 2020. Because of this and other factors, maximum feasible carbon reductions in the economy as a whole are less than what a mere summation of resource potentials suggests.

Results: The Least Cost Scenario

Significant reductions in carbon emissions below 1985 levels would result if Western Europe were to pursue strategies that narrowly focus on providing energy services in a least-cost manner. Figure 2 shows the EC-5 energy bill for 2020 in absolute terms as a function of relative emissions.[8] A further parameter in the figure is the availability of constrained low-carbon resources. A total of four sensitivity cases are shown. We focus first on the high/high and low/low fuel and capital cost combinations. We find the following:

- Carbon reductions under least cost policies would be 15 percent below 1985 levels with high fuel price and technology cost assumptions, and 37 percent with low assumptions.

As shown in Figure 2, these least-cost carbon reductions could be realized at major savings to society:

- A least cost strategy would reduce Western Europe's bill for energy services by about 10 to 25 percent and save 45-70 billion dollars per year in 2020, equivalent to about 1-2 percent of GNP.

---

[8] In purchasing parity terms, one German Mark (DM) is equivalent to about 0.5 dollars or to 0.5 ECU.

301

Figure 2: Sensitivity Analysis of the Cost of Carbon Reductions in EC-5, All Sectors (Minimum Risk Case)

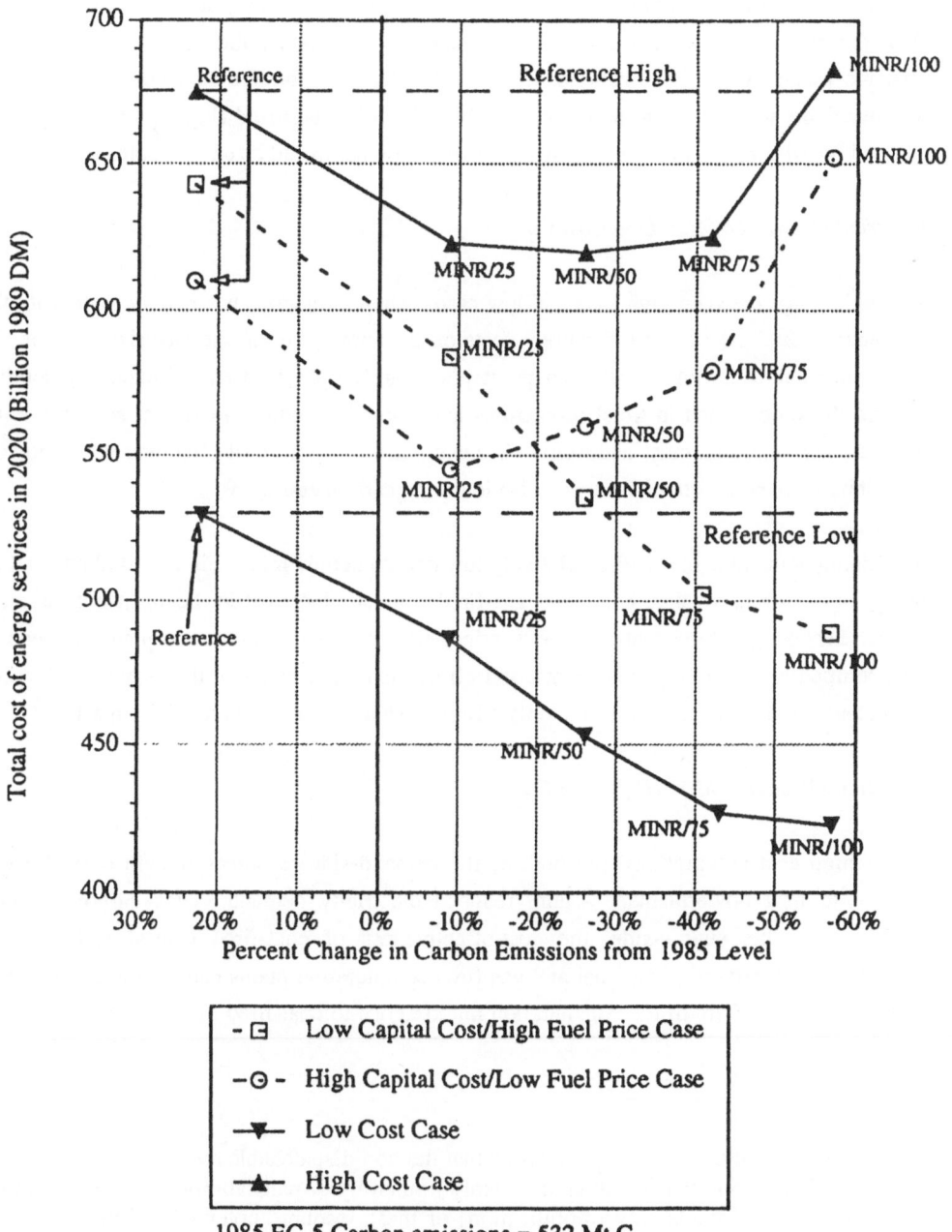

We state our results here for the case where efficiency resource potentials are 100% available (which, in the least-cost case, does not necessarily mean more than 75 percent utilization since the last portion of the DSM resource is relatively expensive). Under ineffective policies and without additional efficiency innovations, these potentials may not be fully available. How the results would then change is indicated by the data points for 25%, 50%, 75% resource availability. For example, if only half the potential for demand-side efficiency improvements were realized, emissions would at least be stabilized (high case) or would be reduced by 10 percent (low case), while economic savings would be comparable in each case.

Electricity Resource Mix: Low Cost Case

The wide range between high case and low case carbon reductions has mainly to do with the influence of fuel prices and technology costs on the supply mix in the electricity sector. In the business-as-usual plan, coal-fired powerplants are emphasized and efficiency potentials beyond those contained in trend projections are ignored. Under a least-cost scenario using low cost assumptions, most additional efficiency resources are cost-effective, which lowers electricity demand substantially while also lowering carbon emissions.

Remaining demand is met with a relatively low carbon supply mix : First, coal plants end up being more expensive than coal- and gas-fired (industrial or district heating) cogeneration plants. Second, gas-fired central stations, wind, biomass waste, and nuclear plants all become cost-competitive with new coal plants, and all of them have to share the remaining market. As a result, coal central stations gain only a limited share of the total electricity market.[9]

Electricity Resource Mix: High Cost Case

Under high cost assumptions, the bulk of the demand-side resources remain cost-effective relative to new powerplants. Again, reduced electricity demand yields carbon savings. However, on the supply-side, the cost-effectiveness of coal-fired central stations now improves substantially. Only coal and gas fired cogeneration plants can compete with them, and a substantial share of the cogeneration market goes to coal-fired units. Gas-fired central

---

[9] We find in our electricity sector analysis that the non-dispatchable resource potentials from wind and solar power generation are mainly limited by dispatchability constraints, and not by the assumed availability (mobilization) of these resource potentials. Similarly, cost-competitive district heating potentials are so large that their contribution to the resource mix is limited more by overall levels of electricity demand and by competition with other supply options than by the size of the cogeneration potential.

stations make only a secondary contribution. Nuclear power and renewables become too expensive to play any substantial role.

As a result, the carbon burden of the supply mix is higher than in the business-as-usual case. This higher carbon burden wipes out some of the emission reductions from increased demand-side efficiency, and the total reduction in carbon emissions is substantially lower than in the low case.

Sensitivity Tests

We also analyze high/low and low/high combinations of cost assumptions on an all-sector basis, and still more detailed sensitivity tests in the electricity sector using various high/low cost combinations for coal, gas, nuclear, renewables, and efficiency. Figure 2 shows these high/low and low/high results on an all-sector basis. Under least cost integration, these sensitivity cases have only a modest effect on the overall range of economic savings and carbon emissions.[10] This limited sensitivity, which is due to countervaling carbon impacts in the electricity sector and in direct fuel applications, indicates that our high/high and low/low results are reasonable proxies for our central findings.

Results: The Minimum Risk Scenario

In a minimum risk strategy, carbon reductions are substantially larger than in a least-cost strategy. Efficiency potentials are fully implemented in all sectors, and wind, biomass wastes, and other renewables are given first priority in electricity supply, followed by low-carbon gas-fired cogeneration. Oil is mainly used for transport, with limited use for heating, for chemical feedstocks, and for meeting peak demand in the elecricity system. Figure 3 shows our key results:

- If demand-side resource potentials are fully mobilized, a minimum risk strategy would reduce carbon dioxide releases in the EC-5 region by 60 percent below present levels by the year 2020, despite the phasing out of nuclear power.

- Relative to the least cost scenario, carbon emissions would be cut by about a further third (low/low cost case) to half (high/high cost case).

[10] In the worst case combination for carbon reductions in the electricity sector - low coal prices and high costs for everything else - (not shown in Figure 2), carbon emissions drop only 2 percent below 1985 levels, again based on full availability (but not necessarily more than 75 percent utilization) of the DSM resource.

Figure 3: Sensitivity Analysis of the Cost of Carbon Reductions in EC-5, All Sectors (Least Cost Case)

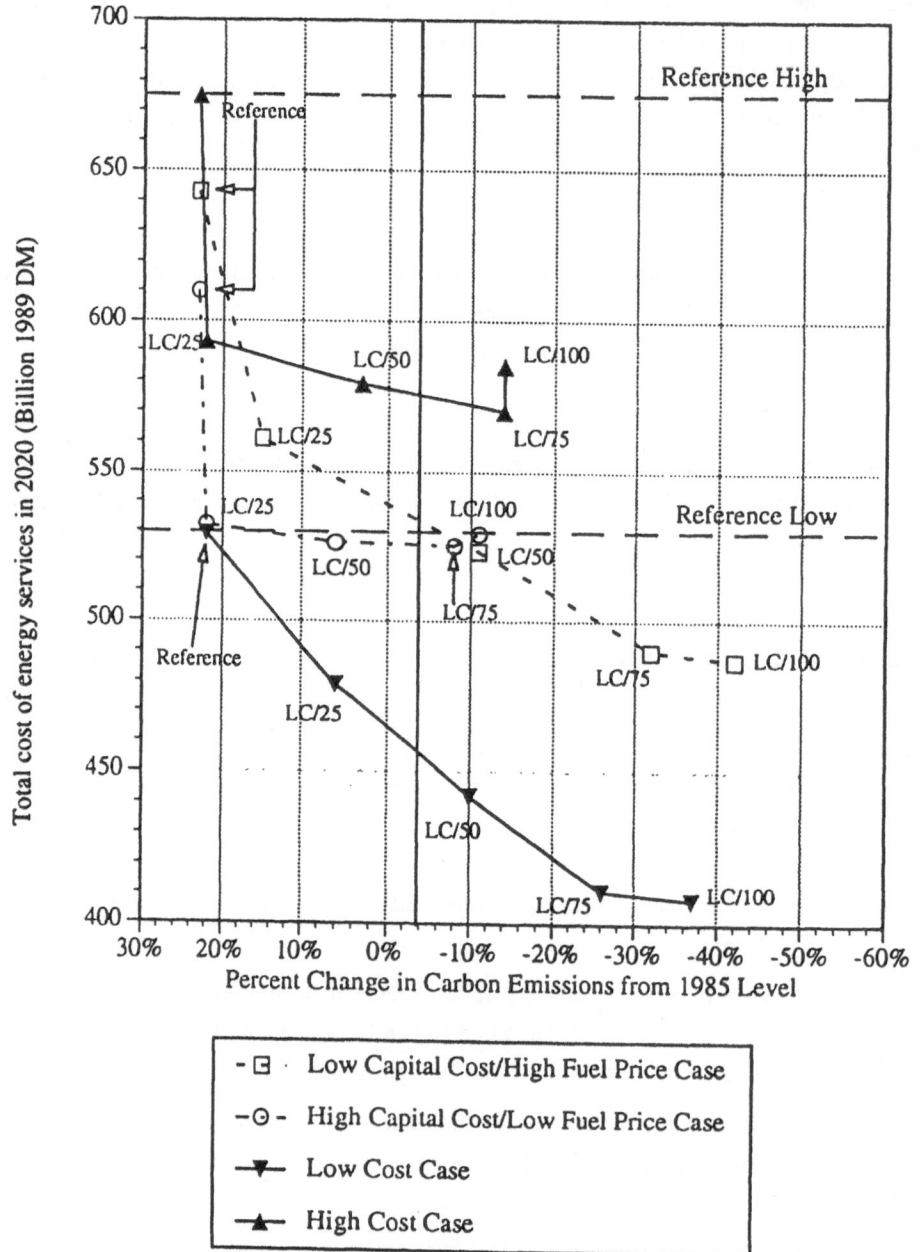

1985 EC-5 Carbon emissions = 532 Mt C

Though a minimum risk strategy would cost more than a least-cost strategy, it would not necessarily cost more than business-as-usual plans.

Sensitivity Tests

Our sensitivity results are also shown in Figure 3, which plots absolute costs as a function of relative emissions. Regardless of the case, emission reductions of up to about 40 percent could still be realized at large savings relative to the corresponding reference case. This is true even in the most unfavorable case where high capital costs are combined with low fossil fuel prices. Only for reductions in excess of 50 percent do costs exceed the energy bill of the business-as-usual case, and here only with high technology cost assumptions.[11]

Gas Requirements

Under DGXVII's business-as-usual trends, gas consumption in Western Europe increases by about 50 percent over the next 30 years.

Both in the least cost scenario and in the risk minimization scenario, gas requirements are highest when only 25 percent of the demand-side resource is available. In that case, gas requirements rise by up to 150 percent relative to the base year. In the 50% case, gas requirements would be 10-50 percent higher than in the business-as-usual case. For resource availabilities of 75-100 percent, which yield large reductions in final energy use and carbon emissions, gas requirements for the least cost mix are comparable to those of the reference case. In the risk minimization scenario, gas requirements drop up to 40 percent below base year levels, due to the larger contribution from renewables.

Though the EC-5 region is surrounded by exporters with ample supplies to handle this range of gas demands, future gas prices are also uncertain. Our low and high gas price assumptions take account of this uncertainty. Our findings suggest that gas supply constraints would be a concern only under modest carbon reduction targets and when demand-side efficiency is pursued without much vigor.

---

[11] In the electricity sector alone, the worst case (low coal and low nuclear costs, high costs for gas, efficiency, and renewables) results in a cost increase of three percent or 4 billion dollars relative to the reference case (100 percent utilization of the demand-side efficiency resource), or 5 percent (75 percent utilization). This cost penalty is overcompensated by the net savings in other sectors except when the most expensive increment of efficiency resources is used in these sectors.

Indirect Cost Savings and Macroeconomic Implications

Based on our above findings, the potential savings in the EC-5 energy bill are equivalent to one to two percent of GNP in 2020. This figure applies before the macroeconomic amplifications of these savings through higher growth and/or lower fossil fuel prices for remaining fossil purchases are taken into account. It is widely recognized that lower expenditures for energy services mean lower inflation and good news for economic growth.

Our cost estimates do not yet take into account a number of indirect economic benefits:

- Expenditures for controlling acid rain emissions and other externalities of current energy use will be greatly reduced.

- On an aggregate basis, Western Europe's fossil fuel imports will decrease. Less money will flow out of the region to pay for imported fuels.

- Lower fossil consumption and imports will put downward pressures on the world prices of coal and oil, and depending on how aggressively efficiency improvements are pursued, even on the regional price of gas. Remaining fossil fuel requirements will thus be met at lower prices than under business-as-usual conditions.

To this list of potential advantages one must add a forth aspect that goes beyond the realm of macroeconomic modeling studies:

- Unilateral action ahead of or beyond international agreements could give Western Europe a head start in technological innovations that will have world-wide markets once least-cost and low carbon policies gain broad international acceptance.

An illustration of this dynamic is found in the fact that during the fifteen years since the mid-1970s, in which the U.S undertook little environmental legislation, foreign producers of air pollution control equipment have gained 70 percent of the U.S. market. West Germany, which implemented the world's toughest acid rain control standards for stationary sources in the early 1980s, has enjoyed a wide lead in patenting and exporting environmental technologies (Porter 1991).

These advantages, together with the above macroeconomic benefits, suggest that a strategy to reduce carbon emissions swiftly could be good industrial policy for Western Europe, irrespective of whether other countries follow suit or not.

What Policies Are Needed?

While carbon reductions in Western Europe could be robustly advantageous in economic terms, realizing this potential requires a suitable mix of policies. Among economists, the favored mechanism are price instruments such as carbon taxes. Unfortunately, carbon reduction targets cannot be achieved mainly or solely through price instruments, on account of two complications: economic feedbacks and market failures.

Limits of Price-Based Policies

When applied in isolation, carbon tax policies are susceptible to various feedback effects both within the EC and in the global economy and energy markets that could undermine carbon reduction goals. These include take-back effects (greater consumption of energy services and fossil fuels on account of higher disposable income and lower than projected fossil fuel prices) within the EC, effects of international trade with regions that have no carbon policies, and take-back effects outside the EC. The longer-term technology goals of developing countries under competitiveness must also be considered.

A detailed discussion of these issues, and of policy measures that could mitigate the effects of feedback, is provided in our full report (Krause et al. 1992a). Briefly, we find that unilateral carbon reduction initiatives by large economic entities such as the EC (and by implication, the U.S. and Japan) could overcome most of these feedbacks provided they use a series of coordinated policies in the fiscal, regulatory, and international assistence arena. The competitiveness issue is more limiting for smaller countries or subnational entities, such as individual states or utilities that lack international market power. A detailed case study analysis of these competitiveness limitations for individual utilities is found in Krause et al. (1992d).

Even in the absence of these dynamics, carbon taxes alone are unsuited for garnering the potential economic benefits identified in our study. Energy and carbon taxes are an important and desirable signal, but they are a blunt sword without specific regulatory policies and incentives programs for individual technologies, end-uses, customer groups, and sectors. Without such measures, pervasive market and regulatory barriers will continue to make price signals as ineffective as they are in most energy service markets today.

Integrated Carbon Reduction Policies: A Checklist

We propose a different and potentially more effective approach, consisting of a number of policy instruments that complement each other:

1) Legally binding reduction national and EC-wide targets and timetables, coupled with market mechanisms for carbon emission trading for suitable end-uses or subsectors;

2) Strict minimum energy efficiency standards for buildings, appliances, lighting systems, vehicles, and other suitable end-uses, with scheduled updates every few years.

3) Complementary extension services, financing, and incentive programs to help industries and consumers invest in cost-effective equipment, vehicles, homes, appliances, etc. whose efficiencies exceed standards or for which standards cannot be implemented.

4) Fee/rebate (feebate) programs that finance rebates on purchases of energy-efficient vehicles or other products by fees on inefficient ones.

5) Least-cost planning reforms in utility sector regulation, including demand-side efficiency programs, fair prices and grid access for independent power producers, and profit incentives for utility companies that effectively implement money-saving efficiency programs.

6) Financial incentives (golden carrots) for manufacturers that increase the energy efficiency of their products beyond best available levels.

7) In special cases, transitional employment subsidies and industrial conversion incentives (golden parachutes) for fossil-fuel producing regions that will lose assets and markets as a result of climate stabilization policies.

8) A reorientation of EC and national research and development programs toward least-cost carbon reduction options, notably high efficiency technologies and modular cogeneration and renewables-based technologies.

9) Combined carbon/energy taxes sufficient to fund the above carbon substitution programs plus an across-the-board shift in the tax structure from labor and investments to pollution.

In this integrated approach, an energy/carbon tax is primarily a direct funding mechanism for the investments and programs that will reduce emissions and other supply-related risks, and only secondarily a market signal.

The plan by the European Commission (CEC 1991c) to pursue both regulatory measures and an energy/carbon tax principally go in the same direction as our proposal. But the list of non-fiscal measures considered by the CEC is grossly incomplete when compared to our above proposal. In particular, regulatory reforms in the utility sector, and here, profit or shared savings incentives for utility demand-side investments, are sorely missing from the CEC proposal. Despite its demonstrated economic and environmental benefits, integrated, least-cost oriented utility resource planning has not yet been widely adopted in Western Europe. The extension of this approach along the lines of our ELCP concept offers a cost-efficient way of implementing the progressive carbon reduction goals of the European Community.

Conclusion: The Greenhouse Dividend

The results from our study challenge the widespread notion that carbon reductions beyond modest reductions would necessarily be expensive. Western Europe is endowed with sufficient low-carbon resource options to cut its carbon emissions by up to 50 percent and more over the next 30 years. Our least-cost oriented analysis shows that with proper implementation, such reductions could yield a large economic dividend. Opportunities for savings in year 2020 energy expenditure alone are equivalent to 1-2 percent of GNP per year, and economic benefits could be even greater if macroeconomic amplifications are taken into account.

These figures suggest economic impacts from internalizing global warming risks into energy policy that are the exact opposite of those found in top-down, macroeconomic modeling studies. We find that the econometric methodologies used in these studies are inherently unsuited for identifying least cost options. They also tend to narrowly steer the attention of policy-makers toward carbon taxes which are not the most effective implementation option.

The most decisive element in garnering the greenhouse dividend is a balanced portfolio of policies that effectively correct market and regulatory failures. A carbon tax is a sensible part of this portfolio to strengthen price signals. In using this tax, it is important to recycle carbon tax revenues into tax relief for capital investments or similarly the macroeconomically efficient measures.

An emphasis on non-fiscal measures is needed to integrate demand-side efficiency resources into a least-cost approach. This integration will by itself yield significant carbon savings in addition to economic savings. These savings can buy contributions from the more expensive low-carbon resources while still realizing net savings in the energy or electricity bill of society. For the same reasons, it is possible to fashion economically advantageous strategies that save large amounts of carbon without relying on nuclear power.

Our favorable results on the direct costs of integrated carbon reduction strategies need to be augmented by considerations of important feedbacks in the international economy. These feedbacks underline that a global approach is desirable in addressing the threat of global warming, but they do not diminish the opportunity for economically sound leadership by the European Community.

Acknowledgement

The research summarized here was conducted by IPSEP with funding from the Dutch Ministry of Environment. Preparation of this article was partially supported by the Energy Foundation.

References

Brinner, R. E., J.M. Yanchar, M. Shelby, R. Shakleton 1992. *"Balanced Policies to Address Global Warming,"* Draft paper presented at the Meeting of the Western Economics Association, San Francisco, California, July.

British Energy Committee 1990. *"The Cost of Nuclear Power. Report together with Proceedings of the Committee and Memoranda of Evidence,"* House of Commons, Fourth Report, Session 1989-90, London, June.

CEC 1990, *Energy for a New Century: The European Perspective,* Special issue of *Energy in Europe,* Commission of the European Communities, Directorate-General for Energy (DGXVII), July.

CEC 1991a, *CO2 Study-Crash programme: Increase of taxes of energy as a way to reduce CO2 emissions: problems and accompanying measures,* Directorate Generale XII, JOULE Program, Commission of the European Communities, Bruxelles, July.

CEC 1991b, *CO2 Study-Crash programme: Increase of taxes of energy as a way to reduce CO2 emissions: problems and accompanying measures,* Directorate Generale XII, JOULE Program, Commission of the European Communities, Bruxelles, July.

CEC 1991c, *"A Community Strategy to Limit Carbon Dioxide Emissions and to Improve Energy Efficiency,"* SEC(91) 1744 Final, Commission of the European Communities, Bruxelles, May.

Cline, W.R. 1992, *The Economics of Global Warming*, Institute for International Economics, Washington, DC, June.

Enquete Kommission 1990, *Schutz der Erde - Eine Bestandsaufnahme mit Vorschlägen zu einer neuen Energiepolitik*, 3. Bericht, Enquete-Kommission zum Schutz der Erdatmosphäre, 11. Deutscher Bundestag, Bonn, FRG.

Friedrich, R., Kallenbach, U., Thoene, E., Voss, A., H.-H. Rogner, H.-D. Fritsche, U., L. Rausch, K.-H. Simon 1989, *"Externe Kosten der Stromerzeugung,"* Vereinigung Deutscher Elektrizitaetswerke (VDEW), Frankfurt.

IPCC 1990, *Climate Change: The IPCC Scientific Assessment*, Intergovernmental Panel on Climate Change, WMO/UNEP, Geneva, August.

Krause, F., and J. Eto 1988, *Least-Cost Utility Planning* - A Handbook for Public Utility Commissioners. Volume Two: The Demand-Side: Conceptual and Methodological Issues, National Association of Regulatory Utility Commissioners (NARUC), Washington, DC, December.

Krause, F., W. Bach, J. Koomey 1989, *Energy Policy in the Greenhouse, Volume I: From Warming Fate to Warming Limit: Benchmarks for a Global Climate Convention*, Prepared for the Air Pollution Directorate, Dutch Ministry of Environment. Also published as *Energy Policy in the Greenhouse* by Earthscan Books, London 1990, and John Wiley&Sons, New York 1992.

Krause, F., J. Koomey, D. Olivier, D. Bleviss, G. Onufrio, P. Radanne, H. Becht 1992a: *Energy Policy in the Greenhouse, Volume Two. Least-Cost Insurance against Climate Risks: The Cost of Cutting Carbon Emissions in Western Europe.* IPSEP, El Cerrito, CA. Prepared for the Air Pollution Directorate, Dutch Ministry of Environment.

Krause, F., J. Koomey, D. Olivier et al. 1992b, *Energy Policy in the Greenhouse, Volume II-A. Demand-Side Resources in the EC-5 Region*, IPSEP, El Cerrito, CA. Prepared for the Air Pollution Directorate, Dutch Ministry of Environment.

Krause, F., J. Koomey, et al. 1992c, *Energy Policy in the Greenhouse, Volume II-B. Low carbon supply options for the EC-5 Power Sector*, IPSEP, El Cerrito, CA. Prepared for the Air Pollution Directorate, Dutch Ministry of Environment.

Krause, F., J. Busch, J. Koomey 1992d, *Incorporating Global Warming Risks in Power Sector Planning: A Case Study of New England,* Draft Final report, LBL-30797, Energy and Environment Division, Lawrence Berkeley Laboratory, Berkeley, California.

Mills, E., D. Wilson, and T.B. Johansson 1991: Getting started. No regrets strategies for reducing greenhouse emissions, *Energy Policy,* July/August, pp. 526-542.

Ministere de l'Industrie 1990, *Production d'Electricité Thermique: Les Couts de Reference,* Direction Generale De l'Energie et des Matieres Premieres, Paris.

NAS 1991, *Policy Implications of Greenhouse Warming - Report of the Mitigation Panel,* National Academy of Sciences, Washington, DC.

OTA 1991, *Changing by Degrees: Steps to Reduce Greenhouse Gases,* Office of Technology Assessment, Congress of the United States, Washington, DC, February.

Porter, Michael E. 1991, *"Green Competitiveness,"* *Scientific American,* April.

Russ, P., H.D. Haasis, O. Rentz 1991, *Elaborations of Strategies to Reduce CO2 Emissions in the European Community by Using a Linear Optimization Approach,* Institute of Industrial Production, University of Karlsruhe, prepared for Directorate Generale XII, Commission of the European Communities, Bruxelles, January.

Sanghi, A.K., J. Wang, and A. L. Joseph 1991, *"State Level Impacts of a Federal Carbon Tax: A New York State Example,"* New York State Energy Office, albany, New York, June.

UCS, ASE, ACEEE, NRDC 1991: *America's Energy Choices, Investing in a Strong Economy and a Clean Environment,* Union of Concerned Scientists, Alliance to Save Energy, American Council for an Energy Efficient Economy, and Natural Resources Defence Council, Washington, DC.

US DOE 1991, *Limiting Net Greenhouse Gas Emissions in the United States.* Executive Summary. U.S. Department of Energy, Washington, DC, September.

Zimmerman, M.B. 1990, *"Assessing the Cost of Climate Change Policies: The Uses and Limits of Models,"* Alliance to Save Energy, Washington, DC, April.

22. Observations on Extending the Set of Externalities to be Quantified

Robert D. Rowe; Carolyn Lang
RCG/Hagler, Bailly, Inc. P.O. Drawer O
Boulder, Colorado 80306-1906

1 Introduction

In the past, consideration of environmental and social impacts of electricity generation were often confined to the Environmental Impact Statement (EIS) and siting processes and focused upon the existence of impacts, mitigation strategies and costs, and whether acceptable levels of externalities could be reached to allow operation of a facility in a selected location. Today, the emphasis has expanded to quantify the social value (damage) of externalities that may remain even after siting regulations are met. Resource selection decisions are then based upon total social costs rather than just financial costs.

Thus far, however, efforts to include well-documented quantified externality damages of electric generation in resource planning decisions have placed the most emphasis upon a subset of ubiquitous well-known air pollutants. These include the U.S. National Ambient Air Pollutant Criteria Standard (NAAQS) pollutants of ozone (O_3), particulate matter (PM_{10} and TSP), sulfur compounds (SO_2, SO_4, and SO_x), lead (Pb) and carbon monoxide (CO). This focus has been motivated by the prevalence of these air pollution emissions from many electric generation alternatives such as fossil fuel-fired facilities, the potential significance of the associated damages, and, perhaps more significantly, due to the extensive volume of applied scientific and economic data on these pollutants developed over the past two decades to support ambient air quality regulations at the federal, state and local level. Examples of studies aimed at the valuation of air pollution injuries and damages using well developed and accepted scientific and economic damage function methods abound (See, for example[1-10]).

New externality valuation studies are continuing to expand and refine the treatment of major air pollution damages (as reported at this conference). Recent past and ongoing efforts have also begun to expand the set of externalities given detailed quantification attention beyond the NAAQS air pollutants. Due to less well developed (or discovered) literature, and time and budget constraints, past efforts to address new externalities have often been made using short cut valuation methods (such as the cost of control approach) or provided damage values based upon insufficiently developed and documented analyses. Several current studies to expand the externalities set are underway, but procedures and results have yet to be released. Just as for criteria air pollutants, considerable time and resources may be required before scientifically defendable damage values are developed and widely accepted.

This paper discusses some of our observations on avenues to expand the set of quantified externality values using the damage function approach, existing literature and moderate amounts of additional resources. The observations are based upon a variety of projects that we have been and are involved in, as well as discussions with principal investigators on other externality projects. Section 2 uses a simplified set of taxonomies of the externality groups associated with electricity generation to place in perspective the need to expand the set of externalities. Because not all externalities may need the same level of assessment, Section 3 discusses alternative strategies for quantifying externalities using a damage function approach. Section 4 discusses selected promising avenues for expanding the externality quantification using a damage function approach. The externalities selected are not intended to be an exhaustive list of where damage functions analysis may be undertaken, but to illustrate a subset of externalities for which literature currently exists for defensible analyses.

2   Externalities Associated with Electricity Generation

The breadth of environmental externalities of potential interest in externality valuation exercises is significant and well exceeds the conventional air pollutants considered in many past analyses. To illustrate this, Table 1 relates selected electricity generation alternatives to environmental burdens associated with the electricity alternative. For many generation alternatives, environmental burdens are associated with each of four production phases:

- Fuel acquisition and processing (mining, oil production, oil refining);
- Fuel transport and storage (truck, railroad, barge);
- Generation (construction, operation, decommissioning); and
- Waste disposal (solid waste), effluent discharge, air emissions.

Table 1 summarizes key environmental burdens for each phase for each technology from a list of selected technology groups. Environmental burdens are grouped by the environmental medium impacted, including outdoor air, indoor air, surface water, solid waste and others under construction and operation. Each cell within the table identifies emissions from the fundamental process itself or from related processes such as air emissions from the combustion of fossil-fuels used in mining equipment or transportation. Tertiary emissions, which refer to emissions from the production of steel used to create boilers and generators,

Table 1:   Electricity Alternatives - Fuel Cycle Emissions

	Coal Steam, Fluidized Bed, IGCC				BioMass Wood-fired			Nuclear			
	Mining	Fuel Proc.	Transport	Generate	Harvest	Transport	Generate	Mining	Fuel Proc.	Transport	Generate
OUTDOOR AIR											
Particulates	X	X	X	X	X	X	X	X	X	X	
SO2	X	X	X	X	X	X	X	X	X	X	
NOx, Nitrate, NO2	X	X	X	X	X	X	X	X	X	X	
Toxics and Metals	X	X		X	X		X	X	X		
CO	X	X	X	X	X	X	X	X	X	X	
Greenhouse Gas/CO2	X	X	X	X	X	X	X	X	X	X	
CFC											
Steam				X			X				X
Radioactive	X			X				X			X
SECONDARY OUTDOOR AIR											
Acid Aerosols	X	X	X	X	X	X	X	X	X	X	
Acid Precip.	X	X	X	X	X	X	X	X	X	X	
Ozone (HC, VOC)	X	X	X	X	X	X	X	X	X	X	
INDOOR AIR											
SURFACE WATER											
Chemicals	X	X	X	X		X	X	X		X	X
Thermal				X			X				X
Impinge/Entrain				X							X
Radioactive	X	X		X				X	X		X
Impoundment											
Consumption				X			X				X
SOLID WASTE											
Transportation							X	X			X
Volume/Land Use				X			X	X			
Hazardous/PCB				X			X				
Toxics in Ash				X							
Radioactive High								X	X		X
Radioactive Low	X			X							
CONSTRUCTION/OPERATION											
Construction				X			X	X			X
Land Use	X			X	X		X	X			X
Transmission – Land				X			X	X			X
Transmission – EMF				X			X	X			X
Explosion/accident											
Nuclear Accident											X
Spills											
Decommissioning											X

Table 1 continued

	RDF/MSW		Natural Gas/Oil–fired CT, CC, Steam				Hydro	Photovoltaic Cells		DSM	
	RDF Proc.	Generate	Production	Refining	Transport	Generate	Generate	Manufacture	Operation	Manufacture	Operation
OUTDOOR AIR											
Particulates	X	X	X	X	X	X		X		X	
SO2	X	X	X	X	X	X		X		X	
NOx, Nitrate, NO2	X	X	X	X	X	X		X		X	
Toxics and Metals	X	X	X	X	X	X		X		X	
CO	X	X	X	X	X	X		X		X	
Greenhouse Gas/CO2	–X	X	X	X	X	X	X	X		X	
CFC											X
Steam		X				X					
Radioactive											
SECONDARY OUTDOOR AIR											
Acid Aerosols	X	X	X	X	X	X		X		X	
Acid Precip.	X	X	X	X	X	X		X		X	
Ozone (HC, VOC)	X	X	X	X	X	X		X		X	X
INDOOR AIR											
SURFACE WATER											
Chemicals	X	X	X	X	X	X	X	X		X	
Thermal	X	X				X	X				
Impinge/Entrain						X	X				
Radioactive											
Impoundment						X	X				
Consumption			X			X					
SOLID WASTE											
Transportation						X			X		
Volume/Land Use	–X								X		
Hazardous/PCB						X					X
Toxics in Ash	X										
Radioactive High											
Radioactive Low											
CONSTRUCTION/OPERATION											
Construction		X				X	X		X		
Land Use		X	X			X	X		X		
Transmission – Land		X				X	X		X		
Transmission – EMF		X	X			X	X		X		
Explosion/accident		X				X	X				
Nuclear Accident						X					
Spills			X								
Decommissioning											

are not considered because impacts from these emissions are expected to be small over the lifetime of the facility.

Table 2 relates the groups of environmental burdens to social and environmental impacts such as mortality, morbidity, visibility aesthetics and the like. Each cell in Table 2 may represent one or many potential externalities. For example, there are a variety of potential human health effects associated with elevated levels of ambient air particulate matter, there are a variety of potential fishery and aquatic impacts associated with surface water emissions, and so forth. In ongoing work we have identified well over 300 specific impacts that may be considered in an externality analysis. While the criteria air pollution health effects may dominate many other air pollution effects (See, for example endnotes[1][2]), and perhaps other non air pollution externalities, Tables 1 and 2 make it clear that more research work is required to expand the set of externalities.

3   Alternative Strategies for Quantifying Externalities

The strategy for quantifying externalities using a damage function approach may vary across the specific impacts to be considered. Not all externalities may have sufficient scientific data for detailed analysis and alternative approaches may be necessary. For example, at least three damage function strategies might be considered:

1. *Generalized Approach*

    This approach specifies functions for the physical effects and economic damages associated with alternative levels of environmental burden, background conditions, and other relevant factors. With location and electricity resource specific information, the functions can be readily applied to all applications, as has been the case for many of the criteria air pollution damage function studies identified above.

2. *Case Study*

    Case study assessments may be used to determine a likely order of magnitude for damages. Case study results are then used to approximate the externality value for similar cases, although the accuracy of the results for other applications and locations may be reduced.

3. *Bounding Analyses*

    A bounding analysis would typically use available literature or defensible assumptions to ascertain the maximum damages that are expected, i.e., determining potential damages

Table 2:   Environmental Damage Groups

Sources of Environmental Impacts	Mortality	Morbidity	Accidental Injury	Materials	Crops/Veg	Forests	Fisheries	Aquatic	Terrestrial	Ground-water	Climate Change	Visibility	Other Aesthetics	Other
OUTDOOR AIR														
Particulates	X	X		X								X		
SO_2	X	X		X	X	X						X		
NO_x, Nitrate, NO_2	X	X			X	X		X				X		
CO	X													
Toxics and Metals	X	X			X	X	X	X	X	X				
CO_2/Greenhouse Gas	X	X			X	X	X	X	X		X			
CFC	X	X		X	X	X	X	X	X		X			
Steam				X							X	X	X	
Radioactive	X	X			X	X	X	X	X					
SECONDARY OUTDOOR AIR														
Acid Aerosols	X	X										X		
Acid Deposition				X	X	X	X	X	X					
Ozone (HC, VOC)		X		X	X	X			X					
INDOOR AIR														
Radon	X	X												
PM, NO_x, Toxics	X	X												
SURFACE WATER DISPOSAL														
Chemical	X	X			X		X	X					X	
Thermal							X	X						
Impinge/Entrain							X	X						
Radioactive	X	X					X	X	X					
Impoundment/Passage							X	X	X				X	X
Consumption										X				X

Table 2 continued

Sources of Environmental Impacts	Human Health			Materials	Ecological	Biologic Resources					Climate	Other		
	Mortality	Morbidity	Accidental Injury		Crops/Veg	Forests	Fisheries	Aquatic	Terrestrial	Ground-water	Climate Change	Visibility	Other Aesthetics	Other
SOLID WASTE														
Transportation			X										X	
Volume/Land Use													X	
Hazardous/PCB	X	X							X					
Toxics in ash	X	X					X	X		X				
Radwaste – high	X	X					X	X		X				
Radwaste – low	X	X					X	X		X				
CONSTRUCTION/OPERATION														
Construction			X		X								X	
Facility: Land Use					X	X			X				X	
Transmission: Land Use						X			X				X	
Transmission – EMF	X	X							X					
Explosion/accident	X		X											
Nuclear accident	X	X	X		X	X	X	X	X	X	X			
Spills	X	X					X	X	X	X			X	
Decommissioning			X						X	X			X	
FUEL ACQUISITION														
Extraction	X	X	X		X	X	X	X	X	X			X	
Processing	X	X	X				X	X	X	X			X	
Transportation/Storage	X	X	X				X	X	X	X			X	

with several orders of magnitude, or a worst case analysis. Threshold analyses might also be applied to determine the likelihood that effect thresholds are exceeded.

The generalized damage function quantification strategy is the most desired but usually requires the highest level of research resources and can require extensive literature that may not be available. Case studies and bounding analyses require less literature and research resources and may adequately address the quantification of damages. For example, assume a detailed generalized damage function estimates damages on the order of $0.001/kWh for externality group E1. If case studies or bounding analyses for externality group E2 result in defensible damage estimates that are several orders of magnitude smaller than for E1, such as $0.0001/kWh, further effort to refine the E2 estimates may not be merited as the added precision may be unlikely to affect any resource decisions because they are dominated by E1. Furthermore, the value for E2 may be in the rounding error of E1 and further research efforts might be better placed on refined estimates of E1 damages. Consideration of these types of evaluation criteria may significantly enhance our ability to efficiently expand the set of externalities that can be quantified with limited research resources.

4   Avenues for Expanding the Set of Externalities to be Quantified

We use Table 2 as a guide for discussing directions for expanding the set of externalities to be quantified as an abstract framework focusing upon environmental burdens and environmental and social impacts related to different types of electric resources. Table 2 indicates that discussions of every externality group (and certainly the specific impacts within a group) cannot be accomplished here. Rather, we provide summary comments about selected externality groups and selected key literature to guide the reader interested in further pursuing quantification for individual impact groups.

4.1   Criteria Air Pollutants - Human Health Impacts

As noted above, the quantification of mortality and morbidity impacts of criteria air pollutants has received extensive examination. For recent state-of-the-art applications for ozone and particulate matters[2 10 11]. Significant expansion of externality values for particulate matter may be addressed in a generalized approach with new work on mortality risks[12-14] and for acid aerosols health impacts using bounding analyses and perhaps a generalized approach based on recent EPA work[15]. While work continues on CO, gaseous SO_2, and NO_x, we suggest the effects are likely to be minimal for power plants, on a $/kWh basis, and that bounding analyses could be accomplished using readily available EPA reports and published literature and would be generally sufficient.

An area of potential expansion is health effects due to lead exposure. Effects and damages studies have been conducted for quite some time[16-20] and new damage valuation studies have drawn upon a significantly improved scientific and economic literature[21][22]. But, these studies have an important limitations in that many existing effects studies are based upon older NHANES II data from 1976-1890, when population blood-lead levels were considerably higher than they are now, which adds considerable uncertainty to the results. Fortunately, the NHANES III data is now being analyzed and, when results are released and published over the new few years, they will provide a basis to significantly improve damage estimation for lead emissions.

4.2   Criteria Air Pollutants - Welfare Impacts

Prior externality damage studies have suggested that welfare impacts of criteria air pollutants are significantly smaller than human health impacts[1][3]. These studies have begun to address damages associated with visibility impacts, crop impacts, materials damages and forest impacts of ozone, PM_{10}, sulfates and other pollutants. While new work continues and will improve the assessments, little of this work suggests dramatic changes to the existing approaches for visibility aesthetics, materials damage or crop damage. Some additional work has emerged on forest injuries, as reported in NAPAP work, and can be used to improve these assessments in terms of commercial timber values[23][24].

4.3   Air Toxics and Mercury

Human health impacts of air toxics and mercury provides an area for incorporation of new externality values. One recent study submitted in Massachusetts[4] (as updated in follow-up submissions) used U.S. EPA IRIS (Integrated Risk Information System) cancer risk factors, economic data, and estimates of inhalation and ingestion rates of air toxics[25] to compute bounding analysis estimates of cancer damage from air toxic emissions. Through improved literature on rates of exposure and uptake this analysis can be expanded.   For human health impacts, mercury is not classified as a carcinogen; rather at low doses it is associated with affect on neuro-behavioral and renal functions. At issue is are whether thresholds exist, what percent of the population exceed the thresholds, and damage functions. Expanded bounding or threshold analyses may be undertaken building upon the work in[4][25-27], and new efforts at EPA, and through use of generally available economic damage literature.

Quantification of damages for other non-cancer health impacts of air toxics are generally elusive, although there is some literature on heavy metal impacts upon crops, forests and other vegetation in locations very near to point sources. Generally this literature suggests that such damages associated with crops, vegetation and forests would be expected to have economic damages much less significant than for human health impacts[28-32]. Considerable work has been conducted on mercury impacts to fisheries in the Great Lakes, off Florida and in other locations. This work may provide the foundation for developing more generalized damage assessments, but more investigations appears necessary on the topic.

4.4 Air and Water Emissions of Nuclear Radiation

Nuclear radiation has long been of concern to nuclear regulatory agencies and to industry and environmental groups, but has not been incorporated into externality valuation studies. Commercial nuclear power reactors, under controlled conditions, release small amounts of radioactive materials to air and water resources during normal operations and during waste storage. Through various exposure and uptake routes, this radiation may cause human cancers (specifically leukemia) in workers and potentially, the general population. Non-cancer damages may also occur, but evidence suggests they may be associated with lower total damages than cancer risks. Similarly, evaluation of other biologic impacts of routine radiation releases suggest impacts will be minimal and of significantly lesser concern than for human health risks[33][34]. Human cancer risks can be addressed for externality damage assessments through case studies and bounding analyses to place them in perspective using data from a variety of recent reports[35-40], which generally suggest that the magnitude of damage from routine releases may be small.

Turning to nuclear accidents, case study damage estimates can be started based on the work of the Nuclear Regulatory Commission's NUREG-1150 analysis of Severe Accident Risks at five nuclear power plants[41][42] for latent fatality/cancer risks. The risk of an accident associated with any particular reactor can be ascertained from the unit's Probability Risk Assessment (PRA). While a Chernobyl type accident is deemed highly improbably in the U.S. due to differences in design and operating characteristics of facilities, data on health impacts and damages from the event may be useful to address damages for high level doses (For example, see[43][44]). Effects on severe accidents upon other biological resources may be significant. While there is evidence to begin investigations to quantify damages, the literature is limited in volume and applicability and the damages are likely to be significantly less than human health impacts.

4.5   Radioactive Solid Waste Disposal

Nuclear power plants can generate both low level and high level radwaste during operation and decommissioning. A worst case quantification of human cancer risks can be developed from a study at West Valley, New York as well as other reported contamination events at other low level waste sites[45]. These case studies may overstate damages is applied in a general case because of improved waste disposal practices and landfill design. These measures are likely to reduce both the likelihood and magnitude of these events. Again, the probably magnitude of damages, and limitations on available literature, suggest that non-cancer human health impacts, as well as other biological impacts would be less fruitful to pursue in a damage quantification study at this time[35][46].

Spent fuel comprises high level radwaste. While there are past and ongoing studies on the impacts and damages of high level radwaste at central depositories, no such depository exists or is likely to exist for quite some time. Therefore, all high level radwaste is stored on site for which, however, research is limited to plant PRA. Further, damage assessment is clouded by society's perceptions of risks versus objective risk measurements. Few measurements of social damages for these perceived risks exist.

4.6   Surface Water Disposal

Some of the most challenging valuation research is on impacts related to surface water discharges. One dimension to this challenge is the site-specific nature of the impacts and the relatively little scientific literature available for general impacts. The general impacts categories related to surface water discharges are human health impacts and aquatic wildlife impacts. Human health impacts are caused by consumption of untreated drinking water, consumption of fish loaded with chemicals or heavy metals, and dermal exposure. Biological impacts to aquatic species and plants can be related to chemicals, heavy metals, or habitat changes. Based on our initial research, we expect that most of the damage values generated in the near term will need to be based on case studies. Given the wide variety of surface water environments (stream, rivers, lakes, oceans) and the wide variety of wildlife species living in these environments, generalizing from these case studies may have significant limitations.

For example, in Victoria, Australia, the Latrobe River Catchment Area is known to have high salinity, which may be partially attributed to coal mining and coal-fired power plants. The State utility has implemented (what some would call) an extreme mitigation measure to pipe the saline waste water from these processes 50 miles to the ocean. In addition, the natural status of these streams and rivers has relatively high salinity so that most of the aquatic species are saline tolerant. Estimating the damages associated with saline waste water from electric generation operation, therefore, is very challenging because the small amount of discharge is not directly attributed to the electric resources (because the saline waste water is pumped to the ocean) and because the impacts of incremental saline on saline tolerant species is unknown. These results will not be generalizable to other less saline surface water resources because of the unique riparian environment and species in the Latrobe Valley.

Other impacts to aquatic wildlife can be caused by thermal discharges from power plants discharging waste heat into surface water bodies or impingement and entrainment on intake screens for power plants with once-through cooling systems. Several large power plants along the Hudson river with once-through cooling systems are attributed with significant impacts on the striped bass populations[47]. These impacts are partially mitigated through the operation of a fish hatchery that replaces the fish in the Hudson. Therefore, a damage value estimated impingement/entrainment for these Hudson river plants will be generalizable only to facilities with similar mitigation.

Hydro-electric generating stations have a unique set of impacts related to the river barrier of a dam for human and fish passage and inundation of riparian and terrestrial resources behind the dam. The level of these impacts is highly site-specific. We are currently considering facilities ranging from Niagara Falls to small run-of-the-river facilities. For existing hydro-electric stations, the damages related to the dam are generally considered to be fixed, in that, it is unlikely that the dam will be removed until the end of its serviceable life and any decisions to implement new or other existing resources in the near term will not change these impacts. Some states, including New York, are requiring that new and relicensed hydro-electric facilities mitigate all impacts[48-50]. The interpretation of mitigation will determine the externality values. A mitigation measure such as a fish ladder or "trap and truck" may directly mitigate aquatic impacts. The trade-off, however, between increased lake recreation and "white-water" recreation will make "mitigation" for loss of free-flowing river difficult to determine[51]. Will two days of flat water recreation be sufficient to mitigate for loss of one day of white-water recreation? Since we are considering only the unmitigated damages in externality values, these mitigation programs must be evaluated.

The primary area of valuation research for hydro-electric plants is related to minimum and maximum flow requirements and the impacts related to tidal conditions along controlled rivers. Since these flows are related to hydro resource operation, they are not "fixed" impacts. In addition, the impacts related to minimum flows and tidal conditions are under consideration for new regulations by many State regulatory agencies, and it is possible that some unmitigated damages may be occurring. The most prominent case currently under consideration are related to the flows through the Grand Canyon as controlled by electric demand in the West and Glen Canyon Dam. Proposed regulations may limit electric production from this facility to protect aquatic species in the Grand Canyon. As with other surface water impacts, the damages found in the Grand Canyon that are attributed to tidal conditions will probably not be entirely useful for a general hydro case.

4.7 Fuel Acquisition

As the most recent externality studies are based on a full fuel cycle analysis, the fuel acquisition stage is a new area of externality valuation. Each fuel can have a unique set of impacts that may vary greatly by acquisition process such as coal mining (surface and underground), natural gas/oil production (land and offshore), and bio-mass (clear-cut and select cut for wood). Early studies sponsored federal agencies such as MERES (Matrix of Environmental Residuals for Energy Systems)[52][53] can be useful for determining the emissions associated with each component of the acquisition process. These early studies, however, use emissions factors that are based on regulatory requirements from the 1970's and are likely to be much higher than current emissions levels. In addition, over half of the oil used in the U.S. is imported and produced in other countries that may have different environmental protection standards. Therefore, determining the unmitigated damages associated with each fuel cycle will require extensive research. Some simplifying assumptions, such as assuming all emissions are located in the area of interest no matter where they are produced, can be used to develop boundary values, but may greatly overstate or understate the actual level of damages.

Another dimension of the fuel cycle is transportation from the mining or processing site to the energy facility. Some recent evidence is showing that fuel transport by truck may be causing externalities related to road wear and other infrastructure impacts. These studies show that user fees are only covering about 60 percent of road maintenance and construction costs, therefore, society may be subsidizing truck transport costs. Evaluating the level of externality related to these impacts will require unraveling the user fees and tax systems to determine if the current arrangement is based on an efficient cost allocation. Therefore, the externality associated with these suspected damages may be smaller than the current evidence suggests.

4.8   Uncertain Evidence

Several suspected externalities do not have enough scientific data to warrant a damage function valuation. These include impacts related to electro-magnetic fields (EMF) and greenhouse gases. The scientific evidence related to EMF impacts has not consistently concluded that human health or other impacts exist[54][55]. Until more conclusive evidence of damage exists, these impacts cannot be evaluated with a damage function approach. Similarly, the impacts of global climate change are uncertain[56-58]. Although some recent analysis related to the potential impacts of doubling carbon dioxide (CO_2) concentrations in the atmosphere exist, the relationship between current greenhouse gas emissions levels and the resultant atmospheric concentrations (the carbon cycle) are not well understood. Furthermore, even when doubling of atmospheric CO_2 is assumed, the general circulation model results are so geographically broad (i.e. global mean surface temperature data) that it is difficult to predict the level of impacts related to global climate change in any particular location.

Some recent economic modeling research by Nordhaus and Cline[59][60] may be useful in estimating damages from global climate change, if a doubling of atmospheric CO_2 and a set of associated damages is assumed during the next century. But given the uncertainty related to the carbon cycle, determining the contribution of CO_2 emissions from energy resources today is not possible without a set of heroic assumptions. One final note, since little or no regulation exists for CO_2 emissions, application of the control cost approach or "revealed preference approach" is also not appropriate because no regulations exist that can reveal society's preference for global climate change impact levels or for marginal control costs.

5   Concluding Comments

The subset of externality values presented in this paper show the wide variety of impacts that must be evaluated and the volumes of data that can be required for full damage function analysis. This is not to suggest that the task at hand is impossible, rather there is a surprisingly large level of scientific and economic data available to support valuation of externalities and further research is warranted. This research will require a serious dedication of time and resources to complete the full set of externalities, especially for all general cases.

While not defended here, our perception is that the existing set of quantified externalities related to air emissions are potentially, but not uniformly, a significant component of total externality damages related to electric generation resources. This may be due to two factors: 1) the relatively large magnitude of damages related to human health in comparison to biological or aesthetic damages, and 2) the emerging scientific evidence that existing air regulations may not be fully protecting human health, in some cases.

The damage function approach is proving to be a flexible and productive method of externality valuation. As new scientific and economic evidence emerges, the damage function approach can incorporate these changes easily. In addition, this method highlights the current status of knowledge and uncertainty at several points in the valuation process so that the basis for the resulting values are clearly understood. Overall, our experience shows that extending the current set of externality to be quantified is feasible and that defendable values can be developed with existing literature for many of these externalities.

References

[1] Rowe, Robert D., Lauraine G. Chestnut, Donald C. Peterson and Craig Miller. 1986. The Benefits of Air Pollution Control in California. (Two Volumes) Prepared for California Air Resource Board by Energy and Resource Consultants, Inc., Boulder, Colorado. Contract No: A2-118-32.

[2] Thayer, M., et al. 1991. Estimating the Air Quality Impacts of Alternative Energy Resources, Phase IV. Prepared by Regional Economic Research for the California Energy Commission, Sacramento, California. December 26.

[3] Thayer, M. 1991. Valuing the Environmental Impacts of Alternative Energy Resources: Phase III Task I Report. Prepared by Regional Economic Research for the California Energy Commission, Sacramento, California. July 15.

[4] Rae, D., R.D. Rowe, J. Murdoch and R. Lula. 1991. Valuation of Other Externalities: Air Toxics, Water Consumption, Wastewater, and Land Use. Prepared by RCG/Hagler, Bailly, Inc. for New England Power Service Company, Westborough, MA.

[5] Harrison, David and Albert L. Nichols. 1990. Benefits of the 1989 Air Quality Management Plan for the South Coast Air Basin: A Reassessment. Prepared for California

Council for Environmental and Economic Balance by National Economic Research Associates, Inc., Cambridge, Massachusetts.

[6] Hall, Jane V. 1989. Economic Assessment of the Health Benefits from Improvement in Air Quality in the South Coast Air Basin. Prepared for South Coast Air Management District by California State University Fullerton Foundation, Fullerton, California. Contract No. 5685.

[7] Krupnick, A. and R. Kopp. 1988. The Health and Agricultural Benefits from Reductions in Ambient Ozone in the United States. Resources for the Future, Washington, DC. Discussion Paper QE88-10.

[8] Krupnick, A.J. and J. Kurland. 1988. An Analysis of Selected Health Benefits from Reductions in Photochemical Oxidants in the Northeast United States. Prepared for the Office of Air Quality Planning and Standards, U.S. Environmental Protection Agency, Research Triangle Park, North Carolina by Resources for the Future. Edited by Brian J. Morton.

[9] Krupnick, A. and P. Portney. 1991. "Controlling Urban Air Pollution: A Benefit-Cost Assessment." Science. Vol. 252(April) pp. 522-528.

[10] Cantor, Robin. 1991. The External Costs of Fuel Cycles: Background Document to the Approach and Issue. Prepared for the U.S. Department of Energy by Resources for the Future, Washington, DC.

[11] Thayer, M. 1990. Estimating the Air Quality Impacts of Alternative Energy Resources: Phase II Report. Prepared by Regional Economic Research for the California Energy Commission, Sacramento, California. July 31.

[12] Schwartz, J. and D.W. Dockery. 1992. "Increased Mortality in Philadelphia Associated with Daily Air Pollution Concentrations." American Review of Respiratory Disease. Vol. 145 pp. 600-604.

[13] Schwartz, Joel and Douglas W. Dockery. 1992. "Particulate Air Pollution and Daily Mortality in Steubenville, Ohio." American Journal of Epidemiology. Vol. 135(1) pp. 12-19.

[14] Schwartz, Joel. 1991. "Particulate Air Pollution and Daily Mortality in Detroit." Environmental Research. Vol. 56 pp. 204-213.

[15] U.S. Environmental Protection Agency. 1989. An Acid Aerosols Issue Paper Health Effects and Aerometrics. Prepared by the Office of Health and Environmental Assessment, Washington, DC. EPA/600/8-88/005F.

[16] U.S. Environmental Protection Agency. 1985. Costs and Benefits of Reducing Lead in Gasoline. Office of Policy Analysis, Washington DC. EPA-230-05-85-006.

[17] Brennan, Kathleen M. and Jerome T. Bentley. 1990. Regulatory Impact Analysis on the National Ambient Air Quality Standards for Lead (Draft). Prepared for Office of Air Quality Planning and Standards, U.S. Environmental Protection Agency by Mathtech, Inc. Princeton, NJ. EPA Contract Number: 68D80094.

[18] Brennan, K.M., R.L. Horst, J.M. Hobart, R.M. Black and K.T. Brown. 1987. Methodology for Valuing Health Risks of Ambient Lead Exposure (Final Report). Prepared by Mathtech, Inc. for the U.S. Environmental Protection Agency, Office of Air Quality Planning and Strategies, Research Triangle Park, North Carolina. EPA Contract No. 68-02-4323.

[19] U.S. Environmental Protection Agency. 1989. Case Study Benefit Analysis of Alternative National Ambient Air Quality Standards for Lead (Draft Final Report). Prepared by the U.S. Environmental Protection Agency, Office of Air Quality Planning and Standards, Ambient Standards Branch, Research Triangle Park, North Carolina.

[20] U.S. Environmental Protection Agency. 1986. Reducing Lead in Drinking Water: A Benefit Analysis (Draft Final Report).

[21] Wade Miller Associates Inc. and Abt Associates Inc. 1991. Final Regulatory Impact Analysis of National Primary Drinking Water Regulations for Lead and Copper. Prepared for U.S. Environmental Protection Agency, Office of Drinking Water, Washington, DC. EPA Contract No. 68-CO-0069.

[22] Centers for Disease Control. 1991. Preventing Lead Poisoning in Young Children. A Statement by the Centers for Disease Control, U.S. Department of Health and Human Services. 108 p.

[23] Shriner, David S. 1990. Responses of Vegetation to Atmospheric Deposition and Air Pollution. NAPAP SOS/T 18. In: National Acid Precipitation Assessment Program, Acidic Deposition: State of Science and Technology, Washington, DC, Vol. III.

[24] U.S. National Acid Precipitation Assessment Program. 1991. 1990 Integrated Assessment Report. Office of the Director, Washington, DC.

[25] U.S. Environmental Protection Agency. 1990. Methodology for Assessing Health Risks Associated with Indirect Exposure to Combustor Emissions - Interim Final. Prepared by the Office of Health and Environmental Assessment, Washington, DC. EPA/600/6-90/003.

[26] U.S. Environmental Protection Agency. 1992. Integrated Risk Information System (IRIS), Inorganic Mercury; Methyl Mercury. Washington, DC.

[27] Clement Associates Inc. 1989. Toxicological Profile for Mercury (Draft). Agency for Toxic Substances and Disease Registry, U.S. Public Health Service.

[28] Adriano, D.C., A.L. Page, A.A. Elseewi, A.C. Chang and I. Straughan. 1980. "Utilization and Disposal of Fly Ash and Other Coal Residues in Terrestrial Ecosystems: A Review." Journal of Environmental Quality. Vol. 9(July-September) pp. 333-344.

[29] Alloway (ed.), B.J. 1990. Heavy Metals in Soils. Blackie. London. 339 p.

[30] Long, R.P. and D.D. Davis. 1989. "Major and Trace Element Concentrations in Surface Organic Layers, Mineral Soil, and White Oak Xylem Downwind from a Coal-Fired Power Plant." Canadian Journal of Forest Research. Vol. 19 pp. 1603-1614.

[31] Arthur, Mary A., Gail Rubin, Robert E. Schneider and Leonard H. Weinstein. 1992. "Uptake and Accumulation of Selenium by Terrestrial Plants Growing on a Coal Fly Ash Landfill. Part I: Corn." Environmental Toxicology and Chemistry. Vol. 11 pp. 541-547.

[32] MacNicol, D. and P.H.T. Beckett. 1985. "Critical Tissue Concentrations of Potentially Toxic Elements." Plant and Soil. Vol. 85 pp. 107-129.

[33] Croke, Kevin, Robert Fabian and Gary Brenniman. 1987. "Estimating the Value of Beach Preservation in an Urban Area." The Environmental Professional. Vol. 9 pp. 42-48.

[34] International Commission on Radiological Protection. 1991. "Recommendations of the Commission." Annals of the ICRP. Vol. 21(1-3)

[35] Becker, David V. 1989. "Concluding Remarks (Low Level Radioactive Waste: How Does Society Respond? As part of a Symposium of Science and Society: Low Level Radioactive Waste. Controversy and Resolution)." Bulletin of the New York Academy of Medicine. Vol. 65(4) pp. 553-554.

[36] National Research Council. 1990. Health Effects of Exposure to Low Levels of Ionizing Radiation: BIER V. National Academy Press, Washington, DC.

[37] Ad Hoc Advisory Committee for the Study of Cancer in Populations Living Near Nuclear Facilities. 1990. Consensus Statement of the Ad Hoc Advisory Committee for the Study of Cancer in Populations Living Near Nuclear Facilities. Co-chairmen of Advisory Committee: P. Correa, Louisiana State University Medical Center, New Orleans, Louisiana; and, M. Szklo, Johns Hopkins University School of Hygiene and Public Health, Baltimore, Maryland.

[38] U.S. Nuclear Regulatory Commission. 1991. Generic Environmental Impact Statement for License Renewal of Nuclear Plants (Main Report/Draft Report for Comment). Prepared by the Office of Nuclear Regulatory Research, Washington, DC. NUREG-1437 Vol. 1.

[39] Baker, D.A. 1990. Population Dose Commitments Due to Radioactive Releases from Nuclear Power Plant Sites in 1987. Prepared for the Office of Information Resources Management, U.S. Nuclear Regulatory Commission, Washington, DC by Pacific Northwest Laboratory, Richland, Washington. NUREG/CR--2850-Vol.9.

[40] Baum, J.W. 1991. "Valuation of Dose Avoided at U.S. Nuclear Power Plants." Nuclear Plant Journal. (March-April) pp. 40-44.

[41] U.S. Nuclear Regulatory Commission. 1989. Severe Accident Risks: An Assessment for Five U.S. Nuclear Power Plants - Appendices (Second Draft for Peer Review). Formerly entitled "Reactor Risk Reference Document". Prepared by Division of Systems Research, Office of Nuclear Regulatory Research, Washington, DC. NUREG-1150 Vol.2.

[42] Evans, John S., Dade W. Moeller and Douglas W. Cooper. 1985. Health Effects Model for Nuclear Power Plant Accident Consequence Analysis - Part I: Introduction, Integration, and Summary - Part II: Scientific Basis for Health Effects Models. Prepared by Sandia National Laboratories for the U.S. Nuclear Regulatory Commission. U.S. Department of

Energy Contract DE-AC04-76DP00789. Reproduced by National Technical Information Service, Springfield, Virginia. NUREG/CR-4214. 234 p.

[43] International Advisory Committee. 1991. The International Chernobyl Project An Overview: Assessment of Radiological Consequences and Evaluation of Protective Measures. Published by the International Atomic Energy Agency. Vienna. 57 p.

[44] Seneviratne, Gamini. 1988. "UNSCEAR Finds Chernobyl Doses Lower Though More Core Got Out." Nucleonics Week. (July 7) pp. 4-5.

[45] Bergeron, M.P., J.L. Smoot, M.L. Kemner and W.E. Cronin. 1991. Hydrogeologic Performance Assessment Analysis of the Commercial Low-Level Radioactive Waste Disposal Facility near West Valley, New York. Prepared by Pacific Northwest Laboratory (operated by Battelle Memorial Institute) for U.S. Nuclear Regulatory Commission, Washington, DC. NUREG/CR-5737, PNL-7688.

[46] International Atomic Energy Agency. 1991. Effects of Ionizing Radiation on Plants and Animals at Levels Implied by Current Radiation Protection Standards (Final Draft). Approved for Publication but not yet edited by IAEA Publications Section. Technical Report Series. Vienna.

[47] RCG/Hagler Bailly Inc. 1990. A Review of Workplan for a Benefit-Cost Analysis of Options to Mitigate Power Plant Impacts on Hudson River Fish Populations. Prepared for Consolidated Edison of New York, Inc., New York, New York. August 27, 1990.

[48] New York State Department of Environmental Conservation. 1991. 6 NYCRR Part 700-705, Water Quality Regulations for Surface Waters and Groundwaters (Effective Regulation). Division of Water, Albany, New York.

[49] New York State Department of Environmental Conservation. 1991. Water Quality Standards and Guidance Values. Division of Water, Albany, New York.

[50] Niagara Mohawk Corporation. 1985. Niagara Mohawk Power Corporation, Lower Raquette River Project. Exhibit E - Environmental Report. Niagra Mohawk Power Corporation, Syracuse, New York. FERC No. 2330.

[51] Boyd, D.W., R.F. Jr. Nease, J.S. Rice and M. Taleb-Ibrahimi. 1990. Evaluating Hydro Relicensing Alternatives: Impacts on Power and Nonpower Values of Water Resources.

Prepared by Decision Focus Inc., Los Altos, CA for Electric Power Research Institute, Palo Alto, CA. Final Report August, EPRI GS-6922.

[52] Council of Environmental Quality. 1976. MERES and the Evaluation of Energy Alternatives. U.S. Government Printing Office, Washington, D.C. Stock Number 041-011-0026-2.

[53] The Science and Public Policy Program. 1975. Energy Alternatives: A Comparative Analysis. University of Okalhoma, Norman, OK. May.

[54] National Institute for Occupational Safety and Health. 1991. *Proceedings of the Scientific* Workshop on the Health Effects of Electric and Magnetic Fields on Workers. Cincinnati, Ohio. 229 pp.

[55] U.S. Environmental Protection Agency. 1990. Evaluation of the Potential Carcinogenicity of Electromagnetic Fields (Review Draft). Prepared by the Office of Research and Development, Washington, DC. EPA/600/6-90/005B. October.

[56] U.S. Environmental Protection Agency. 1989. Review of the Report to Congress: The Potential Effects of Global Climate Change on the United States. Prepared by the Global Climate Change Subcommittee, Office of the Administrator - Science Advisory Board, Washington, DC. EPA-SAB/EC-89-016. 23 p.

[57] Intergovernmental Panel on Climate Change. 1990. Climate Change: The IPCC Scientific Assessment. Cambridge University Press. New York. 364 p.

[58] Intergovernmental Panel on Climate Change. 1992. 1992 IPCC Supplement. Sponsored by the the World Meterological Organization and the United Nations Environmental Program, Washington, DC, February. 70 p.

[59] Nordhaus, William D. 1992. Rolling the Dice: An Optimal Transition Path for Controlling Greenhouse Gases. Paper presented at the Annual Meeting of the American Association for the Advancement of Science, Chicago, Illinois, February, 1992.

[60] Cline, William R. 1992. Global Warming: The Economic Stakes. Institute for International Economics. Washington, DC. 103 p.

23. An Overview of Taxes and Trading as Environmental Control Policies

D. J. Dudek and W.R.Z. Willey
Environmental Defense Fund
257 Park Avenue South, New York, NY 10010

Executive Summary

As the environmental movement has gradually accepted the use of economic incentives as a means of accomplishing environmental goals, the question of which incentives should be employed is arising more frequently. In recent years, we have witnessed controversies -- often counterproductive from an environmental perspective -- concerning which economic instruments constitute correct environmental policy. There is a need for environmentalists to take stock, in a rational and dispassionate manner, of the alternative economic instruments which are available. The intent here is to outline the pluses and minuses of these instruments as regards their ability to deliver environmental protection. The following charts summarize the assets and liabilities of each policy as discussed in the paper.

TAX POLICY ASSETS

- Raise Revenue which can be beneficial if targeted to environmental expenditures
- Readily understood by all
- Widely accepted as a tool of public policy
- Effective if set correctly
- Collection apparatus is in place

TRADING POLICY ASSETS

- Highly effective in hitting the environmental target as long as monitoring and enforcement are adequate
- Creates an environmental asset which can aid in financing pollution control
- Rewards innovation and stimulates R&D
- The political process determines the aggregate control level
- The political process allocates entitlements
- Focuses enforcement resources efficiently
- Applicable to transboundary and international environmental problems
- Built-in incentive for technology transfer

```
TAX POLICY LIABILITIES

• Politically unpopular
• Difficult to set at levels which are effective in reducing emissions
• Difficult to adjust to new environmental information
• Large quantity of information required
• If set on industrial inputs, may not be transmitted to consumers
• Difficult to implement internationally due to sovereignty issues
• Potentially serious distributional effects among firms
• Potentially serious equity impacts among consumers
• Administrations and agencies are likely to be addicted to the revenues


TRADING POLICY LIABILITIES

• Political agreement on allocation can be difficult
• Some environmentalists mistrust
• Public does not readily understand
```

Economic Incentives as Environmental Policies[1]

Across the spectrum of environmental problems, the use of economic incentives is always an option. We consider here the main categories of instruments, according to the emphasis placed on government regulatory requirements. Emphasize that the use of economic incentives always requires government regulation -- the main question is what kind of regulation, and how likely is such regulation to actually achieve environmental goals. Most government regulatory programs which have used economic incentives have heretofore focused either on price regulation directly through administered prices or indirectly using subsidies or taxes.

Government-administered prices are perhaps most notorious in terms of environmental damages in programs to allocate public timber, water supplies, range, electricity, and wastewater disposition. Administered prices have been set at levels so far below economic values that depletion and overexploitation have resulted. Overlapping with many of these programs have been subsidies using taxpayer funds to allow lower administered prices. A common use of taxes as an indirect pricing measure has been the use of tax credits for environmental programs encouraging adoption of energy and water conservation

---

[1] The authors acknowledge contributions to this paper from Kenneth Sewall and Alice LeBlanc. As usual, any errors are ours..

technologies. These uses of taxes as subsidies have been supplemented to a degree by various use-related taxes, or "mil taxes," to provide revenues to help defray government regulatory costs. Pesticide and electricity sales taxes and water pollution discharge permit fees are examples wherein the respective state or federal agency receives these revenues. Such taxes have not, however, been of sufficient magnitude to cause reductions in resource use or to generate substantial revenues. The use of taxes on resource uses as a punitive instrument to discourage such uses and/or to generate significant revenues for environmental programs is relatively new to the active public policy menu. Taxes as punishment and revenue-generators are becoming a popular price regulation device for environmentalists, and therefore merit careful consideration and comparison with other economic instruments available for environmental policy.

In all of these types of price regulation, there have been environmental goals associated with their use or misuse, but those goals have not been accompanied by specific targets. Instead, the link to environmental goals has been of the general sort that "encouragement" of actions consistent with such goals is posited as a result of these price regulations. Examples are taxes or increased administered prices on water, timber, grazing, and highway use intended to encourage conservation and reduce resource depletion; and "green taxes" on uses of gasoline, carbon, and other contributors to atmospheric degradation.

A second general category of environmental policies using economic incentives focuses on the setting of environmental targets and goals directly, while administered pricing is de-emphasized from a regulatory perspective. The fundamental difference of this approach from that of price regulation is that the regulatory and political processes focus directly on the adoption of physical environmental goals rather than indirectly on these goals via pricing. Redistribution of economic wealth and resources is still a political issue, but determining what that distribution ought to be is no longer a direct concern of regulatory politics through such issues as the regressivity of government pricing policies. To be sure, wealth distribution remains an important issue in the setting and implementation of specific environmental goals. But distribution becomes much less of a directly and politically volatile issue.

Examples of programs aimed directly at environmental goals are much more recent and limited. The targeting of critical habitats for direct land acquisition is perhaps the oldest and most frequently used example, with The Nature Conservancy, Ducks Unlimited, and Trust for the Public Land being leaders. Direct acquisitions of water supplies to support specific ecosystems and habitats have recently begun to be utilized. In addition, projects to compensate property owners to manage ecosystems on their land for stewardship of targeted species are beginning to be initiated by both private and public entities. Finally, the approach is now being applied for the first time in a systematic way to the acid rain problem using tradable sulfur dioxide permits whose price is about to be determined by a powerful mix of spot and futures markets. In each of these cases, the environmental target, whether it be additional duck habitat or reduced acid rain, was determined by the political process with the price of achieving these goals to be determined by relevant markets.

These two approaches to the use of economic incentives for environmental policies are fundamentally different in terms of strategies necessary to implement them as well as likelihood of achieving environmental results. They should be viewed as competing approaches for all practical purposes, and compared when formulating agendas for solving specific environmental problems. As most environmental problems, from the regional to the global, have not been solved, the choice between these approaches is of increasing importance. Delineating the strengths and weaknesses of both, then, is central to sensible environmental advocacy.

Alternative Economic Instruments

We have learned that initiating programs for environmental advocacy can lead to large expenditures of resources in terms of time and money. Therefore, choosing which approach to apply deserves some degree of dispassionate, a priori analysis removed as much as possible from the inevitable current political machinations. Can we systematically and deliberately choose, or are we destined to ad hoc choice based on reading the current political "tea leaves?" Without the former, environmentalists are left with checking around in political circles and in the media to determine what is correct policy regardless of the likelihood of actually achieving environmental goals. Given the array of unsolved and intensifying environmental problems, this is a chilling prospect.

Establishing Environmental Limits

The first step in creating any credible environmental policy irrespective of the choice of instrument is the articulation of the environmental goal or objective. Most often such goals are expressed in the form of specific limitations on the discharge of specific pollutants. While the process of determining those limits is often thought of as one of balancing benefits and costs, the practice is much less precise. As a practical matter, public decisions concerning the appropriate level of environmental protection or quality are typically fraught with uncertainties. The environmental benefits of preserving the last remaining individual of a species is an excellent example. What values should be assigned to the potential future demands for its continued existence? What value should be assigned to the willingness of individuals to pay to see that the earth is home to more than homo sapiens?

While environmental benefits are frequently imponderable, environmental costs are commonly believed to be more tangible and certain involving expenditures obligating the current members of society. However, environmental costs themselves are uncertain varying substantially with the policy instrument selected for implementation. Both a Rolls Royce and a Volkswagen get you to the same destination, but at a substantially different cost.

In the face of imprecise quantitative decision methods, it is the function of representative democratic government to make the determination of the appropriate level of environmental risk and protection that society should consume. Examples of this type of environmental limit setting are numerous. Much of the debate over the acid rain provisions of the Clean Air Act of 1990 centered on the appropriate level of reductions in sulphur dioxide emissions. In Oregon, the governor of that state has asked the public to establish appropriate targets for the preservation and maintenance of species of wild salmon. The recent United Nations Conference on Environment and Development which spawned the Rio Convention was the forum for the debate about greenhouse gas limits. **Irrespective of the policy tool selected, there is no substitute for the clear specification of the environmental objective. Without credibly enforceable environmental limits, society cannot expect any environmental policy to function.**

Basis for Policy Choices by Environmentalists

The globe is being swept by a resurgent interest in the power and importance of markets and economic incentives. From the restructuring of formerly communist economies to development in the third world to the ongoing struggle to open markets for international trade, efforts are underway to unshackle decision makers and to present the opportunities and discipline of markets. One of the last social frontiers to feel the winds of these changes is management of the environment. The Environmental Defense Fund has long held the tenet that both economic well-being and environmental quality are best served by informed decision makers facing the full consequence of their choices.

Pollution Taxes -- Informational Burdens and Uncertainties

Taxation is one of several policy instruments available to correct market distortions in the prices of polluting inputs, goods, or production processes. The aim of using taxation in this way is the attainment of environmental goals. tHE price administration required by such taxation stands in contrast to tradable permits or resource use allowances which directly regulate physical environmental targets and rights to use resources within these targets. Other price administrative instruments, such as user fees or effluent charges, can be used in conjunction with such resource use allowance instruments to generate revenues for environmental acquisition funds and other compensation-based programs which are compatible with resource use allowance strategies.

In order to evaluate the effectiveness of imposing taxes as a means to improve environmental quality, there are three general areas we can explore. First, we can try to better understand what the expected environmental results of a tax should be. We can do this by examining why, from an economic point of view, pollution is a problem or why markets don't work in allocating the full cost of pollution, in short why government intervention is necessary at all. Secondly, we need to know what information is required in order to set the level of the tax correctly so as to attain the desired environmental goals. In addition, we need to know the likelihood of obtaining the required information and the implications of not setting the tax at the proper level. Third, it will be helpful to examine two specific examples of potential taxes -- one on the use of gasoline and one to facilitate the control of sulfur dioxide (SO_2) emissions --in terms of the information required to set the tax correctly, how the tax would work to obtain the environmental goal and how the tax compares to other available control measures.

An externality is a type of market failure which refers to a damage or benefit that one individual or firm derives from another without market-based arbitration. For example, someone who enjoys the flowers his neighbor plants pays nothing but receives a benefit. Pollution is an example of a negative externality in that emissions from one source damage resources others use. For example,environmental pollution is estimated to reduce the national output of the Polish economy by 20% (Nelson, 1990).

A key feature of a ideal market system is that valuable assets can be individually owned. Together with competitiveness, this ensures that all costs of production and consumption are borne by the producers and consumers directly involved. Some important, extremely valuable assets, however, do not lend themselves to private ownership. These include the atmosphere, oceans and other waters. Because these assets, referred to as common property resources, are not owned and cannot be exchanged like ordinary commodities, they are one of the major sources of externalities. Without ownership of these environmental commodities, there is no market incentive to reduce emissions or to provide environmental protection. Open and unpriced access to these resources means that the full costs of production and consumption of polluting goods are not reflected in the market prices, resulting in greater than optimal production or consumption levels. In other words, a firm that attempted to produce cleaner air, for example, would not be able to exclude nonpaying consumers from enjoying the benefits of its activities.

The following diagram (Figure 1) illustrates the supply, or marginal costs, and demand for a good that includes a polluting externality. Social marginal costs (SMC) represent the full costs to society including the damage costs produced by pollution. Private marginal costs (PMC) represent the market costs that individual firms and consumers face. The market equilibrium price and quantity are P_1 and Q_1, but the optimal price and quantity when full social costs are taken into account are P_2 and Q_2. The difference between the two prices -- P_2 minus P_1 -- is the level at which the tax should be set. By setting the tax at this level the price facing consumers will reflect the full social costs and result in the economically optimal level of consumption and pollution. The tax thus works indirectly to raise the price of the good and induce a reduction in demand and consequently in the amount of pollution. The difference between the areas under the two marginal cost curves (SMC minus PMC) from 0 to Q_1 represents the total value of environmental damages at the market equilibrium level.

In this example it is necessary to know the demand and marginal cost curves in order to determine the correct tax level. Such a determination is particularly critical when the tax is the sole economic instrument to achieve a stated environmental goal. However, pollution can be produced in a variety of ways. It can be associated with a particular input, for example, coal containing sulfur. Pollution can be produced as a byproduct of particular technology as in the case of chlorofluorocarbons used as aerosol propellants. In each case, the use of the environment as a free disposal resource produces excessive waste and environmental damage. An alternate way of examining this issue is to let the good in question be emissions reductions and ask what information is needed to determine the tax level required to achieve a given reduction in emissions of a given pollutant. The marginal cost curve in this case is the cost of control technology and the marginal benefits curve represents benefits to society from a given level of pollution reduction measured as avoided damages. Without a tax or intervention from outside the market to control the externality, the supply of pollution control is not brought into play. The emissions tax should be set at the price level P* where the two curves cross (Figure 2).

Figure 1:   Private and Social Costs

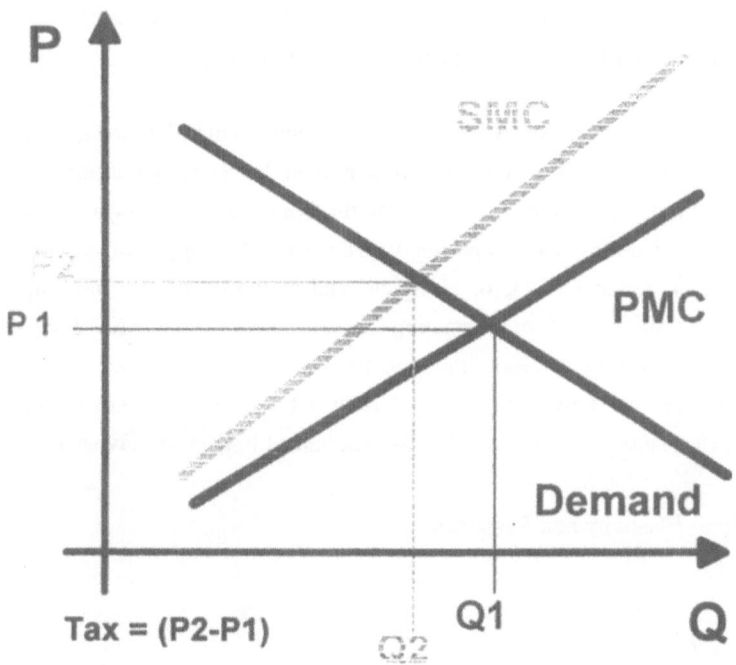

Figure 2:   Costs and Benefits of Reductions

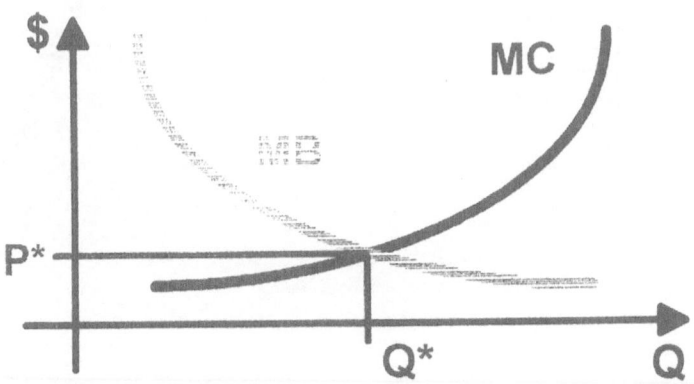

Again, we ask the question what information is necessary to set the correct tax level and under what conditions is a tax the most appropriate policy instrument? In a world of perfect information, the choice of policy tool is unimportant because both optimal price and quantity are known and the desired result can be achieved either by controlling price through a tax or controlling quantity through strict standards or tradable permits.

In the real world of imperfect information and uncertainty, however, it is unlikely that the exact characteristics of these curves will be known. Uncertainty concerning the marginal cost curve of pollution abatement results partially from unknowns regarding future technologic developments, learning curves and the diffusion of technology. Uncertainties in the marginal benefits curve relate to lack of complete understanding of the full impacts of pollution damages and difficulties in estimating costs of some damages. Frequently, measuring environmental damages raises imponderables such as the value of a human life or an endangered species. Indeed, one of the serious deficiencies of the cost-benefit approach is the difficulty of quantifying all of the non-market valued impacts of environmental pollution.

Figure 3:   Elasticity and Response

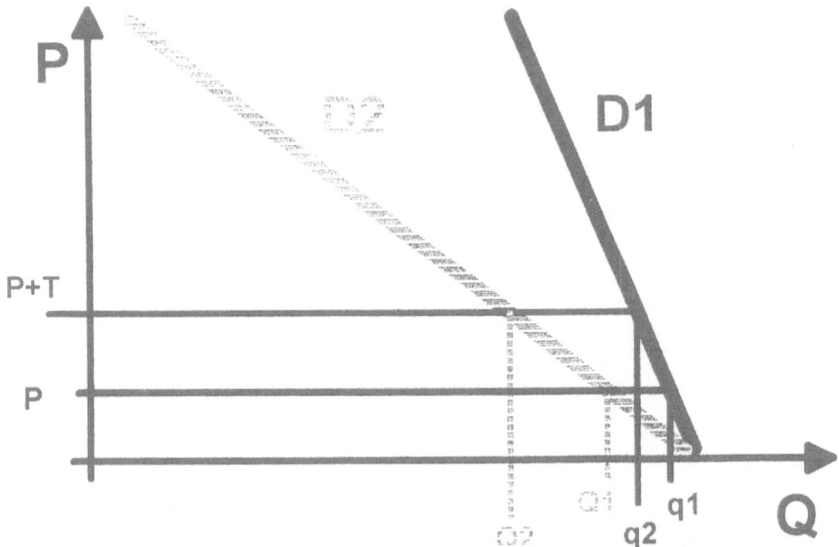

Figure 3 illustrates the importance of knowing the degree of responsiveness of the demand for discharges. If the revenue from a tax is designed to be pledged to an environmental trust fund to fuel expenditures which is the main mechanism to produce the intended environmental outcome, then the administering authority must find a relatively stable source of revenue as illustrated by D1. If the authority picks an emission where polluters have flexible responses as in case D2, then serious revenue shortfalls would result from polluters reducing discharges to avoid taxation. On the other hand, if the main environmental force of the policy is to change behavior and reduce emissions directly, then case D2 is the situation regulators should seek.

As difficult as the problem of identifying appropriate circumstances for taxes as behavioral versus revenue tools may be, it is complicated by the fortunate fact that things change over time, particularly after the application of human ingenuity. While the regulator may begin levying a tax at T under discharge conditions represented by D1, innovation over time can change polluters options to lower treatment costs to D2. When "green taxes" are presumed to be enacted either for the funding of environmental programs or for other fiscal objectives, revenue shortfalls can frustrate those goals.

Figure 4:   Revenue Changes Over Time

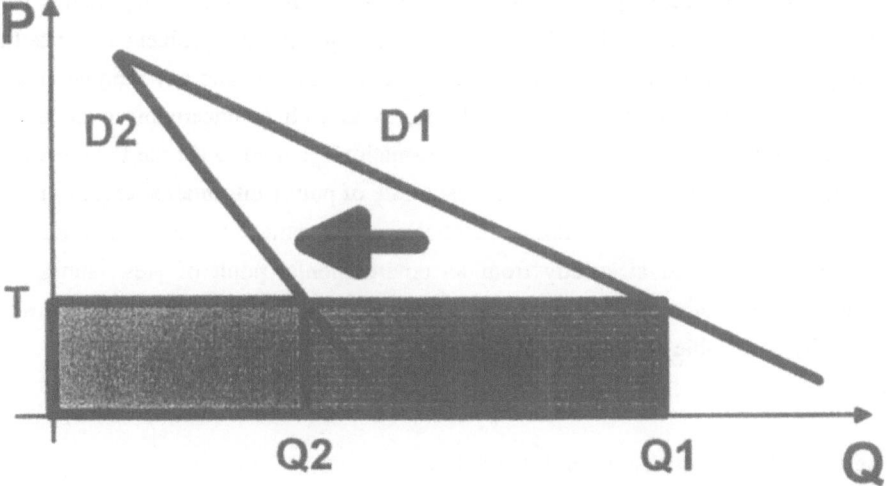

The Case of Acid Rain and SO$_2$ Taxes

Setting the level of a tax incorrectly may produce either little environmental benefit or costly over-control. An example of this problem can be found in some of the legislative proposals which were considered in the search for a solution to the nation's acid rain problem. One proposal would have set fees on industrial sources of SO$_2$ emissions at \$55 per ton while another would initially set the fee at \$45 per ton with a future range of \$30 to \$80 per ton. These fees can be compared to the cost of cost-effective emissions reductions reported by ICF Corporation in an October 1985 study on sulfur dioxide emission reduction alternatives for industrial sources. The 1995 cost per ton of cost-effective reductions of SO$_2$ emissions for industrial boilers reported in that study ranged from \$284 to \$579 depending on different reduction scenarios. If these cost estimates are at all realistic, the fee levels set in these two proposals would clearly fall below most of the range of the relevant cost curves and provide no incentive for industrial emitters to reduce SO$_2$. In order for these emission fees to have an impact on the behavior of polluters, other than raising revenues for utility subsidies, these fee levels would have to be significantly higher.

Given the time and financial requirements for planning and implementing control technologies, an iterative process through which the tax is reset following market response is not feasible. Of course, this strategy completely ignores the problem that taxation is an authority that Congress is unlikely to cede to the bureaucrats and any iterative process must include substantial lags for the political process as well as uncertainty over the ability to sustain political consensus. The cost of information required to set the tax correctly is high even in the acid rain case where a significant base of public information exists. In the face of uncertainty, "prices can be a disastrous choice of instrument far more than quantities can" (Weitzman). Looked at strictly from an environmental point of view, setting limits on quantities may not reach the optimum economic solution, but will at least result in the minimally acceptable environmental solution.

The Case of Gasoline Taxes and Air Pollution

Much research has been conducted on the effects of a gasoline tax on consumer demand. Most of the analysis involves measuring the relationship between change in price of gasoline and change in demand. This relationship, the percent change in quantity consumed divided by the percent change in price, is the price elasticity of demand. It is necessary to estimate this measure in order to have some idea of the extent of reduction in gasoline consumption, and any change in the suite of emissions produced from gasoline combustion from a given level of gasoline tax.

Literature reviews of the research indicate a wide range of elasticity estimates. In addition, there are two different types of price elasticity, the short run, usually defined as within a year after the price change, and the long run, usually meant to extend at least 10 years from the price change or tax. Since gasoline is a derived demand, i.e. derived from the demand for transportation services, equations which model demand often specify the variables that directly influence gasoline consumption, namely vehicle miles traveled and average miles per gallon. Vehicle miles traveled depends in turn on variables such as real income levels, the relative price of gasoline, and other transportation options.

A literature review prepared by the Energy Information Administration indicates a range of short term price elasticities between -.1 and -.3 and long term elasticities between -.3 and -.9. Similarly, the range of estimates in a literature review by Carol Dahl are -.03 to -.52 for short term and -.1 to -1.61 for long term. Differences in the estimates can result from the type of data used. This can be time series (monthly, quarterly or annual) and/or cross sectional (across regions, countries or individuals). It can also result from the scope of the modeling effort, that is from the comprehensiveness of variables and effects taken into consideration. In addition the statistical technique employed will also affect the estimates. Though each elasticity estimate may be valid in its own context, the diversity contributes to uncertainty in determining one estimate of the true price elasticity of gasoline demand.

The elasticities chosen by the EIA for use in their own automobile gasoline model were developed by James Sweeney. In prior research Sweeney estimated elasticities in the short term from -.12 to -.22 and in the long term from -.73 to -.78. For the EIA model his estimates are -.28 for a one year period and -.82 for a 10 year period. However in a short term model that does not account for demand response because of change in vehicle efficiency, the estimate used by the EIA is -.16. Similarly , in discussing the difference between short and long term demand responses, the analysis reported by the EIA in "Price Elasticities of Demand for Motor Gasoline and Other Petroleum Products" concludes that "The greatest proportion of 10 year price response manifests itself in increase in vehicle efficiency." Dahl in a paper in which she estimates the long term price elasticity states "The short run estimated elasticity is -.442 of which half comes from changes in miles per gallon. In the long run, demand is more elastic than -.778 of which nine-tenths comes from changes in miles per gallon."

It seems clear that the immediate demand response to change in gasoline prices is small relative to the price change and takes the form of reduction in vehicle miles traveled through carpooling, use of public transportation, and simply less travel. Most of the overall effect on demand results from the change in auto stock to more fuel efficient cars. While this begins to take effect in the first year following a price change, the full effect is not felt for at least ten years.

To summarize, the estimates for the demand response to a gasoline price change vary significantly. There is some consensus, however, that the response in the short term is not very great and that the long term response occurs over a period of at least ten years through a switch to a more fuel efficient stock of automobiles. However, over a period of ten years, factors such as changes in world oil markets and changes in real income levels can intervene and change the desired tax level, assuming the level could be determined in the first place. The results of the research indicate uncertainty in setting the tax level correctly initially, little effect in the short term, and the possibility of the need to reset the level in the long term in order to maintain the real relative effect of the tax. It is also difficult to directly translate reduced fuel consumption into emissions reductions and environmental improvements. Emissions reductions would depend critically on the polluting characteristics of the fleet as well as fuel consumption.

The Case of Tradable Permits and Atmospheric Pollution

The acid rain title of the U.S. Clean Air Act of 1990 utilizes a market-based strategy to achieve sulfur dioxide (SO_2) reductions. The program begins with a 10 million ton SO2 reduction objective. Each polluter is given an endowment of emission allowances for each plant, much like initial deposits in a checking account. These allowances are freely exchangeable throughout the U.S. In fact, anyone can hold these transferable permits. The allowance allocations simultaneously prescribe permissible emissions and the reduction responsibilities required at each location. Through the artifice of the transferability of these permits, the U.S. Congress created a valuable asset, one which can be utilized as part of the financing strategy to produce compliance with the acid rain program.

The flexibility of the SO_2 emissions market will produce innovative solutions for both control and financing. Trading itself will not only reduce compliance costs but will also create new revenue flows for some polluters. Public and private auctions are part of the design of the program with no restrictions upon who may participate. A futures market has also been announced by the Chicago Board of Trade (CBOT) and several firms are developing insurance services based upon allowance pooling. Electronic bulletin boards and computerized tracking systems have been developed to facilitate trading.

While these ancillary institutions have seemingly occurred relatively rapidly, markets for environmental discharges have been under development for over a decade. As such, marketable permits for atmospheric emissions can be seen as a natural outcome of the evolution of market-based environmental policy experiences in the U.S. The transition from emissions trading in nonattainment areas to national SO_2 trading through lead rights trading among refiners and transferable production permits for CFCs is very clear. This substantial experience stands in stark relief to the paucity of practical experience with emission fees as an incentive tool in the U.S.

This national emissions market will lead to a successful and durable acid rain program because it is founded on an inextricable linkage between environmental protection and powerful economic efficiencies. The crucial elements are: 1) substantial required reductions in sulfur dioxide emissions and emissions of oxides of nitrogen, 2) strict liability for severe, unavoidable penalties for sources that fail to meet their reduction requirements, 3) a nationwide cap on total emissions once required emissions reductions have been achieved and 4) complete flexibility for sources in the means they chose for compliance with their reduction requirements. Such flexibility includes the ability of sources to "trade" emissions reduction credits, which would be earned by any source that reduced its emissions beyond its reduction requirements, to another source that could use such credits, in lieu of on-site reductions, to meet its reduction obligations.

Due to the dynamics of a multi-state sulfur dioxide emissions trading market, emissions trading may provide a form of cost sharing for those areas, such as the Midwest, which at one and the same time will have the largest emissions reductions and the greatest opportunities to produce cost-effectively "excess" emissions reductions that could be sold to sources in other regions of the country. Finally, by making sources strictly liable for severe penalties when they fail to achieve required emissions reductions, the program can relieve both sources and the Environmental Protection Agency of overly cumbersome compliance planning obligations. As a result, sources enjoy a greater margin for innovation and enforcement resources can be concentrated on demonstrated wrongdoers subject to virtually automatic sanctions.

The value of encouraging innovation is one of the critical advantages of using an economic method to solve the acid rain problem. Since emissions trading is performance based, polluters are not forced to use specific technologies. As a result, institutional innovations like emissions banking and technical innovations like sorbent injection are allowed to develop and penetrate the emissions control market. To evaluate the importance of allowing innovations to be adopted, a sensitivity analysis of ICF's CEUM linear programming model was conducted. Basic program costs, already reduced due to the efficiencies of the trading system, would likely be reduced further. Allowing emissions banking, cost-effective industrial trading, and sorbent injection (conservatively available in 1997) would reduce basic program costs by a further 30%. While some environmentalists remain unconcerned about potential cost savings, cost savings lower political resistance to otherwise more expensive environmental goals and free up resources to be applied to the solution of other environmental problems. As the debate at the Earth Summit in Rio last year demonstrate, this lesson has not been widely learned.

Criteria for Evaluation

While the nation is gripped in the throes of debating new policy strategies and approaches, it is important to clearly establish the criteria against which alternative policies should be evaluated. As the din from competing interests becomes increasingly confused between goals and means, it is critical for environmentalists, in particular, to be focused on objectives as clearly distinguished from tools. The following are sets of criteria which are germane to this choice between strategies based on regulation of prices versus environmental targets.

Environmentalism is dedicated to a wide array of objectives -- all unabashedly environmental. Thus, it should be no surprise that this first criteria heading the list is environmental performance. How well does the policy deliver the environmental bacon?

With environmental taxes, price levels are subject to direct controls and the resulting impacts on actions, and consequent environmental effects, are uncertain since they are only indirectly produced by behavioral changes induced by the tax. Of course, the most critical factors in determining the effectiveness of any tax are its level and the underlying elasticities, i.e. the degree of flexibility in response.

With trading systems, environmental goals are directly articulated and set; however price (cost) levels are subject to uncertainty and lack of control. In effect, the environmental goal is guaranteed if the trading system has been properly designed to include credible monitoring, enforcement, and penalties. In reality costs are not completely unknown as policy analyses are routinely conducted by the government and other interested parties which frequently bracket expected costs. The dynamics of markets, explicitly harnessed in this approach, usually imply falling costs over time as learning and innovation occur.

A second criteria to consider when selecting among policy choices is the financing of the strategy. Environmental taxes are generally set so low as not to affect behavior and to yield revenue. The revenues may either accrue to the general treasury or be committed to special purposes. In the later case, the environmental tax is known as a user fee. In either case, the availability of additional revenues provides the opportunity for the government to pick winners and losers, bestowing the revenues upon favored industries or strategies.

Policies using marketable permits begin from a superior position financially as the policy creates tangible, fungible assets of value. The creation of environmental assets allows many self-financing opportunities. For example, under the acid rain program futures markets, auctions, leases, insurance pools are all being developed. Each will ease the financing problem either by providing a direct source of new revenue or by reducing the costs that need to be financed.

Another important criteria, one that underlies this entire document, is the economic performance of the policy tool selected. Certainly, both the theoretical and empirical literature support the idea that economic incentives generally offer at least the same environmental performance but with significant cost savings. As between economic policy tools, the potential economic impact is not so clear. Some argue that emission fees limit the cost of an environmental policy by fixing an upper limit on the amount per unit pollution levied. Whereas under marketable permit systems, there is no upper limit to cost except through operation of the market. In this latter regard, the dynamics of the behavior of the costs of alternatives is critical. Here again, we find the evidence compelling in favor of the stimulus to innovate provided by a fixed resource limit.

Conclusions

The taxing power of the Federal governmental is a potentially powerful policy tool for the solution of environmental problems. In the simplest case, user fees or taxes can be levied and those revenues dedicated to specific environmental purposes. However, it is significantly more complex to design and assess taxes whose function is to cause polluters to take account of their discharges and to adopt pollution abatement strategies in order to reach pre-specified environmental targets. In such cases, alternative economic instruments such as marketable permits are preferred for their certainty in attaining environmental goals. Nonetheless, taxes could be used in concert with other economic and regulatory tools in a suite of policies to achieve environmental goals if the tax revenues are pledged for environmental purposes and allocated efficiently. Lastly, the tax code itself contains provisions which induce environmental pollution through subsidies for natural resource extraction and use. Deficit reduction with the revenues generated by taxing undesirable economic activities like pollution can also be accomplished by the reduction of federal spending to encourage activities which produce environmental degradation.

References

Adar, Zvi and Griffin, James M. "Uncertainty and the Choice of Pollution Control Instruments" Journal of Environmental Economics and Management. Vol. 3 1976 pp.178-188.

Fischelson, Gideon. "Emission Control Policies under Uncertainty." Journal of Environmental Economics and Management. Vol. 3 1976 pp.189-197.

Dahl, Carol A. "Gasoline Demand Survey." The Energy Journal, Vol.7, No.1. 1986 pp. 67-81.

Dahl, Carol A. "Consumer Adjustment to a Gasoline Tax." The Review of Economics and Statistics. 1978 pp 427-432.

U.S. Department of Energy, Energy Information Administration. May 1981. Price Elasticities of Demand for Motor Gasoline and Other Petroleum Products. AR/IA/EUA/81-03.

Weitzman, Martin L. "Prices vs. Quantities." Review of Economic Studies. Vol.41 1974 pp. 477-491.

24. Utility Externalities and Emissions Trading: California is Developing a Better Way

David Harrison, Jr.
National Economic Research Associates
One Main Street, Cambridge, MA 02142

1 Introduction

Regulators in the Los Angeles Air Basin in California are developing an emissions trading program that provides a far better means of internalizing utility externalities in utility resource planning than the approach now being developed by Public Utility Commissions (PUCs) in many states. PUCs in at least 29 states have reportedly either adopted or considered requiring "environmental adders" to reflect the social costs from power plants that remain after environmental controls are in place.

That "adders" approach has many shortcomings, both in concept and in practice (see Joskow 1992). Emissions trading programs such as the one being developed in the Los Angeles Basin avoid most of these difficulties. Utility rate payers and the public in general would be better served if environmental agencies and state Public Utility Commissions worked together to develop comprehensive emissions trading programs.[1]

This paper summarizes the Los Angeles Basin trading program and explains its implications for electric utility planning. We show how emissions trading programs internalize utility residual emissions regardless of whether the utility actually buys emission allowances for these emissions. The paper also deals with some of the complications that arise when emissions trading programs are actually implemented. We begin with a brief overview of emission trading.

---

[1] Emissions trading programs may not be justified in all air basins given their administrative costs. Such costs may not be justified in many areas with minimal air pollution cases. In those areas, PUC's should determine whether utility emissions pose a sufficient problem to justify accounting for residual emissions, given the limitations of the "adder" approach. If "adders" are deemed appropriate, the proper approach is to estimate the *damages* that residual emissions cause. See Harrison, Nichols, Evans and Zona (1992) for an illustration of the damage-based approach applied to emissions in the Los Angeles Air Basin and other California air basins.

2 Concept of Emissions Trading

The basic concept behind emissions trading is simple. A trading scheme sets overall limits on emissions, allocates the initial allowances (i.e., the right to emit a ton of the pollutant), and then allows firms to trade emissions allocations. Trading reduces the cost of meeting emissions targets and establishes results in a market price for emissions allowances.[2]

2.1 Cost Savings from Emissions Trading

Figure 1 illustrates how trading works to lower the cost of emissions reductions and how both buyers and sellers gain from trades. The figure shows hypothetical marginal costs for two sources faced with the same emission standard. Since emission standards are typically set on the basis of the availability or affordability of control equipment—rather than on the cost-effectiveness of controls—the same standard can lead to vastly different costs per ton. In Figure 1, Source A incurs a marginal cost of $5,000 per ton controlled, while Source B spends $30,000 for the last ton.

Clearly we could achieve the same overall reduction in emissions at lower cost by tightening controls at Source A and relaxing them at Source B. Loosening controls at B by one ton saves $30,000, while controlling that added ton at Source A raises costs by only $5,000, for a net savings in compliance cost of $25,000 per ton. Achieving that savings with standard command-and-control regulations is unlikely—even if there were the will to do so—because the government is not likely to have such precise information on source-specific control costs.

Allowing firms to trade emissions for a price leads to the cost savings illustrated in Figure 1. Figure 2 shows how control costs would be reduced—and how both the buyer and the seller would gain—if there were an emissions trading market. Assuming a market price of $20,000 per ton, the seller (Source A) gains by paying $5,000 to control one more ton and getting $20,000, for a net profit of $15,000. The buyer (Source B) gains by reducing control costs by $30,000 and paying only $20,000 for the ton of emissions, for a net savings of $10,000. Thus, the total savings of $25,000 is split between the buyer and the seller, with both gaining from the trading program.

---

[2] The following discussion relies upon Harrison and Nichols (1990) and Nichols and Harrison (1991), which also discuss previous experience with emissions trading programs.

Figure 1:   Standards Often Lead to Large Differences in Marginal Costs

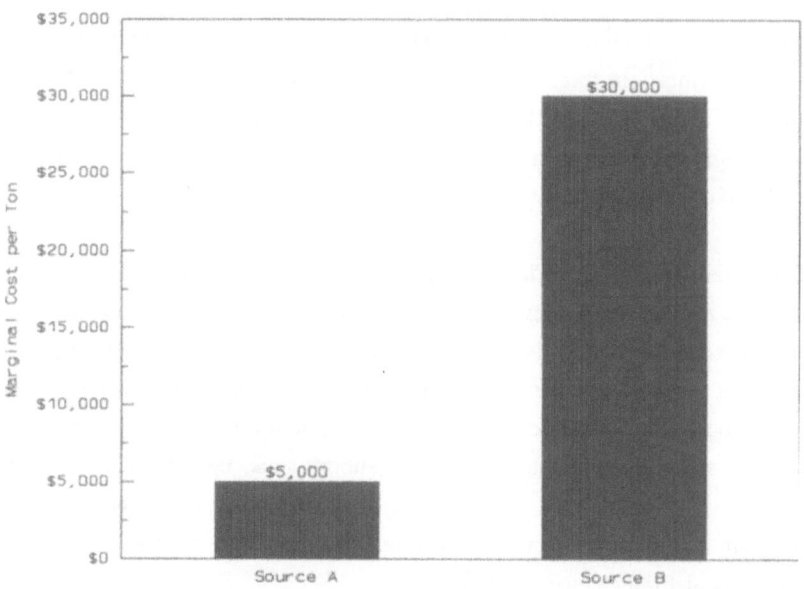

Figure 2:   Trading Can Reduce Overall Control Costs

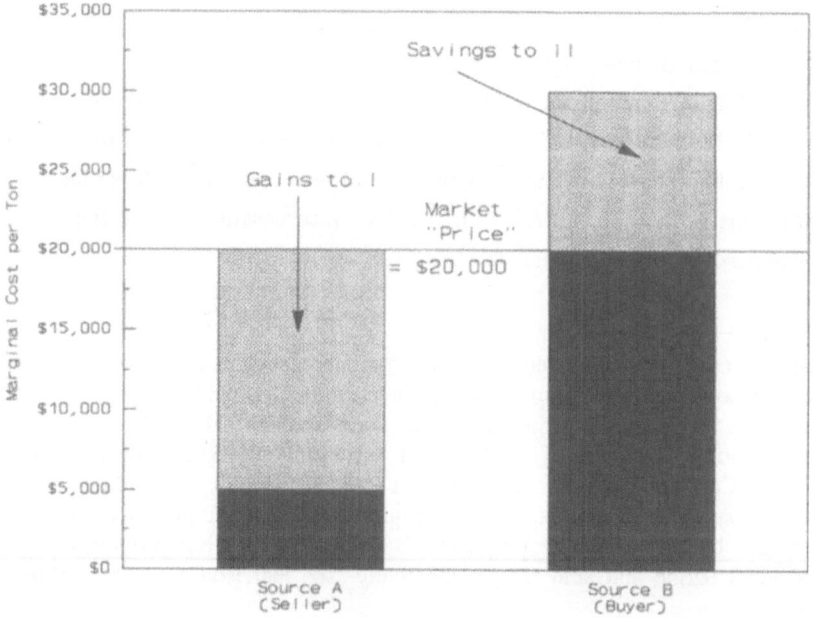

2.2   Emissions Trading in the 1990 Clean Air Act

Emissions trading is not a new concept. What is new is the pattern of increasing applications and proposals for emissions trading in the United States and elsewhere.[3] The 1990 Clean Air Act provided a major impetus for the greater application of economic incentives and emissions trading, most prominently through the Title IV acid rain trading program but also through a host of other provisions.

The 1990 Amendments to the Clean Air Act have shifted environmental policy from an almost exclusive reliance on command-and-control regulation toward a much greater reliance on the use of economic incentives. All seven of the major titles of the act include opportunities for economic incentives. These provisions include mandatory as well as discretionary programs. Economic instruments are required for extreme ozone nonattainment areas that fail to meet an applicable milestone or submit an acceptable plan. (The instrument can include a trading program, a fee program, or other similar measures.)   The acid rain trading program (discussed below) is also mandated as the method for reducing SO_2 emissions from coal-fired electric utility boilers.

The range of *possible* trading programs in the Clean Air Act is wide. Any State may choose to adopt a trading program as part of the State Implementation Program (SIP) it submits to demonstrate compliance with federal requirements. EPA is currently developing rules for mandatory economic incentives that will serve as policy guidance for these discretionary programs. Options include expanded use of EPA's three established (but seldom used) Emissions Trading (ET) programs, the bubble, offset and netting programs. (The use of these programs might be increased under potential new rules EPA is considering.) Trading also can apply to mobile sources through options for refiners and others to participate in credit trading programs—much like EPA's earlier and very successful lead trading program—for reformulated gasoline or oxygenated gasoline.[4]

---

3   Emissions trading and other "economic instruments" (such as emissions taxes or deposit-refund systems) are becoming increasingly prominent outside the U.S. as well. The Organisation for Economic Cooperation and Development (OECD) has recommended the use of economic instruments to deal with environmental problems (Organisation for Economic Cooperation and Development 1991).   Many individual countries are developing specific proposals. For example, the Canadian provinces of Ontario and Alberta are both considering proposals for emissions trading programs to deal with ground-level ozone and acid rain, respectively (see Harrison forthcoming and Nichols 1992).

4   See Harrison and Nichols (1990) for a summary of experience with ET programs and the lead-in-gasoline trading program.

Economic incentives may also play a significant role in the air toxics program. Major emitters of hazardous air pollutants are subject to Maximum Available Control Technology (MACT) standards. Economic instruments can be used in several areas under this new program, including the early reduction program, the compliance with individual MACT standards, and an emissions offsetting provision.

The emissions trading program designed to control acid rain is the most prominent program. It establishes a cap on total SO_2 emissions from coal-fired power plants of 8.9 million tons by 2000, roughly a 50 percent reduction from current levels. Electric utilities are allocated allowances to emit SO_2, based primarily upon historical emissions. They are allowed to trade SO_2 allowances among themselves, and industrial sources can "opt in" to the trading program as well.

The estimated cost savings from the program are substantial. Studies sponsored by EPA predict that the trading approach can reduce utility compliance costs by as much as 20 to 50 percent compared to the traditional command-and-control approach. The cost savings could be even greater if industrial sources decide to opt into the trading program.

The eventual scope of emission trading programs under the 1990 Clean Air Act will depend upon how the various provisions are implemented as well as on the initiative that States and local air quality agencies exercise to take advantage of this greater flexibility. Nowhere is the initiative being seized with greater vigor than in the South Coast Basin in California.

3   Emissions Trading in California's South Coast Air Basin

The South Coast Air Quality Management District (SCAQMD) is developing an ambitious · emissions trading program for the Los Angeles Air Basin that would dwarf even the national acid rain program in scope and complexity. The program is called the Regional Clean Air Incentives Market (RECLAIM), emphasizing its objective of reclaiming both clean air and a healthy regional economy. Following a year-long feasibility study and earlier scoping studies—and with the support of EPA and the California Air Resources Board—the SCAQMD Governing Board voted on March 6, 1992 to authorize a full-scale rule development process.[5] The SCAQMD hopes to adopt rules for the program by early 1993.

---

[5]   The earlier scoping studies were Harrison and Nichols (1990) and Noll (1990), both of which were presented to a workshop on trading programs organized by the SCAQMD in January 1991. The SCAQMD's feasibility study, which included five working papers, culminated in the set of final recommendations reported in SCAQMD (1992).

RECLAIM is being developed as a response both to the need to reduce air pollution in the most heavily polluted region of the U.S. and to the enormous cost of the traditional "command-and-control" approach. The maximum concentration for ozone is almost three times the federal ambient air quality standard. The SCAQMD estimates that the Basin needs to reduce overall precursor emissions of ROG and NO_x by about 85 percent from current levels, which already reflect the most stringent controls in the country. The SCAQMD developed a detailed Air Quality Management Plan containing more than 130 control measures ranging from controls on outdoor barbecue lighter fluid to requirements for large numbers of electric vehicles by the year 2000. The SCAQMD did not estimate the cost of the entire AQMP, but rough estimates based upon the existing cost data put the annual cost of the entire AQMP at about $13.5 billion per year, or more than $2,200 per household per year (see Harrison 1989 and Harrison and Nichols 1990).

RECLAIM promises to reduce the cost of meeting emission reduction targets substantially. Trading among sources in the same industry shifts emission reductions from high-cost to lower-cost options, as the simple example given above illustrated. Trading between facilities in different industries extends the cost savings. Tentative estimates put the cost savings from RECLAIM at 40 percent of the costs of overall cost of the standards approved (Harrison and Nichols 1992). Moreover, creating a market for emission allowances tends to provide greater incentives for technological innovation that could further reduce control costs as well as improve future air quality.

RECLAIM as currently proposed would be limited to large stationary sources—those with emissions of more than four tons per year—but the SCAQMD intends to extend the program to smaller stationary sources and to mobile sources. Markets would be set up for three major pollutants, nitrogen oxides (NO_x), reactive organic gases (ROG) and sulfur dioxide (SO_2). The initial markets are expected to include about 2,000 ROG facilities, 700 NO_x facilities and 70 SO_2 facilities. (These account for about 85 percent, 95 percent and 80 percent of the total emissions of the respective pollutants.) RECLAIM would set overall caps from emissions from these sources that would decline over time in order to meet federal ambient air quality standards by the year 2010.

The SCAQMD is developing the specific elements of RECLAIM based in part on contributions from an Advisory Committee set up to advise on the program.[6] Many questions arise in turning theory into practice. How should the allowances be allocated initially? What restrictions, if any, should be placed on trades? Should separate markets be established for

---

[6] The author is a member of the Advisory Committee and is serving as a consultant to the SCAQMD.

emissions in different subareas of the South Coast Air Basin or in different seasons of the year? What monitoring requirements should be established? What reporting requirements should be established? What role should the SCAQMD play in overseeing trading? What protection, if any should be developed for small businesses and others that might be unfamiliar with the program? What impacts will the trading program have on employment and the regional economy?

4 Implications of Emissions Trading for Electric Utilities

The development of emissions trading programs such as RECLAIM have two major implications for electric utilities:

1. Trading will *reduce* the cost of meeting environmental targets for electric utility plants; and

2. The market price for emissions means that the social cost of the emissions that remain will be *internalized* into the private cost of resource options.

The internalization of the social cost of residual emissions means that creating a system of "adders" is not necessary and, indeed, would be inefficient double-counting.

4.1 Cost Savings from Trading

The cost savings to utilities from emissions trading can be substantial. Data provided by the SCAQMD allows the calculation of a lower-bound estimate of the savings due to switching from a conventional emission standard to an emissions trading program (see Harrison and Nichols 1990). In developing a rule on NO_x emissions from utility generating stations in the Los Angeles Basin, the SCAQMD identified three different potential NO_x emission levels for the standard. Two were based on emissions per unit of heat *input*: 0.03 pounds per million Btu (lbs/MMBtu) and 0.01 lbs/MMBtu. The third potential standard was expressed in terms of emissions per unit of electricity *output*: 0.25 pounds per megawatt hour (MWHr). Overall, this last standard is roughly in between the others.

The SCAQMD identified 54 active utility boilers in the South Coast Basin. For each, they had estimated the cost of meeting three increasingly stringent emission standards. These data allow us to estimate the overall cost of meeting each one of these standards as well as the total tons of emissions that would be reduced under each standard. For example, achieving the intermediate standard of 0.25 lbs/MWHr results in costs of about $63 million per year for these 54 boilers and reductions in NO_x emissions of about 21 tons per day. These overall figures mask enormous variation in the cost per ton of controls at the various facilities. (This variation is due in large part to large differences in projected capacity utilization of the different boilers which, when coupled with relatively large fix costs of retrofitting boilers with controls, leads to very different cost per ton estimates.) Figure 3 plots the distribution of costs per ton to meet the 0.25 lbs/MWHr standard. For the entire group of 54 boilers, the average cost to control a ton of NO_x ranges from $3,000 to more than $130,000.

Figure 3: Variation in Cost per Ton of Control for Electric Utility Boilers

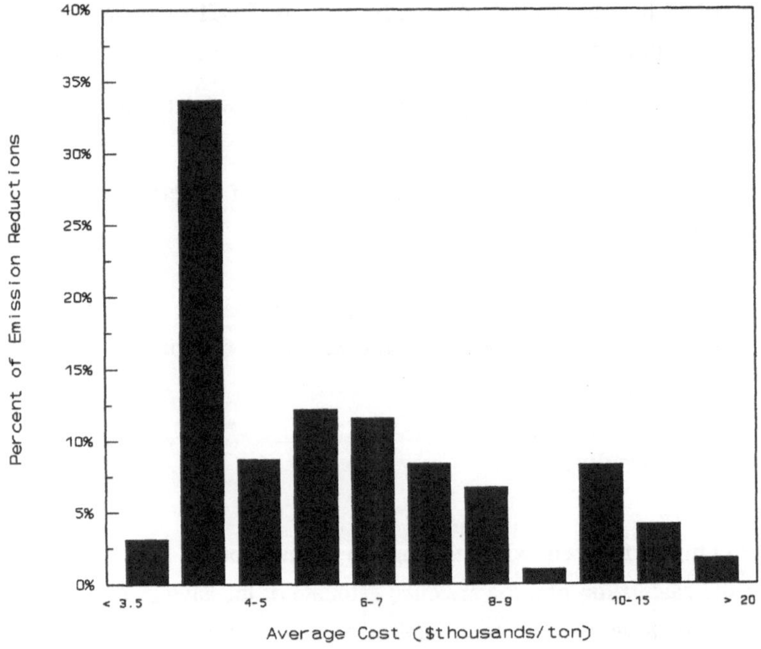

Source: Harrison and Nichols 1992, p 13

If utilities were allowed to trade emissions reductions among themselves, the high-cost compliers would look for opportunities to buy emission reductions and the low-cost compliers would look for ways to control more in order to sell allowances at a profit. The market would tend to establish a market price, with buyers having costs above that price and sellers having costs (at the margin) below that price. Using the SCAQMD data, it is possible to rank the options by their marginal cost and plot the results of computing the lowest cost means of achieving increasing emission reduction targets. Figure 4 shows the results for the 54 utility boilers. The horizontal axis shows total emissions reductions, while the vertical axis measures costs. The three potential standards are plotted as individual points connected by lines. The bottom curve shows the minimum cost of achieving any given level of emission reductions; it was constructed by ranking the various plant-control combinations in ascending order of their incremental cost per ton reduced.

Note that the least-cost line lies well below the points representing the individual standards. That means that substantial cost savings could be reaped with trading. If we focus on the intermediate standard of 0.25 lbs/MWHr, the cost savings with the least-cost outcome obtained with trading would be about $16 million per year, or about 25 percent of the cost of the uniform emission standard. The cost savings, of course, depend upon the level of the emission standard; savings from trading would be about 43 percent if the standard were set at the less stringent level of 0.03 lbs/MMBtu.

Figure 4:   Cost Savings to Electric Utility Boilers Under Trading

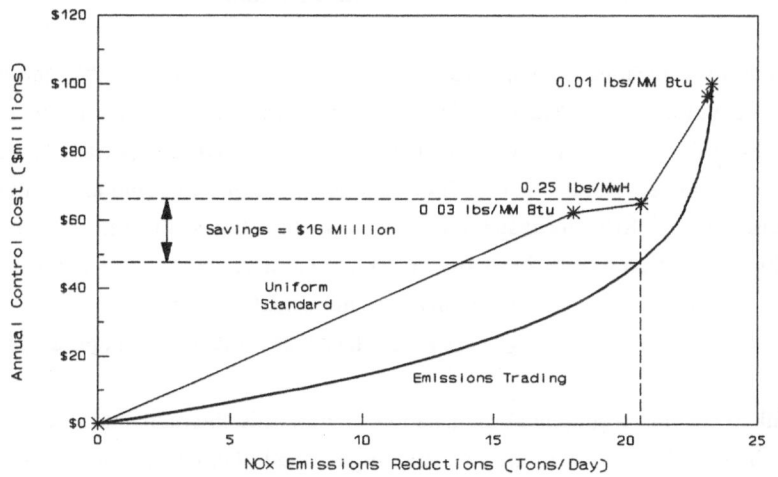

Source: Harrison and Nichols 1992, p. 12.

4.2   Internalization of Residual Emissions Damages

The interest in "externality adders" among PUC's is based upon concerns for the emissions that remain after controls such as the ones discussed above are in place. For example, the 54 utility boilers would have combined residual emissions of about three tons per day if emission reduction targets were based upon the intermediate standard.   Even new boilers, which typically face much stricter standards, would have residual emissions. Moreover, the level of residual emissions will differ a great deal depending upon the boiler technology and fuel used. That means that alternative resource plans could lead to very different levels of residual emissions.

These considerations have recently led PUC's to require that utilities take these residual emissions into account in determining their resource plans. Utilities have long been required to take into account the future costs of emission controls in deciding among alternative supply sources. In this context, the interest in "adders" for residual emissions is a logical extension although, as noted in the Introduction, there are many conceptual and practical problems in developing "adders" only for electric utility residual emissions.

Participation in emissions trading programs like RECLAIM means that these "adders" programs are not needed. Such participation does *not* mean that there are no residual emissions from the new plants. What is different about the emissions trading program—in contrast to emission standards—is that these residual emissions will tend to be internalized in resource decisions. Why? Because the utility will include the cost of the allowances needed to cover the residual emissions in its own calculations of the cost of the facility. This cost calculation will take place even if the utility does not actual pay for the emissions allowances, because it is allocated enough allowances at the start to cover its residual emissions.

It is useful to provide a graphical explanation of how this internalization occurs. Figure 5 illustrates the situation facing a utility considering the cost of emission control for a proposed plant with a trading program such as RECLAIM in place. The plant is assumed to have uncontrolled emissions of 10,000 tons of NO_X per year.[7] The marginal cost of control schedule shows the technical options the utility has to control emissions. The slope shows that the added costs of controlling a ton of emissions tend to increase as controls become tighter; the initial reductions are reasonably cheap on a per ton basis but the costs per ton increase substantially as emissions decrease.

---

[7]   Although the example describes emissions trading for NO_X, the same concepts would apply to any pollutant.

RECLAIM will tend to result in a common market price for a ton of emissions. In Figure 5, the price is assumed to be \$20,000 per ton of NO_x.[8] That means that any firm could buy or sell the right to emit a ton of NO_x for \$20,000. Like other markets, there may be some differences among the prices in different transactions, but they will tend to center on a common price at one point in time.

Figure 5 shows that utilities participating in RECLAIM will incur two types of costs for their emissions:

1. Control costs for emissions reductions from uncontrolled levels; and

2. Allowance costs to cover residual emissions (area B).

Figure 5:   The Social Cost of Emissions Equals Control Costs Plus the Cost of Allowances

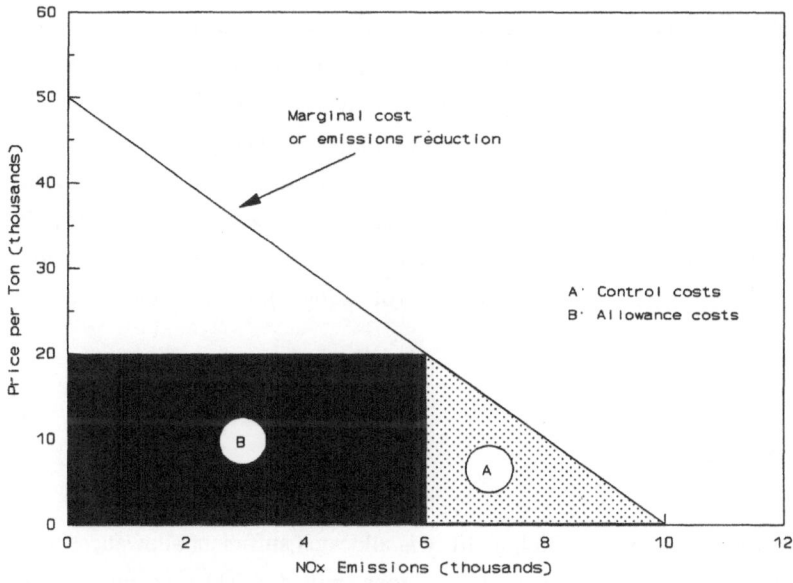

Assuming that PUC's give utilities the proper incentives to choose the least-cost strategy, the utility will control emissions that cost less than \$20,000 per ton. The utility in Figure 5 would reduce emissions from 10,000 per year to 6,000 per year, a reduction of 4,000 tons per year. The cost of controlling those 4,000 tons is shown as area A, equal to \$40 million. For

---

8   The figure of \$20,000 per ton is illustrative, based on assumed conditions in the Los Angeles Air Basin.

the 6,000 tons that remain as residual emissions, the utility would purchase allowances equal to the area B, or $120 million. Thus, the total "internalized" cost of NO_X emissions from that plant would be $160 million, the sum of cost-effective controls plus the cost of allowances for the 6,000 tons that remain.

4.3 The "Opportunity Cost" of Using Allowances

Allowances are typically distributed initially to market participants, rather than auctioned by the government. Under the acid rain trading program, for example, SO_2 allowances are distributed to utilities primarily on the basis of the historical production levels of their participating power plants. The same general strategy is likely to be followed for RECLAIM. That means that utilities will typically not pay for allowances to cover all of their emissions.

Even though some allowances are acquired at "zero cost," the utility should include the market price for *all* its residual emissions in determining the full cost of emissions for that plant. Why? Because the use of the allowances involves a very real economic cost. That cost is based upon the revenues that the utility forgoes by using the allowance to cover its own emissions rather than selling the allowance to another utility. If the utility adds a resource that emits more pollution, it must retain more allowances. The utility thus would have fewer allowances to sell to others, and a result would sacrifice revenues. These foregone revenues are the *opportunity costs* of using allowances the utility is initially allocated.

4.3.1 Allocated Allowances

Figure 6 illustrates the case in which a utility is allocated sufficient allowances to cover *part* of its residual emissions. Of the total of 6,000 tons emitted, 4,000 are covered by the utility's initial allocation and 2,000 are assumed to be purchased at the market price of $20,000 per ton. The "out-of-pocket" costs are therefore only $40 million for allowances, resulting in a total "out-of-pocket" cost of $80 million taking into account control costs. But the full cost includes the "opportunity cost" of using the 4,000 allowances. That opportunity cost—the money the utility could obtain if it sold the allowances—is equal to an additional $80 million. Thus, the total cost of emissions from the plant is equal to $160 million, the same figure as when no allowances are allocated initially.

Figure 6: The Social Cost of Emissions Does Not Depend on the Allocation of Allowances to the Utility

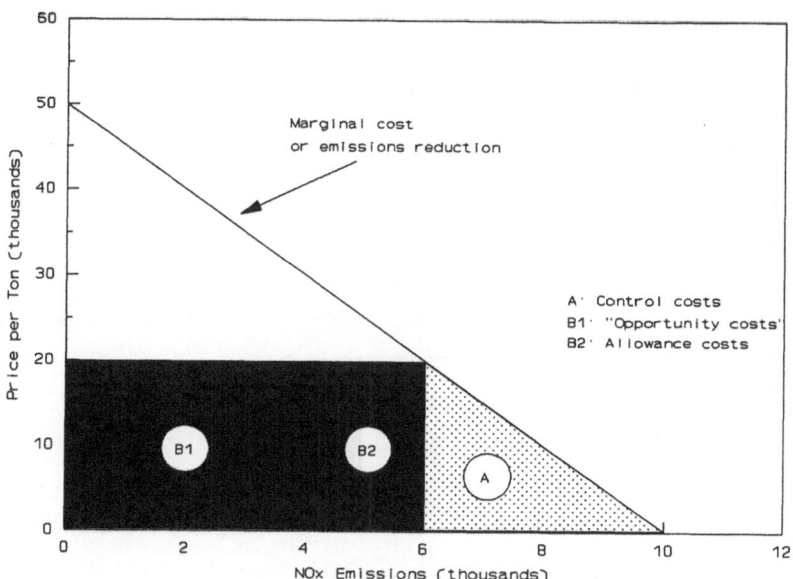

NOx Emissions (thousands)

A· Control costs
B1· "Opportunity costs"
B2· Allowance costs

Whether a utility purchases allowances or uses its allocated allowances does *not* affect the internalized cost of a plant's NO_X emissions. However, the initial allocation of NO_X emissions does affect the financial implications of the trading program; greater initial allocations mean that the utility (and its ratepayers) will bear smaller financial burdens.

4.3.2 Allocated Allowances Exceed Residual Emissions

The cases discussed above assume that the utility is a net purchaser of allowances. The RECLAIM trading market will of course have utilities that are net sellers as well, i.e., utilities whose initial allocation of allowances is greater than residual emissions.

Figure 7 shows the case in which the utility initially receives 8,000 tons of allowances. A profit-maximizing utility would still control emissions from the plant to the same level (6,000 tons) given the market price of $20,000 per ton. It would then sell the excess 2,000 tons of allowances at the market price of $20,000. Why? Because the cost of reducing emissions from 8,000 tons to 6,000 tons is less than $20,000 per ton, the utility gains profits from selling the excess allowances rather than using them to cover emissions from its plant.

Figure 7: The Allocation of Allowances Affects the Utility's Revenues but Not the Social Cost of Emissions

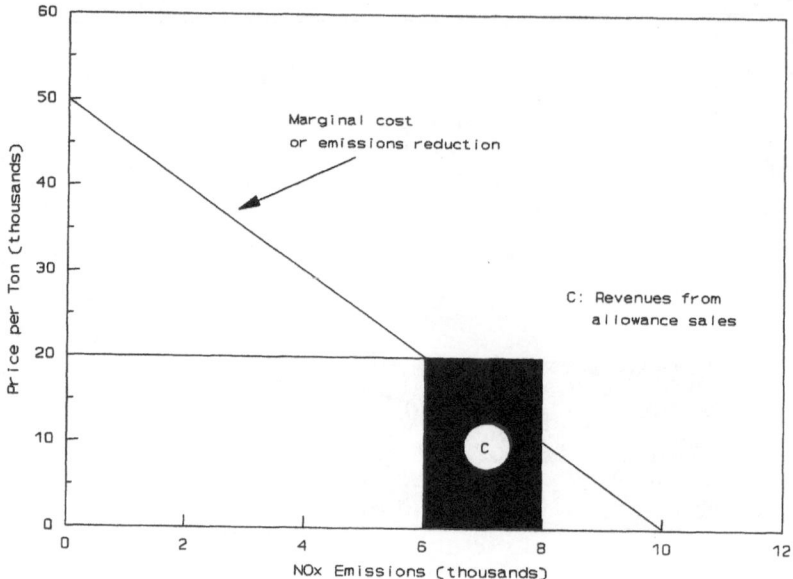

In the specific case shown in Figure 7, the net profit from selling the 2,000 excess allowances just equals the utility's cost of reducing emissions from 10,000 tons to 8,000 tons. That means that the trading program reduces the net cost of NO_x controls to zero. But although the *financial* impact of the plant's emissions is zero, the *social* cost of those emissions remains $160 million. The social cost is still equal to the control cost plus the opportunity cost of the allowances needed to cover residual emissions.

4.4 Other Complications

Beyond the need to include opportunity cost, there are several complications involved in translating the concept of emissions trading into practice.

4.4.1 Purchased Power

Many utilities purchase power from utilities in other jurisdictions. Should a utility add estimates of the cost of allowances to its estimates of purchased power? Such a calculation is implicit in many states' concerns for calculating residual values for out-of-state power.

The answer depends both on whether out-of-state utilities are covered by emissions trading programs and on the deference paid to residents of those other states with regard to air quality issues. If emissions are covered by a trading program, such as with SO_2 emissions under the national acid rain trading program, the price quoted for out-of-state power should include the cost of purchasing the necessary number of allowances to cover residual emissions. That means that, for example, the price a California utility pays for power from the Southwest should include the value of the SO_2 allowances needed in producing the power.

If no emissions trading is in place for the out-of-state emissions, the answer is less clear. One answer is to determine a separate set of "adders" for emissions from that state, on the theory that damages out-of-state are as important as those in state. There are several reasons, however, to exclude out-of-state emissions from the "adders" process (in addition to the conceptual and practical limitations when applied to in-state emissions). For one thing, it may be virtually impossible to identify which out-of-state resources supply the power, and it would thus be impossible to determine the nature and location of the relevant emissions. Moreover, state policies are typically made on the basis of effects on in-state residents, on the theory that other states can take care of their own problems. That would argue for leaving to other states the responsibility for adopting policies on the treatment of residual emissions from their own power plants.

4.4.2 External Costs Before Trading Programs Apply

Emissions trading programs are likely to have a transition period between when the program is mandated and when it actually takes effect. For example, an acid rain trading program was instituted in the 1990 Clean Air Act, but does not actually begin to apply until 1995 for some utilities and in 2000 for others. How should emissions before those dates be treated? The RECLAIM program will also have some transition period before the program is in force. What costs should be attributed to a planned unit's emissions before the trading program is in place?

The answer is that these emissions should be treated the same as those for which no trading program is in place. The external costs are equal to the damages the emissions cause. For example, external costs from SO_2 emissions in California before 2000 (when the acid rain trading program first applies) would be equal to the health and other damages from SO_2 directly and indirectly through contributions to ambient levels of small particulate matter.[9]

4.4.3 Types of Damages Internalized

The SO_2 trading program appears to have been motivated by concerns about the acid rain effects of the pollutant on rivers, lakes, forests and soils. A single national market was established because of the importance of the long-range transport of SO_2 to sensitive areas.

But SO_2 is also a localized pollutant, linked to health damages, visibility impairment and other effects. Does the trading program internalize these damages as well? There is no clear answer to this question, partly because there is no clear indication of what Congress intended when it set the national targets. But one way of putting an upper bound on the value would be to estimate the local damages from a ton of SO_2. The sum of the allowance price and these local damages would put an upper bound on the social cost for planning purposes.

It is important to remember, however, that the national cap means that any "adder" will shift SO_2 emissions around rather than reduce the national total. Because of the possibility of long-range transfer, some shifts may well lead to worse conditions for the very states that set "adders."

4.4.4 Optimality of the Trading Program

Environmental "adders" could in theory be useful complements to an emissions trading program. If the program did not set the optimum number of allowances or otherwise did not function properly, environmental adders could be used to correct the inadequacy.

However, if sufficient information is available to determine that the market is not optimal, it would seem more appropriate to attempt to improve the trading program rather than to add a

---

[9] See Harrison, Nichols, Evans and Zona (1992) for an example of the use of the damage assessment methodology to estimate the damages from emissions in air basins in California.

second (and inherently inferior) element. Utility "adders" would still suffer from the "piecemeal" problem even if they were "correct" for utility emissions.

4.4.5 Projections of Prices for Emissions Allowances

An emissions trading program means that residual emissions are no longer externalities. A program like RECLAIM leads utilities to internalize the costs of residual emissions in the private costs they calculate for alternative resources plans.

This does not, however, mean that internalization is automatic in resource planning. The utility must project the future prices of emissions allowances, just as it must project future energy prices and other input prices. In addition, like these other inputs, the PUC must provide the appropriate incentives for the utilities to make cost-effective choices given the projected prices and the opportunities they have to control emissions from the various resource alternatives.

5 Concluding Remarks

The rise of emissions trading requires that utility regulators rethink their approach to including environmental considerations in electric utility planning. Indeed, the rise of emissions trading can be seen as leading to a new phase in the evolution of PUCs' concerns for environmental concerns. In the first phase, in the 1970's and early 1980's, PUC's focused on insuring that utilities factor in the future costs of *required emission controls* in their resource planning. Thus, if Congress required scrubbers on new coal-fired plants, the PUC wanted to make sure that the utility included the initial and ongoing costs of that scrubber in their planning.

In a second phase, beginning in the late 1980's, PUC's began to concentrate on the residual emissions that remained after required control were in place. This has led to the proposals to include *"externality adders"* to reflect the damages from residual emissions. Although this approach appears consistent with proper pricing principles for environmental externalities, success has been limited both by the "piecemeal" problem inherent in only covering emissions from new electric utility facilities as well as by some major flaws in implementation. Because the conceptually-correct measure of damages has been seen as difficult to estimate, some states have relied upon other methodologies to estimate the "adders." A growing number of states have, however, recently turned to the correct damage-based approach to calculate "adders."

A third phase is evident in the 1990's as emissions trading programs expand in importance. These programs shift the PUC's efforts from predicting the dollar damages from utility emissions to predicting the market prices in emissions markets. Air pollution thus becomes like fuel and other inputs in constructing and operating electric utility resources: the utility has to project how much it will buy and at what price. Moreover, because utilities operate in the same market as other emitters, emissions trading does not have the "piecemeal" problem that the adder approach creates. It is significant that the California Energy Commission has recently endorsed the development of emissions trading programs as the most effective way to determine the value of residual emissions for purposes of resource planning.[10] This view is likely to become more prominent as more State agencies consider the issue.

In sum, emissions trading programs like that being developed in California promise to bring utility emissions into a truly coordinated least-cost framework. Encouraging the adoption of emissions trading programs would allow PUC's and utilities to contribute to environmental policies that are both coordinated and efficient.

References

California Energy Commission. 1992. *Draft 1992 Electricity Report*. Sacramento, CA: CEC, September.

Harrison, David Jr. Forthcoming. *The Distributive Effects of Economic Instruments for Environmental Policy*. Prepared for the U.S. Environmental Protection Agency and the Organisation for Economic Cooperation and Development. Paris: Organisation for Economic Cooperation and Development.

Harrison, David Jr. and Albert L. Nichols. 1992. *An Economic Analysis of the RECLAIM Trading Program for the South Coast Air Basin*. Prepared for the Regulatory Flexibility Group and the California Council for Environmental and Economic Balance. Cambridge, MA: National Economic Research Associates, Inc., March.

---

[10] Pending the development of emissions trading programs, the California Energy Commission recommended that residual emissions be valued using the damage function approach (California Energy Commission 1992, p. 1-6).

Harrison, David Jr. and Albert L. Nichols. 1990. *Market-Based Approaches to Reduce the Cost of Clean Air in California's South Coast Basin.* Final Report prepared for the California Council for Environmental and Economic Balance. Cambridge, MA: National Economic Research Associates, Inc., November.

Harrison, David Jr. and Albert L. Nichols. Harrison, David Jr. and Albert L. Nichols. 1989. *Preliminary Comments on Economic Assessment of the Health Benefits from Improvements in Air Quality in the South Coast Air Basin.* Prepared for the California Council for Environmental and Economic Balance. Cambridge, MA: National Economic Research Associates, Inc., August.

Harrison, David Jr., Albert L. Nichols, John S. Evans, and J. Douglas Zona. 1992. *Valuation of Air Pollution Damages.* Prepared for Southern California Edison Company. Cambridge, MA: National Economic Research Associates, Inc., March.

Joskow, Paul. 1992. "Weighing Environmental Externalities: Let's Do It Right!" *Electricity Journal,* May.

Nichols, Albert L. 1992. *Emission Trading Program for Stationary Sources of NO_X in Ontario.* Prepared for the Advisory Group on Emissions Trading, Toronto, Ontario, with assistance from Goodfellow Consultants, Inc. and VHB·HICKLING, October.

Nichols, Albert L. and David Harrison, Jr. 1991. *Market-Based Approaches to Managing Air Emissions in Alberta.* Prepared for Alberta Energy, Alberta Environment, and the Canadian Petroleum Association. Cambridge, MA: National Economic Research Associates, Inc., February.

Noll, Roger G. 1990. "Marketable Emissions Permits in Los Angeles." Stanford, CA: Stanford University, December.

Organisation for Economic Cooperation and Development. 1991. "Recommendation of the Council on the Use of Economic Instruments in Environmental Policy." Paris: OECD, January.

Regulatory Flexibility Group. 1991. *A Marketable Permits Program for the South Coast.* Los Angeles, CA: Regulatory Flexibility Group, December.

South Coast Air Quality Management District. 1992. *RECLAIM: Marketable Permits Program Summary Recommendations.* Los Angeles, CA: SCAQMD, Spring.

SUBJECT AREA 4:

**SOCIAL COSTS AND SUSTAINABLE
DEVELOPMENT**

25. From Social Costing to Sustainable Development: Beyond the Economic Paradigm[1]

Stephen Bernow, Bruce Biewald and Paul Raskin
Tellus Institute
89 Broad Street, Boston, MA 02110-3542
and Stockholm Environment Institute-Boston Center

The time has come, the walrus said,
to talk of many things
Lewis Carroll

Introduction

We are witnessing a major transition in public policy regarding economics, the environment and human well-being. Appropriately enough, it has come at the onset of a new century, indeed millennium. Despite the groundwork laid before us, and that being laid today, we expect that our counterparts and the world's citizens decades from now will recognize some but not all of the methods and approaches developed today, and hope they will look back with some kindness upon our modest beginnings.

In the current period we are undergoing a double movement -- the application of economic principles to environmental policy, and the insertion of ecological principles (and other extra-economic concerns) into economic affairs. The books by Daly and Cobb and Gro Harlem Brundtland are hallmarks of this important transition, which may be seen as the reintegration of economics and ecological science, and with moral philosophy.

Examples of these developments abound. States in the U.S. have begun to account for environmental externalities in energy choices; the World Bank and some European countries are exploring the modification of national income accounts to include environmental impacts and natural resource use (the degradation and depletion of stocks of natural capital), and nations and municipalities are implementing or considering pollution taxes. At the same time, national and international protocols for limits on CFCs, greenhouse gases, and acid rain precursors have been developed.

---

[1] Much of the discussion in this paper was prepared by the authors for the U.S. Office of Technology Assessment.

While we are confident that this transition will be successfully effected, it would be unwise, if we wish to help bring it about, to ignore its evolutionary character and the contention in which it is often embroiled (for example, the debates over externalities valuation, climate change, and forest policy in the U.S.). Recognizing the nascent state of both the professional and public policy discourses, on social costing and sustainable development, we hope that we can strive for both methodological clarity and practical pluralism.

From Energy Externalities to Social Costing

The quote from Lewis Carroll is more than mere whimsy for energy planning in the current context. Through today's perspective, the past two decades appear to be a tumultuous and almost surreal series of twists and turns for the energy industries generally, and for the electric sector in particular. The most obvious analogy to the story of Alice has been the in role of demand growth, with the electric utilities in the U.S. first eating the cookie of rapid expansion, then tasting the fruit of excess capacity, and, finally, reducing their exposure to risk by swallowing prior construction ambitions almost whole.

The similarities between the two stories cut even more deeply and broadly: from nuclear power "too cheap to meter", to electricity costs no one wants to bear, and radioactive waste with no place to go; from displacing oil with coal for fuel supply and cost security, to backing down coal for public health reasons; from natural gas shortages and usage restrictions, to its abundance and encouragement for new power supply; from low to high oil prices, and back again; from tall stacks as part of the solution, to tall stacks as part of the problem, as concern for afflicted lungs is complemented by concern for afflicted environments; finally, from fossil fuels as a resource for development, a heritage millennia in formation, to fossil fuels as the bane of sustainable development, while global climate change threatens the integrity of the biosphere in the next millennium.

At the center of these changes has been the way we use energy, and the relationship of energy use to human goals and the environmental conditions that underlie those goals. Until recently, growth in energy use was itself, with some good reason, equated or associated with economic and social development. However, it has long since been recognized that energy use is not an end in itself, but rather a means to satisfy certain human ends such as the provision of light, heat, food, motive power, built environments, and objects for human use and enjoyment. Traditional sources of energy in the industrial age are not deployed with impunity. As with other raw materials, land, potable water, clean air, species and habitats, energy resources are not inexhaustible. Moreover, these resources are distributed unevenly across the globe, and with increasing use they become increasingly difficult to extract and deliver. Economic, social and political consequences are thus intimately bound up with energy use. Finally, the extraction, processing, transport and conversion of the earth's energy resources -- whether from fossil fuels or river systems -- the construction of energy delivery systems, and the disposal of the waste products of these processes, cause often irreversible harm to human and natural environments.

Thus, energy use imposes a variety of costs on society that are not reflected in energy prices. Most notable among these costs are the damages to the environment, human health, and socio-economic systems that can result from significant impacts on air, water, soils, biota, habitats, and built environments. These occur on local, regional and global scales, and are often quite complex and difficult to predict. The release of pollutants and the use of local resources by energy-related activities can directly affect human health and amenity and the integrity of natural systems in the immediate vicinity of the facilities, thereby raising important siting issues. At the regional impact level, discussions over the last decade in North America and Europe have sought the best means of reducing the release of acid gases that could seriously damage habitats, agricultural production, and a variety of other human and ecological systems. At the global level, attention has been drawn to the accumulation of those greenhouse gases in the upper atmosphere that are released from anthropogenic sources, especially from energy and land use. This accumulation threatens global climate change, with potentially severe environmental, social and economic consequences by the middle of the next century. Thus, energy development decisions are taken under conditions of great uncertainty, with the risk of deleterious and potentially catastrophic outcomes.

This experience suggests that the purpose to which energy use has been put -- the creation of social wealth -- can be undermined by energy use itself. Not only are the sources and means of delivering energy now being called into question as the twenty-first century looms, but some of the ends served by energy use as well. The spatial configuration of society, the relationship of built environments to their natural surroundings, the organization of production, the fabrication of new substances, agriculture, water use, etc. will need to be re-examined from the perspective of ecological principles and the paradigm of sustainable development. Since so many of the world's peoples do not enjoy the living standards brought by industrialization and resource exploitation to so few, deep questions regarding equity, international stability and the preservation of democracy are posed.

Finally, since energy use is but one of several sources of scarce resource depletion and environmental assault on local, regional and global scales, the approach taken to energy issues in the coming period will have to be coordinated with the other essential aspects of sustainable development in order that the broad goals and values of humanity can be realized.[2] The reference to Lewis Carroll, then, is appropriate to the issue at hand. From out of the nineteenth century, whence arose the straightforward commitment to industrial growth, and the emergence of the U.S. and Germany as exemplars of this ethos, comes also an interest in paradox. As with Alice eating the cookie, more may be less, and less more, in a finite world that is incapable of absorbing all manner and magnitude of assault upon it inflicted by traditional economic development. This is indeed a time, when considering energy issues, to "talk of many things" -- of technologies and policies, of human values and natural processes, of local institutions and international protocols -- "of ships and sails and sealing wax, and cabbages and kings".

Externalities: The Economic Paradigm

In recent thinking about pollution reduction policies, the Marshallian supply-demand cross has been employed in a modified form, with pollution *reduction* plotted on the horizontal axis as a commodity for which there are supply and demand curves. The supply curve for pollution reduction is the marginal cost of abatement or control options arranged in order of increasing cost. The demand curve for pollution reduction is the "marginal damage cost" resulting from varying levels of pollution. This is illustrated in Figure 1.

Figure 1:   Economic "Textbook" Approach to Pollution Reduction

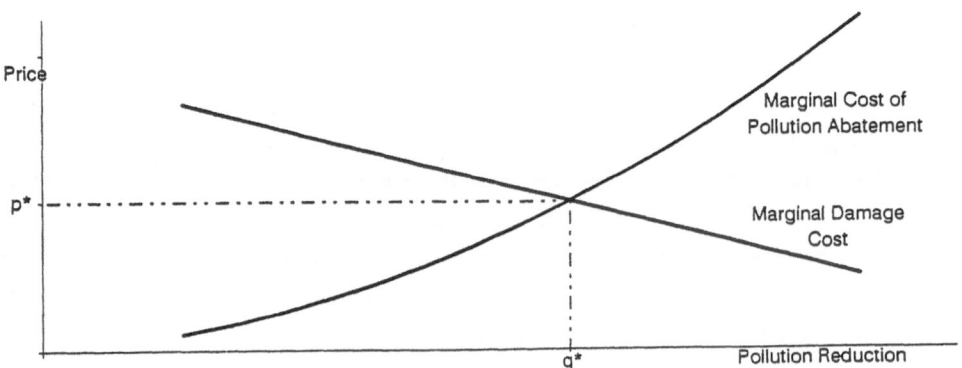

---

2   The concept of sustainable development, reflecting the interplay of ecological, economic, socio-political, and cultural values, has been crystallized in the "Brundtland Report," *Our Common Future* (World Commission on Environment and Development. Oxford University Press. 1987.)

Within this idealized framework, the "optimal" amount of pollution reduction, q^*, is determined as the point on the horizontal axis where the supply and demand curves cross. The standard textbook policy instrument in this situation is to implement a Pigouvian tax set at p^*. The market would then produce q^* of pollution, and the externality could be declared internalized. Other policy instruments are also available for the internalization of the externality, including "command and control" regulation embodied in technology and facility-specific requirements. Setting an aggregate cap on emissions (at q^*) with a permit trading system would have many of the economic efficiency advantages that would be attained with a tax (at p^*). Here, the point is not the mechanism used to address the pollution, but rather, that within this economic approach, the policy is *based upon* a notion of marginal costs and marginal benefits (reduced damages), and is aimed to hit the target point at the intersection of these curves.

There are many simplifications and assumptions embodied in this framework. For example, the graph considers just one pollutant whereas pollutants are usually both emitted and controlled in combination with others in production processes. Another factor is the scope of the regulation; when the regulation only applies to one jurisdiction then the regulation itself can have "external effects" upon other sectors of the economy and other geographic regions. Also, the construction of a single, monotonically increasing curve to represent the marginal cost of abatement is an oversimplification, given uncertainties about the availability, costs and effectiveness of pollution control options, and the pace and character of technological innovation. Markets for pollution reduction are generally "imperfect", owing to lack of information, high transaction costs and lack of competitive forces. Indeed, for many pollutants markets are non-existent.

Markets, however imperfect, can be created through political and regulatory processes. The construction of the marginal damage cost curve is a different sort of problem. It may be impossible as a practical matter in many cases to develop this curve adequately, given the inherent complexities in the ecological processes and the controversies surrounding monetary valuation. The shape of an idealized marginal damage curve may be reasonable for certain damages for small and continuous changes, but severely simplified in many other instances. Indeed, it may be inappropriate as a theoretical and political matter for some environmental impacts.

Valuation of Environmental Resources

If the theory described above is to be put into practice, then marginal damage curves (or at least some points on the marginal damage curves) must be estimated. This involves estimation of the physical impacts on ecosystems and humans, and valuation of these impacts (to the degree that they are not already internalized in direct costs) in monetary terms.

A variety of challenges render the idealized marginal damage cost curve unattainable in practice, if not in principle. Estimating the direct *physical* impacts or damages of environmental loadings (air emissions, water emissions, etc.) can be a large and exceedingly complex task, fraught with uncertainty. These impacts depend on: the *emissions* loadings into various media (air, water, soils); the *transport* of pollutants through those media, based on the physical characteristics of the emissions (e.g., velocity) and climatological (wind speed/direction) and topographical conditions; the *exposure* of receptor area or populations; and the *dose/response* relationship of those populations. Beyond the immediate environmental impacts of energy production and use, there can be a series of subsequent non-physical (e.g., socio-economic) as well as physical impacts that ensue in both the near and long term. The local, regional, and global consequences of air emissions, for example, depend upon atmospheric, biological, chemical, geophysical, ecological, and physiological relationships across time, space, and socio-economic and cultural conditions. These various pathways involve feedbacks and interactions with one another, over differing time scales, with both reversible and irreversible consequences. In modelling such relationships, elements of scientific uncertainty, availability of accurate data, the ecological response to perturbation (including non-linearities and feedback effects, among other factors) must be taken into account.[3]

A National Research Council report addressed the major sources of uncertainty in our understanding of ecosystem response to stress, and identified each as either rectifiable or inherent. For *rectifiable* sources of uncertainty, "it is possible to plan research that could considerably reduce the uncertainty." For *inherent* sources, "it is difficult even to conceive of a research program that could significantly reduce the associated uncertainties" (NRC 1980). The inherent sources are reproduced here as Table 1.

---

[3] For example: "There are strong links between the issues of greenhouse gas induced climatic change, stratospheric ozone depletion, ground level oxidant generation, forest and watershed degradation by acid rain and regional air pollution, and eutrophication of coastal and inland waters". *The Full Range of Responses to Anticipated Climate Change*. (United Nations Environment Programme and Beijer Institute. p. xii.)

Table 1:   Inherent Sources of Uncertainty in Predicting the Response of Ecosystems to
            Stress

Inherent Sources of Uncertainty in Predicting the Response of Ecosystems to Stress
1. Lack of data on, and understanding of, effects of long-term, low-level effluents on ecosystems
2. Difficulty in performing controlled, replicable experiments that provide *in situ* information about ecosystems
3. Lack of models allowing the use of measurable data to predict detailed ecological responses to stress

From Table 1-1 (NRC 1980, page 5)

The list of "rectifiable sources of uncertainty" for which "research is imaginable that could reduce the uncertainty to a considerable extent" is much longer, and is not reproduced here.

If the external environmental impacts can be reasonably estimated, various techniques can then be used to determine their monetary values to construct the marginal damage cost curve. Some of these damages are themselves market goods, for example cattle, fish, or commercial crops affected by industrial or urban pollution. The market value of these damages can be estimated *directly* by assessing lost revenues or remediation costs, although the estimation of the physical impacts, and the potential responses of economic activity and technological developments to those impacts, can be quite complex.

Typically, many of the damage costs are associated with non-market resources, such as natural ecosystems, human health and amenity. For these, *indirect* valuation techniques are available. These techniques involve either: 1) the examination of behavioral responses that are or might be influenced by an externality (e.g., hedonic pricing); 2) the assumption or creation of a fictitious market in order to elicit the value that individuals might assign to an externality (contingent valuation); or 3) an analysis of the implicit value placed on pollution abatement by society through the actions of its regulatory agencies (e.g. regulators' revealed preferences). Each of these has practical and/or inherent limitations that could circumscribe, albeit not entirely obviate, its usefulness. Their most important difficulties concern the lack of information, potential biases, and limitations of perspective that inhere in the actions or hypothetical actions of the individuals studied. These are, at best, indirect techniques to determine the *chimera* of "actual damages" -- the monetized values of environmental impacts, that is close to the heart of the economic approach to externalities. They are perhaps more accurately seen as instruments to move resource decisions in a direction that takes account of the environment and towards meeting concrete environmental goals.

The idealized economic theory of intersecting marginal cost curves, discussed earlier, generally rests on a severe lack of relevant information when applied to environmental policy decisions. Neither the emissions reduction supply curve nor the damage cost curve have been generally accurately established by regulators. While in practice the "control cost" curve is possible to construct with sufficient study, the damage cost curve will likely be more like a cloud of points at best, reflecting the inherent complexities, uncertainties, and variety of valuation perspectives that are entailed in such estimates. Indeed, for certain environmental resources, such valuation may be inappropriate.

Beyond the Economic Paradigm

> *Programming sticks upon the shoals*
> *Of incommensurable multiple goals*
> *And where the tops are no one knows*
> *When all our peaks become plateaus*
> *The top is anything we think*
> *When measuring makes the mountain shrink*
>
> *The upshot is, we cannot tailor*
> *Policy by a single scalar*
> *Unless we know the priceless price*
> *Of honor, justice, pride, and vice*
> *This means a crisis is arising*
> *For simple-minded maximizing.*[4]
>
> Kenneth Boulding

Consumers and Citizens

The role of individuals in microeconomics is typically represented by very abstract simplifications of the complex biological and social beings that we are. *Homo economicus* is a species of utility-maximizing individuals, deciding how to allocate limited disposable

---

[4] We are grateful to Harold Glasser of the Program in Ecology at the University of California-Davis for bringing this poem to our attention.

income to various combinations of commodities. This model emphasizes people acting as individuals and consumers. But we are also members of a society and inhabitants of ecosystems. As social beings, we participate in decisions on society's goals and values. As part of the fabric of life on the planet, we have an interest in maintaining and enhancing the quality of the environment and protecting it for future generations.

Contingent valuation techniques attempt to determine the value of environmental resources by posing questions to individuals, while indirect approaches assess behavior in marketplace decisions in which environmental conditions are often at best implicit. These approaches emphasize the individual as a consumer, and treat environmental protection as a commodity. This is not a substitute for a public discourse on society's values and goals. Individuals acting as citizens in such a process may express very different values than they do when acting as consumers.[5]

For example, individuals may make decisions as consumers that imply a time value of money of as much as 20 percent (e.g., credit card purchases) while the same individuals may support government policies that imply discount rates closer to 2 percent. Recent proposals for "green pricing" of electricity are an interesting example. This pricing approach would offer consumers an option to purchase "clean" power, generated by renewable resources, at a higher price. The logic of charging a premium price for a premium product is attractive, as is the emphasis upon "individual freedom to choose." However, in recognizing the individual as a consumer, this approach fails to recognize the individual as a citizen in society. It would be perfectly reasonable to decline "clean electricity" when offered a small portion as an individual consumer, but welcome a more costly approach in which all consumers of electricity participate in the "subsidization" of clean sources of electric power.

Societal goals and values may be expressed in myriad forms including pollution taxes, emission constraints, command and control regulation, and wilderness set asides. Discussion of these policies may itself imply a certain "willingness to pay" but this would be one constructed quite apart from markets and simulated markets.

The "tragedy of the commons" is the archetypal example of the existence of an externality, wherein individuals maximizing their own utility cause a sub-optimal or disastrous outcome for society. In this situation, the "invisible hand" of the market fails to produce an acceptable

---

[5]   A system based on targets rather than externalities may also open the door for lay people to participate in environmental decision making in a more effective and meaningful way. The average citizen has an important contribution to make in deciding what a "sustainable" environment means and in deciding what type of environment we want to leave for future generations.

result. Society, then, must look for remedies. An economic incentive system (e.g., a tax) may be used to optimize the use of the resource, perhaps based upon an ecological constraint (e.g., grazing not to exceed the rate at which grass can survive and reproduce effectively). Other "commons" may simply be set aside to be bequeathed to future generations. The conventional view of such regulation appears to be that it is generally undesirable, to be implemented and tolerated to the extent that corrections to the market system may be required. One might also look at examples of protection of ecosystems and natural endowments as the best of human behavior. The successful cooperation of individuals, whether voluntary or self-imposed through law and regulation, should be examined, acknowledged and celebrated as a "glory of the commons." In sum, we need to shift our focus from individuals as consumers, to individuals as citizens.

Monetization, Discounting, Intergenerational Equity and Sustainability

"Frank Ramsey, for instance, argued that it was ethically indefensible for society to discount future utilities. Individuals might do so, either because they lack imagination. . . or because they are all too conscious that life is short. In social decision-making, however, there is no excuse for treating generations unequally, and the time-horizon is, or should be, very long. In solemn conclave assembled, so to speak, we ought to act as if the social rate of time preference were zero. . ."
Robert Solow (1974)

Resource allocation decisions often involve tradeoffs over time. In certain cases, e.g., nuclear waste and global warming, the impacts of our decisions can span centuries. Today's CO_2 emissions will likely contribute to climate change for decades, while some of the ecological, economic and demographic impacts global warming may occur over centuries. The usual tool in cost-benefit analysis for combining economic effects spread over time into a single measure of total cost is the "discount rate."

Ordinarily, discount rates are used to express time preference, reflecting willingness to forego certain consumption now by investing to achieve expected consumption later, and the associated risk of that investment decision. Whether undertaken by the individual, the firm, or society, resource allocation must take this into account, as costs and benefits of different resource strategies will have different temporal patterns. Discounting costs enables the comparison between costs that occur now and costs that occur in the future by establishing an *indifference factor* (or equivalence) between consuming/paying now or paying/consuming in the future.[6] Ideally, by establishing this indifference factor, economically efficient resource allocation over time can be ensured.

There are many important considerations in the determination of a discount rate for any particular analysis (see, for example, Lind et. al. 1982), and some analysts have questioned the compatibility of discounting and the maintenance of ecological sustainability (Howarth 1991; Norgaard and Howarth 1992; Bernow and Marron 1991; and Bernow, Biewald and Wolcott 1992).

It is interesting to consider that within a context where only economic losses are considered, and with a discount rate that is higher than the rate of economic growth, the entire value of the planet (i.e., the world's economy) after any particular point is finite, since the annual discounted contribution can rapidly approach zero. Thus, within this economic framework, if one were faced with the policy decision between spending a small annual percentage of GNP in the near term in order to avoid major economic collapse a century or more from now, or not investing the money and suffering the long-term economic loss, one would be compelled to select the latter. This result is, *prima facie*, unacceptable. The limitations of discounting over very long time periods becomes even more problematic when human and ecological consequences are considered. The use of discounting applies best to relatively near-term cash flow analysis, and "marginal" impacts. It may be inappropriate to apply real discount rates to impacts on the time, impact, and geographic scales involved in global warming. Non-linearities in the damage function, including possible threshold effects, also raise problems for the discounting approach, even for more localized environments.

While discounting aims to ensure *efficiency* in resource allocation over time, the use of discounting for long time frames, over which large and irreversible impacts are possible, is problematic. Economic efficiency is not the only societal goal, and the objectives of environmental sustainability and intergenerational equity are distinct from that of economic efficiency.[7] To sustain its existence, society must take steps to guarantee that the welfare of its future generations is not compromised by the short-term decisions of its current generations. *Intergenerational equity* requires that we bequeath sufficient resources so that future generations are at least as well off as current generations. This is at the center of the

---

[6] Every consumer may have a different discount rate, expressing a different willingness to trade future expenditures and goods for current expenditures and goods.

[7] To first order, so too is intragenerational equity, whether amongst groups within nations or amongst nations. If efficiency is defined as Pareto optimality, then certain solutions are precluded which might increase overall welfare. Improved social and economic equity could enhance the prospects for economic efficiency as well as long-term sustainability. We do not address these issues further here.

emerging concept of *sustainable development*. In its weaker sense, it would require that a combination of environmental and productive conditions be maintained at some level or exceeded; in its stronger sense, it would require that environmental conditions and the conditions of production are separately sustained at or above minimum acceptable levels. Finally, either of these two senses of sustainability could be further strengthened by applying them to local as well as more global ecological and economic systems.

Fairness between generations is determined by comparing the welfare of each generation. The value of each generation's welfare is relative to its own time period; that is, a comparison is made between the value a person today has on her situation now and the value a person fifty years from now has on his situation then. It is *not* the value a person today has for the situation fifty years from now. Yet that is what is reflected in conventional discounting from today's perspective. For that reason, discounting is not sufficient, nor are variations in discount rates appropriate, to ensure that the intergenerational equity is reflected in resource allocation decisions.[8]

Indeed, it may be impossible to place an economic value on irreversible changes in certain environmental resources. Conventional economic analysis and discounting assume a reasonable degree of continuity in underlying conditions. The loss of a species, the elimination of a unique habitat, or the potential global and regional impacts of climate change, are discontinuous changes in the environment that are irrevocable, whether local or global. As Cline indicates with a quote from Mishan, we cannot know the value that future generations will place on their environmental resources:

> Whenever intergenerational comparisons are involved. . . it is as well to recognize that there is no satisfactory way of determining social worth at different points in time (Cline, 1992 p. 239).

Discount rate analysis lies strictly in the realm of economic efficiency. The use of a positive discount rate does not necessarily result in resource allocations that are equitable to future generations. Instead, the issues of equity must be decided separately from those of efficiency. Society today must consider the legacy of its resource decisions. To the extent that the resource allocation decisions made today include the possibility of irreversible changes in the environment now or in the future, society must decide whether such change is justifiable. When such change occurs, future generations will not have the option of recovering the resource in exchange for some measure of their own consumption. As Lind observes, the tools of economic efficiency inform the debate, but "other rules or judgement must be evoked

---

8   See, for example, (Dodds and Lesser, 1992, p. 112) and (Howarth 1991).

in deciding whether to incur the risks for the future by creating and storing nuclear waste or by not investing heavily in energy R&D" (Lind, 1986, p. 457).

Society may decide that certain decisions are unacceptable, owing to the consequences that affect future generations, regardless of the values that economic analysis might place on them. Dr. William Nordhaus acknowledges the limitations of cost-benefit analysis in such policy decisions:

> Cost-benefit analyses are a useful starting point for considering government policies, but they raise several issues that must be addressed before making policy recommendations. To begin with, many values cannot be incorporated in a quantitative cost-benefit analysis. For example, climate change may threaten a society's cultural heritage in ways that are not possible to evaluate in an economic framework but which are nonetheless unacceptable. While being unable to put a price tag on Venice, we might decide that it is unacceptable to take actions that threaten Venice's existence. There is not much economic science can say about this issue except to identify such trade-offs (Nordhaus, 1991a, p. 55).

Certain environmental resources would thus not be assigned monetary values within a market-like context for use in a conventional economic analysis. While most of our argument thus far has focussed on long term and/or global impacts, the same considerations could also apply to environmental resources that are more local either spatially or temporally. Instead of assigning monetary values to such "resources" as Venice, the James Bay ecosystem, the Baltic Sea, or particular forest and watershed ecosystems, and adjusting discount rates and costs in an effort to address intergenerational equity in an efficiency analysis, the alternative is to establish constraints or targets based on sustainability considerations, and proceed with the efficiency analysis of resource allocation within those boundaries.

Figure 2 shows how a sustainability target approach to environmental policy might be presented in the graphical form used earlier to present the microeconomic approach. Here, the fictional marginal damage cost curve has been eliminated; a target pollution level, q^*, is determined, based upon physical or ecological considerations; and a marginal cost of pollution reduction is implied by the target and the abatement cost curve.

Avoiding intergenerational inequity may strictly prohibit certain resource decisions, or require additional investments to minimize the risk of future impacts if certain decisions are allowed. Policies establishing targets could rely upon the use of market instruments such as taxes or tradeable permits as long as those instruments are effective at achieving the identified goals. The costs that are incurred by the instruments used to achieve a particular set of goals can then be examined, and used to inform the debate of whether society should continue to pursue those goals given their costs, or to identify alternative instruments that are more cost-effective. The proper means of combining economic and ecological sustainability considerations is to have a societal discourse on its values and goals, the uses and risks to which it subjects its human and environmental "resources", and its "willingness-to-pay" to preserve certain resources for future generations. Environmental and health risk, ecological sustainability, economic cost, and social equity have important places in such a discourse.

Figure 2:   Environmental Target Approach to Pollution Reduction

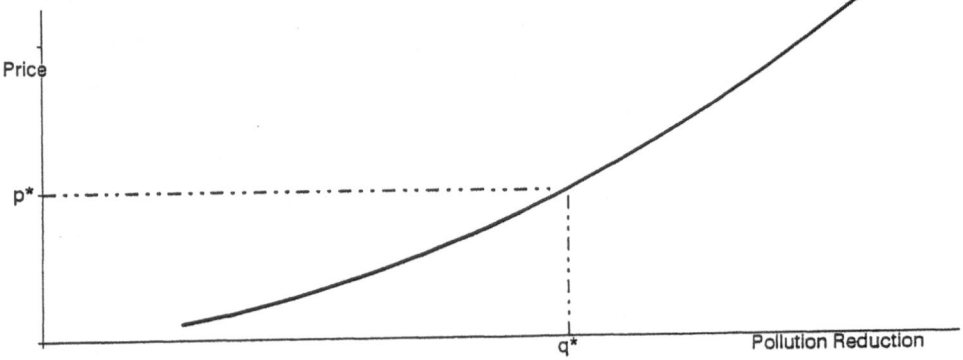

Acid Precipitation Policy in Europe

One example of the environmental standard or target approach is an analysis of continent-wide acid deposition abatement strategies for Europe that has been undertaken by the Stockholm Environment Institute Center at York, U.K. (Chadwick and Kuylensterna 1990).[9] Here, the concept of "critical load" is used for establishing environmental targets based on local area characteristics. Critical loads are the maximum deposition of a pollutant that will not cause significant adverse effects upon particular ecosystems. Taking account of emissions, transport, deposition, and area-specific responses and sensitivities to acid gas depositions, modelling exercises are being carried out to determine cost-effective abatement strategies for Europe. The transport modelling is performed at the level of 150 square kilometer grid sizes, while area sensitivities and critical loads were built up from information at a much finer grid size.

---

[9]   The Boston Center of SEI at Tellus Institute is participating in the modelling efforts on this project.

The abatement strategy analysis involves finding the least costly mix of options affecting the emitters of acid gases that will satisfy the critical load targets or, alternatively, the minimum overall impact based on relative sensitivities of receptor areas for a given level of expenditures. These techniques can be used to explore the use of environmental standards as a means to effect acid gas abatement. They could also be used to inform other policy instruments such as tax-based systems.

Greenhouse Warming Policy in the U.S.

The global warming problem is largely a matter of addressing large and uncertain impacts with incomplete knowledge. In such a context, developing policies to respond to the threat involves a decision that is analogous to deciding whether or not to purchase insurance. That is, is it worthwhile to spend a relatively well-defined sum in the near-term in order to avoid possible large damages (possibly with unknown probabilities of occurring) in the long-term? A deterministic cost-benefit analysis that reflects the environmental consequences of climate change in solely economic terms is not a sound basis for climate change policy.

In his paper "To Slow or Not To Slow . . ." (William Nordhaus, 1991b), Dr. Nordhaus is quite candid about the uncertainties in such a cost-benefits analysis that he pursued:

> We now move from the *terra infirma* of climate change to the *terra incognita* of the social and economic impacts of climate change. Studies of the impacts of climate change are in their infancy, and at this stage we can only hope to obtain an order-of-magnitude estimate of the impact of greenhouse warming upon the global economy. (page 930)

Later in the same paper Dr. Nordhaus points out that:

> A wide variety of non-marketed goods and services escape the net of the national income accounts and might affect the calculations. Among the areas of importance are human health, biological diversity, amenity values of everyday life and leisure, and environmental quality. (page 932)

Of course, under conditions of uncertainty and incomplete knowledge we might be tempted to wait for the global experiment to unfold. Even if the likelihood of significant climate change from existing and projected future carbon dioxide concentrations were small, staking the future of the planet and its peoples on a role of the dice would be unwise social policy.

The observations of Dr. Nordhaus quoted earlier, about the limits of cost-benefit analysis and the appropriate role of economics, suggest that for global warming, as for the case of Venice given by Dr. Nordhaus in this discussion, the appropriate role for economic analysis is limited to the emissions reduction side of the problem, and that the damage side might best be left for scientists, politicians and the public acting through political processes.

Economic analysis may be informative in estimating the costs of achieving specified reductions in carbon emissions. It may also have a role in developing the least-cost combinations of measures to achieve specified reductions. Economic analysis in its current state of development may simply fail, however, at any attempt to determine an "optimal" level of CO_2 emissions. Whether this deficiency can be overcome, or whether it is an inherent limitation in the applicability of economic methods, is a question that is beyond the scope of this paper.

Nevertheless, global warming policies must be evaluated and implemented. An approach that avoids the most problematic aspects of the cost-benefit approach is the setting of targets based upon long-term sustainability criteria. For example, the Stockholm Environment Institute has proposed that "an absolute temperature limit of 2.0 degrees C can be viewed as an upper limit beyond which the risks of grave damage to ecosystems, and of non-linear responses, are expected to increase rapidly" (SEI 1990).

Some believe that in order to achieve climate stabilization there would need to be a reduction in emissions of greenhouse gases of about 40 percent by the year 2010, and 70 percent by 2030, relative to today's levels (UCS 1991). If "targets" are set at these levels, then taxes, planning methods (e.g., externality adders), tradable permit systems, equipment efficiency standards, and other policies can be developed in order to achieve the targets. In the case of economic instruments (e.g., a carbon tax) the level of the tax would not be based upon an estimate of the "marginal global warming damages from carbon dioxide emissions." Rather, it would be an instrument to achieve stabilization, with its magnitude based upon estimate of the level of tax required in order that the goal be realized. With this approach, one might expect periodic adjustment of the tax depending upon the progress toward the goal, and upon the development of technologies to reduce carbon emissions.

Policy Basis and Policy Instrument

The use of environmental targets or monetary values for externalities should be viewed in two ways: as the basis for establishing policy, or as the instrument for implementing policy. As a policy basis, targets and monetary values can be seen to be theoretically equivalent within the economic paradigm. That is, if one believes that acceptable damage and cost curves can be established (as shown in Figure 1 earlier), the optimal level of pollution occurs where the marginal cost and marginal damage curves intersect; there, the appropriate pollution target is directly related to the optimum value (willingness to pay) at the intersection of the two curves. As policy instruments, the Pigouvian tax set at p* and the quantity constraint set at q* are theoretically equivalent if the quantity constraint is associated with a permit auction/ trading system.

However, notwithstanding such theoretical and analytical equivalence, there may be differences in practice owing to technological, institutional, economic and behavioral factors. Perhaps the most important difference is the policy response to uncertainty. If the pollution damages are potentially very high and irreversible, and if there appears be a threshold beyond which these damages grow precipitously, policy-makers may wish to adopt a cap to ensure that a disastrous, albeit uncertain, outcome is avoided. The risk that would be taken instead is of unexpectedly high costs to meet the standard. The pollution tax alternative, perhaps better suited to conditions of greater certainty and smoothness of the physical relationships, would ensure the cost but not the environmental outcome.

But environmental goals and pollution targets can be set *outside of the economic paradigm* and embodied in policies that override and constrain economic activity. Here, the policy basis is sustainability, with targets, critical loads, or safe minimum standards established, based on scientific and policy discourse, and not a damage cost curve, as represented in Figure 1. To be sure, the potential costs and economic impacts of implementing such targets would be of interest to policy-makers, but the point of departure is the sustainability target.

The instrument to implement such policy could be either taxes or targets with tradeable permits, to ensure economically efficient realization. With taxes, however, since institutional and behavioral responses may not ensure that the target is reached, it might be necessary to modify the tax level periodically. For this reason, and the arguments earlier in favor of targets/caps as the best policy response to uncertain and highly risky physical outcomes, targets (with trading) may be preferred for some environmental goals. Moreover, while market-based instruments such as pollution taxes or caps with tradeable permits are desirable to achieve economic efficiency, these might well be supplemented by more direct means such as restrictions on production, or use of certain substances (e.g., CFCs), or on releases of pollutants within certain areas.

Toward Sustainable Development

In the past, global models tended to predict ecological crisis and/or economic stagnation or devolution unless major social, political and/or technological changes were effected. Such models, limited by lack of data, incomplete knowledge, inability to capture all the important physical relationships, and difficulty in representing the interaction between the physical and social-institutional, might best be seen as heuristic and indicative, rather than predictive and prescriptive. The Polestar project of the Stockholm Environment Institute was developed to envision and represent alternative futures in which both environmental and development goals could be realized, and to make these complex issues accessible to policy-makers and the public.

The Polestar Project

Introduction

Economic development and environmental preservation often have been viewed as competing or incompatible values. Recent thinking on global futures reflects a revised interpretation of the interplay between environment and development which holds that the relationship is one of interacting and mutual objectives of "sustainable development". The Bruntland Commission elaborated this concept, suggesting further that sustainable development requires the eradication of extreme poverty, and a social process which promotes equity and fairness between and among the world's peoples.

The emerging sustainable development debate draws attention to the following broad issues:

- If world averages approach western industrial styles of production and consumption, material flows and environmental loadings would dramatically increase, perhaps by factors of 10 to 20. How can a world of over 10 billion people during the 21st century achieve development, sustainability, and equity?

- Sustainability requires that depletion rates of natural stocks and burdens on environmental systems remain below critical values, which if exceeded would significantly disadvantage future generations. What constraints does this put on the levels and types of commodity flows? What kinds of lifestyles are compatible with these constraints?

- The Bruntland Commission concludes that there are no inherent environmental limits to economic growth, putting its faith in the rapid deployment of new technology to elude the environment/development dilemma. What technological innovations to reduce and recycle material flows and to mitigate environmental damage can one prudently assume?

While we have considerable information on various elements of the environment and development nexus, we have no good answers to these basic questions. A major challenge today is to transcend the broadbrush character of the current debate in order to establish concrete guidelines for evaluating which modes of development are compatible with a sustainable, more equitable, higher population world, and which are not. New concepts, methods and tools are needed to practically address the complex social, economic, and ecological interactions that will underpin human development in the 21st century.

In late 1990, Stockholm Environment Institute (SEI) initiated the POLESTAR project to address this challenge. POLESTAR has three dimensions: *research* on global change, *policy assessment* to link long-term sustainability considerations to near-term decisions, and *education* to heighten public appreciation of the issues. The Boston Center of SEI at Tellus Institute is developing the computer modelling system and undertaking the quantitative analysis in which the data, research results, and policy scenarios are being organized, in collaboration with an international team of experts. POLESTAR project activities and outputs are oriented toward these three corresponding audiences -- the scientific community, decision-makers, and the general public.

Approach

At the most general level, it is useful to view the environment and development nexus as a *socio-ecological* system with three major interacting subsystems, as illustrated in Figure 3.[10] The societal system includes the structure of human populations, lifestyles, culture and social organization. The economic subsystem is the realm of production, itself a complex of interacting components. The environmental subsystem -- the biosphere -- is comprised of the great bio-physical process through air, water, land and biotic cycles. Numerous links and influences between these subsystems shape the socio-ecological pattern into a unified whole. Some of the main categories of interaction are shown in the figure.

---

[10] Systems concepts are developed in Shaw, R. et. al., *Sustainable Development: A Systems Approach*, IIASA, January, 1992.

Figure 3:   The Socio-Ecological System

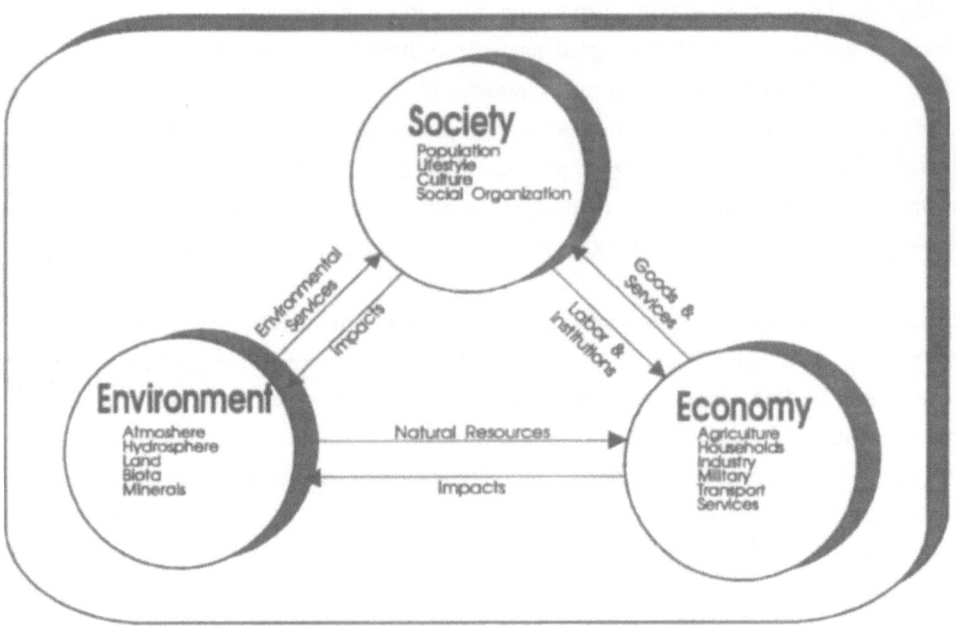

POLESTAR analyzes socio-ecological systems in terms of the major components of the three subsystems and their interactions. At any point in time, the state of the socio-ecological system can be represented as a set of material, economic, and environmental accounts for each of these components. The accounts vary both in space and in time. Analysis of global processes requires that regions which vary significantly in development, natural resources, and trade patterns be considered separately. POLESTAR views the global system as a set of mutually interacting regional socio-ecological systems as illustrated in Figure 4. Regions influence one another through causal pathways such as transboundary pollution, international markets, migration, and cultural and political influence.[11]

To represent alternative global futures, POLESTAR relies on *scenario analysis*, a powerful method for both analysis and pedagogy. Scenario analysis consists of combining the best available data, information, and simulation techniques to paint a picture of the current situation, and future "what if" possibilities. What if current trends persist and where are they leading us? What if alternative population and economic patterns emerge? What if various

---

[11] In principle, each global region can be further decomposed into a nested set of constituent systems at, e.g., national, district, community and even household scales.

policies, programs, and technologies are adopted? What actions are needed at the local, national and global levels? What are the environmental risks and uncertainties and what if the improbable should occur?

Figure 4:   Interregional Linkages

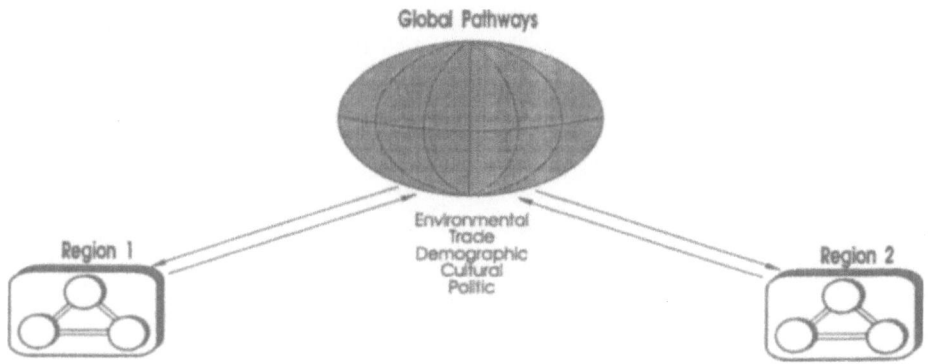

The scenario simulation process is depicted in Figure 5. The current accounts at the regional level (from Figure 4) provides the point of departure. A scenario is characterized by a set of assumptions which alter the current accounts and produce the scenario accounts for a future year.   The influence of assumed changes (e.g., in demographics, lifestyle, industrial technology, agricultural practices, resources, and energy patterns), produces the scenario accounts. Finally, scenarios can be evaluated and compared on environmental and human development criteria.   In this manner, scenario analysis can reveal the implications of alternative approaches to resource and economic development.   By permitting us to "experiment" with the future, scenario analysis provides a laboratory for rational policy formulation today.

The Computational System

A centerpiece of the POLESTAR project is a computer-based framework which provides a formal representation of the approach just described. An "appropriate technology" is required which allows adequate representation of current and future accounts under a wide range of assumptions in order to satisfy scientific objectives, while providing a structure and presentation which is both useful and enticing for a wide range of nonspecialists. A prototype system is now operational.

Figure 5:   POLESTAR Simulation Flow

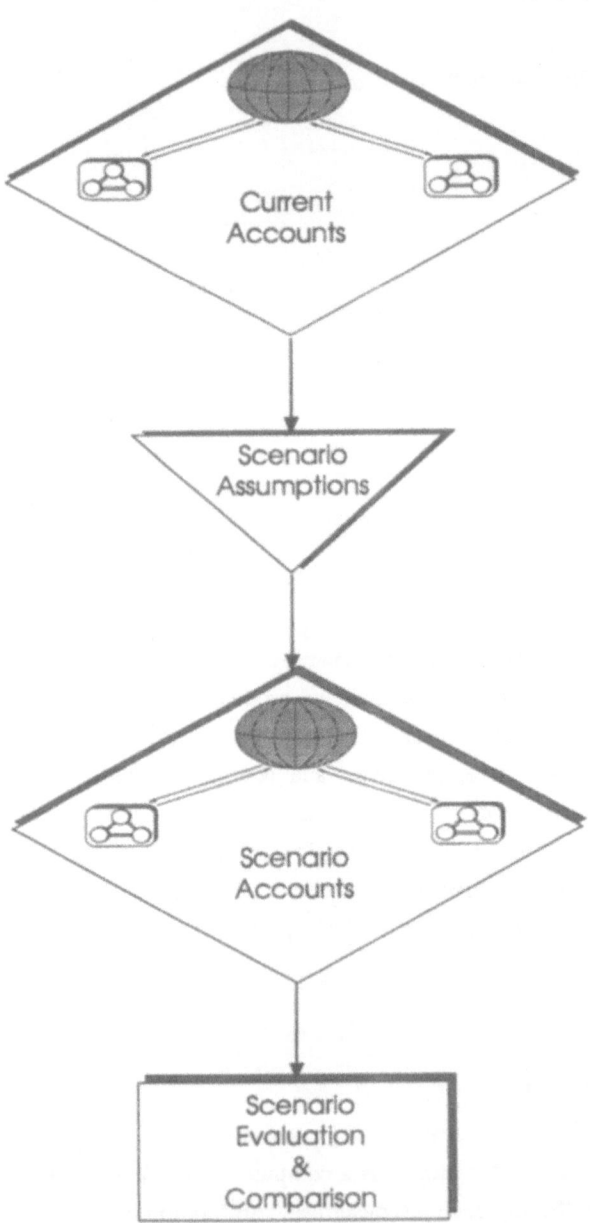

Specific objectives are to:

- Organize a vast amount of data pertaining to sustainability in a compact and accessible fashion.

- Quantitatively represent a broad range of conceivable development scenarios.

- Evaluate scenarios against sustainability indicators, including physical factors (e.g., rates of mineral stock depletion, land degradation, and loads on ecological systems) and development measures.

- Illuminate the issue of sustainable development for a broad audience.

Finally, POLESTAR can play a useful role in guiding the research agenda on sustainability. By integrating sectorally-organized (energy, water, land) and issue-oriented environmental research (global warming, acid rain, waste, deforestation) within a context of socially-defined visions of development, interactions can better be understood. The framework can help set priorities for sustainability research by identifying essential areas where data and knowledge must improve to sharpen our picture of the current socio-ecological condition and the consequences of alternative scenarios for the future.

The approach used in POLESTAR can be contrasted to the first generation of world models of the 1970s which attempted to *predict* the unfolding of global development in terms of fixed feedbacks and dynamics. The price was extreme complexity, opacity, and controversy. More modestly, POLESTAR offers a flexible framework for considering a wide range of perspectives. The microcomputer POLESTAR system builds on SEI's considerable experience in applying scenario-based computer analysis and data bases to resource and environmental policy analysis. Using the latest micro-computer programming techniques which allow for rapid upgrade of software, POLESTAR will be iteratively refined as data expands, knowledge improves, and policy issues evolve.

The Prototype POLESTAR Scenario System

SEI has designed and implemented a *prototype* of the POLESTAR system. The purpose of the exercise was to establish a detailed structure plan for the full POLESTAR system, to develop an initial data base, and to perform preliminary scenario exercises.

Table 2:   Prototype POLESTAR Structure

Module	Level of Detail (For 5 regions: Industrialized (OECD); East Europe and former Soviet republics; Asia; Africa; Latin America)
Demographics	• Total population • Age structure
Lifestyle	• Settlement pattern • Household size • Shelter floorspace • Built environment • GDP and expenditures pattern • Diet (12 categories) • End-use amenities ownership (8 amenities) • Personal travel pattern
Agriculture & Fisheries	• Crops (8 categories) and managed pasture on 7 land classes (rainfed and irrigated) • Livestock products (6 categories) • Inputs: water, fertilizers, and pesticides by crops and land classes; labor, energy and machinery for agriculture as a whole • Impacts: land use and deforestation; soil erosion; livestock and rice cultivation methane emission; CO_2 emission from disturbed soils; energy-related GHG emissions; water pollution; soil salinization and toxification
Households	• Inputs: water and energy by end-use amenities; • Impacts: energy-related emissions; water pollution; solid wastes generation; methane emissions from landfills
Transport	• 8 passenger and 5 freight types; • Energy use and related emissions by type
Industry & Services	• Interlinks of industry, services, and other economic macro sectors • 8 industrial sub-sectors • Reuse and recycling • Inputs: labor, water, energy, 6 metals, wood products • Impacts: air and water pollution; solid and hazardous waste
Forestry	• Waste fraction • Inputs: labor and energy • Impacts: deforestation; energy related emissions
Mining	• Inputs: labor, water, and energy • Impacts: resources depletion; water and air pollution
Energy System	• 15 fuels • Efficiencies and fuel mixes of electricity, heat, hydrogen generation, and refineries • Efficiencies and biomass requirement of charcoal and ethanol production • Distribution and transportation losses • Impacts: air and water pollution, solid and hazardous waste
Water System	• Distribution and transportation losses • Inputs: labor and energy • Impacts: flooding; energy related emissions
Trade	• Agriculture and fishery products • Fuels and electricity • Non-fuel minerals • Wood products • Industrial goods and services
Natural Resources	• Land (arable, pasture, forests, other) • Fossil fuels and other minerals • Fresh water

Figure 6:   Sample Computational Flow: Food and Agriculture

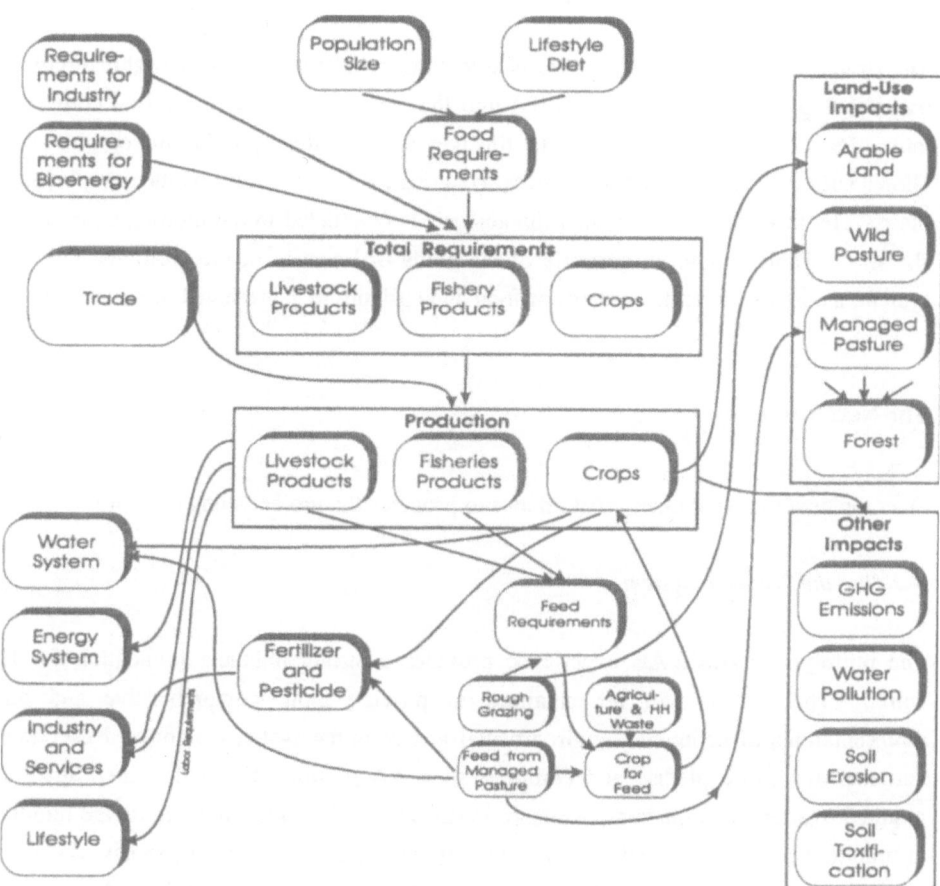

This "proof-of-concept" phase of the POLESTAR project is now complete and has included:

- discussions with internationally recognized global change analysts on concept and approach;

- design of prototype computational framework;

- microcomputer implementation of prototype system with user-interaction features (menus, graphics, prompts, etc.);

- collection of standard data sources and compilation of initial global data base; and

- design of simple development scenarios (e.g., global attainment of variations on current industrial country lifestyles and production patterns).

The structure of the prototype system is summarized in Table 2. For five global regions, the socio-ecological system is decomposed into the "modules" listed in the first column of the table. Data on these modules and the linkages between them provide the current accounts. Global scenarios are built out of separate scenarios for each module and the interplay between them. The level of detail at which the analysis is conducted in the prototype system is also shown on Table 2. As an example of the kinds of linkages in the prototype POLESTAR system, a computational flow chart for food and agriculture is presented in Figure 6.[12]

The Next Phase

Over the next two year phase, SEI intends to progress the project along several lines.

Building the Scenario System

The prototype POLESTAR experience provides a strong practical foundation for further work. The system will be enhanced to provide more comprehensive and detailed representations of economic, environmental and resource systems; improved user interface features; and more precise and complete data. A number of detailed investigations are required to strengthen the components of POLESTAR. Notable among these are building the information base on alternative agriculture strategies, on the material-process-product flows in industrial production and restructuring options, on macroeconomic and inter-industry linkages, on the interactions among environmental variables, and on land and resource accounts. SEI will collaborate with leading experts in researching these areas and linking results to the POLESTAR framework.

Workshops

SEI intends to actively involve an expanding spectrum of participants corresponding to its tripartite purpose -- research project, educational project, and policy project. During the next two-year phase of the project, SEI will organize three series of workshops: *expert workshops* in which technical analysts, scientists, and modelers would examine conceptual, data, and computer system issues; *vision workshops* in which social philosophers, ethicists, and artists

---

[12] A complete description will be found in a companion document: *The Prototype POLE-STAR Scenario System: Technical Description.*

would explore qualitative aspects of scenario alternatives; and *policy workshops* in which decision-makers, social scientist, and policy analysts would consider the implications of the long-term POLESTAR perspective for near-term policy. The workshops would both inform the evolution of the POLESTAR scenario framework and use that framework to shape discussions.

Dissemination

The computerized POLESTAR scenario framework and data base itself will be a major output of the project. The computerized system will be made available to international organizations, research institutes, and national governments. It will provide the basis for further enhancements, for example, the development of POLESTAR applications at the national level for sustainable development planning.

Selected papers from each of the workshops and an overview analysis would be issued as SEI reports and given wide distribution. The POLESTAR research results, per se, would be documented in both technical and popular articles. Based on the foregoing, a book will be prepared on issues, methods, and findings, and targeted toward an international audience of educated laymen. It would discuss the current crossroads of human development, appraise the status of environment and development today, introduce alternative development trajectories and their consequences, and advance directions for sustainable development.

The process envisioned for POLESTAR is an ongoing iteration between scenario analysis and dissemination activities, as illustrated in Figure 7.

The POLESTAR analytic framework, data base, and scenario analysis will develop along with the evolution of scientific understanding, data, and recipient needs. The concepts, perspectives and information generated by dissemination efforts will stimulate revisions of POLESTAR simulations. In turn, the POLESTAR system will provide an objective tool, source of information, and quantitative framework for the debates, instruction, and research that are so vitally needed in this period of global transition.

Figure 7:   POLESTAR: An Iterative Process

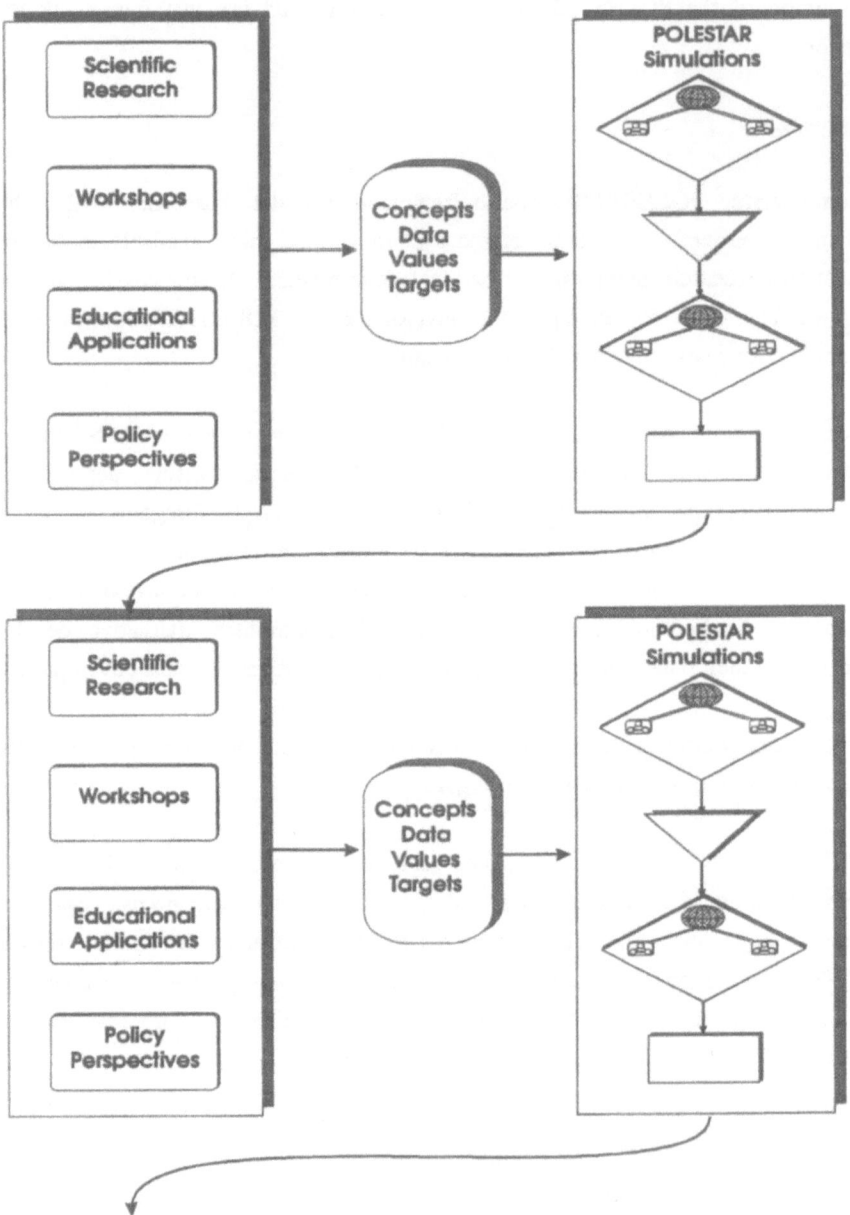

Conclusion

For the most part, the scale of human economic activity has been limited by the constraints of socio-political organization and the level of development of technology. To be sure, local depletion and degradation of the environment can occur, and has occurred, at levels of population and production far below today's, and in those contexts new social forms and technologies have often provided the basis for avoiding or superseding such crises. Today, however, the scale and range of human activity, and the role of technology, are more problematical. The condition of the biosphere continues to affect the opportunities for human activity generally and economic development in particular. But the condition of the biosphere has in turn been affected by such activity which today has reached a magnitude that threatens further development at local, regional, national and global levels. While in the past the amount of food harvested from the land and water bodies was limited by available tools, today it is increasingly being limited by the amount and condition of the resources themselves.

This is an underlying theme of sustainable development: how can we produce today without undermining the conditions for production in the future? Moreover, if the world of resources is being diminished rather than expanded by the deployment of modern industrial technology, a second theme emerges: how can the earth support closure of the gap between the rich and poor nations (and rich and poor generally)? It is noteworthy that in both of the foregoing questions, both economic and ecological conditions are regarded as essential to sustainable development. In its simplest, most emblematic formulation, sustainability entails economic development, environmental preservation, and equity (both intergenerational and international). The goals and policies for sustainable development will necessarily have elements appropriate to global, regional, national, and local contexts. Finally, the purely quantitative aspects of the problem -- embodied in sustainability indicators and targets, and economic activity -- will have to be supplemented by institutions that preserve, and indeed expand, democracy in the face of restrictions on use of the commons.

We are not obliged to complete the work,
but neither are we free to desist from it.

Book of Enoch

References

Ackerman, Frank. 1992. "The Natural Interest Rate of the Forest." Presented to the Second Conference of the International Society for Ecological Economics. Stockholm, Sweden. August 3-6.

Ahmad, Yusuf J., Salah El Serafy, and Ernst Lutz, eds. 1989. "Environmental Accounting for Sustainable Development." The World Bank. Washington, D.C.

Batie, Sandra S. and Shugart, Herman H. 1989. "The Biological Consequences of Climate Changes: An Ecological and Economic Assessment." Chapter 9 in *Greenhouse Warming: Abatement and Adaptation*. Norman Rosenberg, William Easterling, Pierre Crosson, and Joel Darmstadter, editors. Resources for the Future. Washington, D.C.

Bernow, Stephen, Bruce Biewald and David Wolcott. 1992. "Modelling Fuel Cycle and Site-Dependent Environmental Impacts in Electric Resource Planning." Paper presented at the OECD/IEA Expert Workshop on the Environmental Impacts of Energy Systems, Life-Cycle Analysis: Methods and Experience. Paris. May 18 and 19.

Bernow, Stephen and Donald Marron. 1991. "The Inclusion of Environmental Goals in Electric Resource Evaluation: A Case Study in Vermont," Chapter 13 in *Energy Efficiency and the Environment: Forging the Link*, Edward Vine, Drury Crawley and Paul Centolella, eds. American Council for an Energy Efficient Economy, in conjunction with the University-wide Energy Research Group, University of California, Berkeley.

Biewald, Bruce and Stephen Bernow, 1992. "Climate Change and the U.S. Electric Sector," presented at the DOE-NARUC Fourth National Conference on Integrated Resource Planning, Burlington, Vermont. September 13-16.

Chadwick, M.J. and Kuylenstierna, J.C. 1990. "The Relative Sensitivity of Ecosystems in Europue to Acidic Depositions." A preliminary assessment of the sensitivity of aquatic and terrestrial ecosystems. Stockholm Environment Institute at York, University of York, U.K.

Cline, William. 1992. "The Economics of Global Warming." Institute for International Economics, Washington, D.C. June.

Daly, Herman E. 1990. "Towards and Environmental Macroeconomics."

Dodds, D. and Lesser, J. "Appropriate Use of Numeric and Monetary Values for Environmental Impacts of Energy Resource Development and Use Decisions." Washington State Energy Office. Olympia, WA.

Howarth, Richard. 1991. "Economic Efficiency, Intergenerational Equity, and Uncertainty: An Application to Climate Change." Presented to the Peder Sather Symposium on Global Climate Change, Berkeley, California, October 16-18, 1991.

Howarth, Richard, and Richard Norgaard. 1992. "Intergenerational Transfers and the Social Discount Rate," presented to the Second Conference of the International Society for Ecological Economics. Stockholm, Sweden. August 3-6.

Howarth, Richard B. 1991. "Intergenerational Competitive Equilibria under Technological Uncertainty and an Exhaustible Resource Constraint." *Journal of Environmental Economics and Management.* 21, 225-243.

Johansson, Per-Olav. 1987. "The Economic Theory and Measurement of Environmental Benefits." Cambridge University Press.

Kuik, Onno and Harmen Verbruggen, eds. 1991. "In Search of Indicators of Sustainable Development." Kluwer Academic Publishers. Boston/Dordrecht/London.

Lind, Robert, et. al. 1982. "Discounting for Time and Risk in Energy Policy." Resources for the Future. Washington, D.C.

Mills, Evan, Deborah Wilson and Thomas B. Johansson, 1991. "Getting Started: No-regrets Strategies for Reducing Greenhouse Gas Emissions." *Energy Policy,* July/August.

Mitchell, R. and R. Carson. 1987. "How Far Along the Learning Curve is the Contingent Valuation Method?," (discussion paper QE87-07). Resources for the Future. Washington, D.C.

National Research Council. 1980. "Energy and the Fate of Ecosystems." The Report of the Ecosystems Impacts Resource Group, Risk and Impact Panel of the Committee on Nuclear and Alternative Energy Systems, National Academy Press.

National Academy of Sciences, National Academy of Engineering, and Institute of Medicine. 1991. "Policy Implications of Greenhouse Warming." Synthesis Panel, Committee on Science, Engineering, and Public Policy. National Academy Press, Washington, D.C.

Nordhaus, William. 1991b. "To Slow or Not to Slow: The Economics of the Greenhouse Effect." *The Economics Journal*, 101. July.

Nordhaus, William. 1991a. "Economic Approaches to Global Warming" in *Global Warming: Economic Policy Responses*, Rudiger Dornbush and James Poterba, editors. MIT Press.

Norgaard, Richard and Richard Howarth. 1991. "Sustainability and Discounting the Future" in *Ecological Economics: The Science and Management of Sustainability*, R. Costanza, ed. New York: Columbia University Press.

Organisation for Economic Co-operation and Development, 1989. "Environmental Policy Benefits: Monetary Valuation." OECD, Paris.

Ortolano, Leonard. 1984. "Environmental Planning and Decision Making." John Wiley & Sons, Inc.

Ottinger, Richard. 1990. "The True Cost of Electric Power," in *the Electricity Journal*. Vol. 3, No. 6. July.

Shllyakhter, ALexander I., and Daniel Kammen. 1992. "Estimating the Range of Uncertainty in Future Development From Trends in Physical Constants and Predictions of Global Change." Department of Physics and Northeast Regional Center for Global Environmental Change, Harvard University, Cambridge, Massachusetts.

Solow, Robert. 1974. "The Economics of Resources or the Resources of Economics." Richard T. Ely Lecture, American Economic Review 64, No. 2. May.

Stockholm Environment Institute. 1990. "Targets and Indicators of Climatic Change." Edited by F.R. Rijsberman and R.J. Swart.

Stockholm Environment Institute. 1991. "An Outline of the Stockholm Environment Institute's Coordinated Abatement Strategy Model (CASM)." SEI, Stockholm, Sweden. November.

UCS. 1991. "America's Energy Choices: Investing in a Strong Economy and a Clean Environment." Union of Concerned Scientists, Natural Resources Defense Council, American Council, American Council for an Energy-Efficient-Economy, and Alliance to Save Energy, in consultation with Tellus Institute. Published by Union of Concerned Scientists, Cambridge, Massachusetts.

United Nations Environment Programme and Beijer Institute. "The Full Range of Responses to Anticipated Climate Change." Nairobi. 1989.

26. Beyond External Costs - A Simple Way to Achieve a Sustainable Energy Future, International and Intergenerational Equity by a Straightforward Reinvestment Surcharge Regime

Olav Hohmeyer
Fraunhofer Institute for Systems and Innovation Research
Breslauer Str. 48, D-7500 Karlsruhe

1 Sustainability, Intergenerational Equity and Energy Markets

Beside the environmental and health damages caused by the use of conventional energy sources, which have been discussed under the heading of external or social costs at length during the conference so far, there seems to be another fundamental problem, which most likely is not dealt with adequately by the energy markets. It is the problem of long term scarcity of non renewable resources of the planet earth being exhausted at an ever increasing rate.

Presently these resources are valued mostly at the cost of digging them out of the ground and making them available for our present use. The fact that we are dealing with an inherited stock of resource capital is practically ignored. National accounting systems do not deduct any value for using up such natural resources in our GNP and capital stock accounting. Prevailing neoclassical economic theory is only looking for the optimal point of final depletion of these resources assuming that these are non essential resources (easy to substitute by any type of financial investment) and that enough financial resources will be set aside by the present resource owners to secure an ever growing future economy. Under these two assumptions it does not seem to be necessary to worry about any future problems of resource availability. Questions of international or intergenerational equity in energy resource availability do not enter into the neoclassical discussion of the allocation of non renewable energy resources.

To the author present knowledge seems to suggest that energy can not simply be substituted by other production inputs. To a very substantial extend energy is an essential production input which cannot be substituted. Due to the laws of thermodynamics in many applications non renewable energy resources can only be replaced by renewable energy sources like solar energy. Secondly the present income derived from diminishing our energy capital stock seems to be used mostly for increased consumption in the energy exporting and the industrialized countries. No mechanism appears to secure an adequate functional reinvestment of these incomes into a future energy capital stock based on technologies for the use of renewable energy sources.

It looks as if the basic assumptions are wrong on which our conviction is based that the energy markets will adequately handle long term scarcity problems. To the author it seems highly likely that we are fuelling very high levels of present consumption in the industrialized world and some oil exporting countries by simply wasting away our inherited resource capital. If this is true and the actual reinvestment of income from our present resource depletion is not adequate to secure a comparable energy resource stock for future generations, our use of energy resources will not be sustainable in the sense of the Bruntland Commission. As discussed by the author before (see Hohmeyer 1988, 1989, and 1992) intergenerational equity will not be secured as well.

If this is true and we agree on the idea that it is a duty of each generation to secure the essentials for the survival of the following generations (**sustainability**) and that all generations should generally be treated as equal (**intergenerational equity**) we need to devise a scheme which allows to secure the necessary reinvestment for keeping the energy resource stock functionally constant (see Pearce and Turner 1990, pp. 48ff). The following paper gives a short outline of one simple strategy to secure a constant functional energy resource stock indefinitely.

2   International Equity and Energy Markets

Once we agree that the different generations should have equal access to an energy resource stock over time, an other equity problem concerning the access to the worlds energy resource stock comes into view. This is the availability of energy resources to the different people of the world today. Due to the vast differences in buying power a few rich countries of the world can consume the major share of our energy resources at a very low price. Most countries of the world lagging behind in industrial development can afford only very little consumption of these cheap energy resources today. By the time these countries may have advanced further in their own development needing cheap energy to fuel industrialization, energy will probably be far more expensive due to the rather wasteful use by the rich countries of today.

As a result the poor countries will be caught in a poverty trap set up unconsciously by the rich. Because they are poor today, the necessary economic development to improve the situation of these countries in the future will be far more expensive than it has been for the presently rich nations. In line with the demand for intergenerational equity follows the demand for **international equity** in the availability of energy resources. The proposed strategy to secure sustainability and intergenerational equity can be expanded to take care of international equity in energy resource availability as well.

3 Securing a Sustainable Energy Future Through a Simple Reinvestment Strategy

The strategy developed is based on a special case of the so called 'max-min' approach (see Heal 1980, p. 51), which is geared towards securing even resource access for all generations. Based on a special case of 'Hartwick's-Rule', which shows the existence of a constant consumption path for total reinvestment of the resource rents (see Hartwick 1977 and 1978), it can be shown that the reinvestment of these rents into renewable energy resources as backstop technologies can secure even accesses to energy services derived from the energy resource stock for all generations (see Ströbele 1987, p. 34).

To secure a functionally constant energy resource stock it is essential to set aside the necessary funds to make up for each unit of non renewable energy consumed through future reinvestment into technologies for the use of renewable energy sources. The simplest type of such a reinvestment surcharge (s_{TT}) is the price of the unit of energy supplied by a renewable backstop technology at the point of exhaustion of the non renewable energy resource (T), the time for reinvestment. As it is not necessary to invest the surcharge at the time of collection, the funds can be invested in other options in the meantime. As the funds can earn interest (r) in between, the necessary sum for the reinvestment at point (T) can be discounted for this fact. We get a discounted reinvestment surcharge (w) to be collected today:

$$w_t = s_{TT} * (1 + r)^{-(T-t)} \qquad (1)$$

If (s_{TT}) covers the investment costs for setting up the new energy resource stock based on backstop technologies it secures the functional continuity of the energy resource stock. For each unit of non renewable energy taken from the resource stock a new unit of the resource stock is created by investment into the backstop technology. The requirement for sustainability is met.

However, the requirements for intergenerational equity are not necessarily met by this simple reinvestment regime. If there are substantial mining or drilling costs as well as reprocessing and transportation costs for using the present non renewable energy sources, the generation using these energy resources at (t) are put at a disadvantage by the amount that these costs $(m_t + c_t)$ exceed the non investment costs for utilizing the backstop technology (e_T). Intergenerational equity demands that this cost difference is included in our calculation as well, we get:

$$w_t = (s_{IT} + e_T - (m_t + c_t)) * (1 + r)^{-(T-t)} \qquad (2)$$

If we combine the investment costs (s_{IT}) and the non investment costs (e_T) of the backstop technology into the total backstop costs (s_T) we can write:

$$w_t = (s_T - (m_t + c_t)) * (1 + r)^{-(T-t)}. \qquad (3)$$

As non of the elements of this calculation should contain any profits, which are appropriated by each generation itself, this reinvestment surcharge secures intergenerational equity by putting the same financial burdens for each unit of energy on every generation.

Welsch (1989) has shown that external costs of the use of non renewable energy sources (q) need to be taken into account to that extent to which these external costs are already being internalized. The final reinvestment surcharge takes the form of:

$$w_t = (s_T - (m_t + c_t + q_t)) * (1 + r)^{-(T-t)}. \qquad (4)$$

If we are able to estimate the magnitude of this reinvestment surcharge, collect it and secure its final use for the investment into the backstop technology, we can achieve a sustainable energy future as well as intergenerational justice.

The approach outlined above is based on the global availability and use of non renewable energy sources. The point of reinvestment is determined by the global exhaustion of these resources. The time span until (T) can be calculated by dividing the estimated non renewable resource stock through the present annual global consumption of the resource. Two modifications seem to be necessary to make this a meaningful approach. First, it does not seem to be likely that rather scarce fuels like mineral oil or gas can be substituted by the far more abundant coal in the future due to restrictions in the permissible amount of global CO_2 emissions. Thus, each type of fossil energy will have to be substituted by non CO_2 emitting renewable backstop technologies. Due to this restriction for the more scarce fuels (T) moves much towards the present. Second, we are faced with fundamental international disparities in the use on non renewable energy sources today. If we want to take care of international equity in the use of non renewable energy sources, we need to use national annual per capita consumption rates to calculate the point of reinvestment (T). This leads to far lower reinvestment surcharges for the countries with low energy consumption and to substantially higher surcharges for the energy guzzling industrialized countries, which need to switch to renewable energy sources as soon as they have used up their fair share of the global non renewable energy resources. In the following empirical estimation of the reinvestment surcharges these calcualtions will first be based on the average global per capita energy consumption and then be recalculated on the basis of some prominent examples of national per capita energy consumptions.

4 Empirical Estimation of the Suggested Reinvestment Surcharge

For a first empirical estimation of the reinvestment surcharge data for the following parameters are necessary:

- costs of a probable backstop technology at the time of reinvestment

- mining, drilling, reprocessing and transportation costs of non renewable energy sources of the base year

- estimated world resources as well as global and national per capita consumption rates of non renewable energy resources in the base year

- interest rates of long term safe financial investment for the interim use of the reinvestment funds

- development of technical change with respect to the rational use of energy and

- size and development of the population in the countries investigated.

The United Nations assume that the world population will grow to somewhere between ten and fifteen Billion people by the year 2100 (see Hohmeyer 1989, p. 123ff). This would mean a very substantial shortening of the remaining time until reinvestment in backstop technologies will become necessary. On the other hand we know that substantial improvements in energy efficiency can be achieved if energy will become more expensive. According to Jochem and Schäfer (1991) it is possible to reduce the specific energy use of the Federal Republic of Germany by as much as 30-70% as compared to the 1990 situation in a country already using rather high efficiency equipment. Neither of the two developments can be well foreseen over the next one hundred years. As both are more or less counterbalancing each other neither effect will be included in the following calculations.

4.1 Costs of a Probable Backstop Technology

For all following calculations photovoltaics is used as the backstop technology which can potentially cover all global energy needs. The photovoltaic generation of electricity is used in combination with a partial conversion to hydrogen, substantial hydrogen storage capacity, and partial reconversion of some hydrogen back to electricity by fuel cells. Table 1 gives the cost figures derived for the different types of backstop energy supplied on this basis.

Table 1: Backstop costs for electricity directly supplied by photovolatics, hydrogen produced from it, and electricity supplied by reconversion of such hydrogen to electricity

Year	PV invest. costs $/kWp	PV efficiency %	PV costs $/kWh	H2 invest. costs $/kWp	H2 O & M costs $/kWp	H2 efficiency electro- lysis %	H2 from PV total costs $/kWh	Electricity from fuel cells PV and H2 total costs $/kWh
1982	20269	8	2.2	647	46	64	3.43	6.23
1990	11321	9.6	1.0	611	46	66	1.58	2.88
2000	6072	13.5	0.5	574	46	68	0.83	1.50
2010	3829	14.4	0.3	537	46	70	0.53	0.96
2020	2635	15.3	0.2	469	41	73	0.36	0.65
2050	1922	15.3	0.2	400	36	76	0.26	0.48
2065	1667	15.3	0.1	400	36	76	0.24	0.43
2080	1543	15.3	0.1	400	36	76	0.22	0.40

Assumptions:
- all prices 1988
- solar radioation 1100 kwh/m2 per year
- real intrest rate 5%/a
- duration of use of installations:
 -- PV 20 years
 -- electrolysis 30 years
- annuity: PV 8.024 real, electrolysis 6.505% real
- O & M costs PV 1.5% of investment costs per year
- no costs for fuel cells assumed
- fuel cell efficiency 55%
- prices originally calculated in DM, conversion factor used 1.5 DM/$

Sources: Hohmeyer 1989, Pálz and Schmid 1990, Prognos and ISI 1991

4.2   Present Market Prices of Non Renewable Energy

The reinvestment surcharge demands that only the exploration, exploitation, reprocessing and transportation costs of the present use of non renewable energy sources are included in the calculations. As it is not possible to calculate the profit margins incorporated in the market prices, we will use total market prices as approximations of the cost elements which should be included. This leads to overestimating these cost elements and low estimates for the necessary reinvestment surcharges putting the present generation at an advantage over future generations. Table 2 shows the costs for different non renewable fuels in the FR Germany in 1988. A price of 70 $/toe for imported hard coal shows that even very dramatic reductions in backstop costs will not make energy available to future generations at costs anywhere near to such price level. By the year 2050 hydrogen costs of about 2 500 $/toe are expected even after substantial cost reductions.

4.3   Estimated Present Resources of Non Renewable Energy

To calculate the remaining time until a switch to backstop technologies will be necessary the estimated world resources are of non renewable energy are used. This contrasts with many other approaches where only reasonably assured reserves are used for such calculations. The estimated resources for oil and gas are more than double the assured reserves, while in the case of coal the resources are larger by factor five. Table 3 shows the estimated resources in billion tons of oil equivalent (1 000 Mtoe) and the portion of the overall resources used up by 1988. Taking the global annual depletion rates of the different fuels of 1988 the remaining time spans until the exhaustion of these resources are calculated.

4.4   National Differences in Energy Consumption

As pointed out before there are great differences in the international per capita consumption of energy. Table 4 shows that an average US citizen used about 7.319 toe of energy while a citizen of Tanzania used only 0.026 toe in 1988. Thus, the US citizen reduced the worlds non renewable energy reserves by more than the energy consumption of 282 citizens of Tanzania in the same year. A German citizen used as much energy as almost 162 people in Tanzania in 1988. We see that the different nations contributed very disparate amounts to the depletion of our global non renewable energy resource stocks. Table 4 shows the national per capita consumption rates and the resulting national time spans until the resources would be exhausted if the whole world would have the per capita energy use of the country listed. Figure 1 exhibits the vast differences in national per capita energy consumptions.

Table 2:   Energy costs for different types of non renewable energy in the Federal Republic of Germany in 1988 (conversion rate used is 1.5 DM/$)

Fuel prices		
Crude oil import price	$/toe	135.6
Heavy fuel oil excl. tax	$/toe	124.5
Light fuel oil excl. tax	$/toe	216.5
Automotive diesel oil	$/toe	597.3
Automotive premium	$/toe	845.3
Natural gas import price	$/toe	103.8
Natural gas industy price excl. tax	$/toe	184.0
Imported coal import price	$/toe	70.2
Domestic steam coal excl. tax	$/toe	273.6
Lignite excl. tax	$/toe	210.4
Internal electricity costs		
Based on importe hard coal	$/kWh	0.07
Based on domestic hard coal	$/kWh	0.11
Based on domestic lignite	$/kWh	0.07
Based on uranium 235	$/kWh	0.07
Based on heavy fuel oil	$/kWh	0.15
Based on natural gas	$/kWh	0.22

Assumptions:
- all quoted prices FRG 1988
- all prices without tax
- lignite only household price available
- annual hours of operation for power plants:
  — coal and nuclear 5000 h/a (intermediate load)
  — natural gas 1680 h/a (peak load)
  — heavy fuel oil 435 h/a (peak load)
- prices originally given in DM, conversion factor used 1.5 DM/$

Sources: IEA 1992, Brand 1986 and Schmitt 1988

Table 3:   Estimated resources of different types of non renewable energy, depletion until
            1988, and remaining time until the exhaustion of the resources, if future
            consumption remains at the level of 1988

	Crude oil 1000 Mtoe	Natural gas 1000 Mtoe	Hard coal 1000 Mtoe	Lignite 1000 Mtoe	Uranium 1000 Mtoe
Estimated max. resources	230.2	198.7	8087.5	1787.6	319.2
Consumption until 1988	80.6	30.8	60.1	5.9	44.3
Share already consumed in %	35.0%	15.5%	0.7%	0.3%	13.9%
Remaining resources by the end of 1988	218.8	193.1	8078.0	1786.6	312.2
Static resource life remaining 1988 (in years)	73.4	130.4	2286.5	1438.5	177.7

Sources : Statistische Jahrbücher 1990 and 1991, Institut de'Economie 1989
          OECD/NEA according to atw different volumes

Table 4: National consumption of non renewable energy resources and the time spans until resource exhaustion based on national per capita consumptions

	Liquid fuels Mtoe	Natural gas Mtoe	Solid fuels Mtoe	Uranium Mtoe
Federal Republic of Germany				
Consumption in 1988	104,420	44,723	74,521	37,181
Population in Mill.	61,990	61,990	61,990	61,990
Toe/capita per year	1,684	0,721	1,202	0,600
Extrapolated world consumption	8611,019	3688,078	6145,335	3066,115
Remaining static resource life	25,0	51,5	1578,5	100,1
France				
Consumption in 1988	77,638	26,382	14,960	70,605
Population in Mill.	55,846	55,846	55,846	55,846
Toe/capita per year	1,390	0,472	0,268	1,264
Extrapolated world consumption	7106,821	2414,968	1369,373	6463,024
Remaining static resource life	30,3	78,6	7083,9	47,5
USA				
Consumption in 1988	755,166	435,102	469,537	143,148
Population in Mill.	246,329	246,329	246,329	246,329
Toe/capita per year	3,066	1,766	1,906	0,581
Extrapolated world consumption	15671,750	9029,552	9744,174	2970,706
Remaining static resource life	13,7	21,0	995,5	103,4
Japan				
Consumption in 1988	191,278	41,794	79,266	45,785
Population in Mill.	123,098	123,098	123,098	123,098
Toe/capita per year	1,554	0,340	0,644	0,372
Extrapolated world consumption	7943,363	1735,596	3291,745	1901,364
Remaining static resource life	27,1	109,4	2946,9	161,5
Former USSR				
Consumption in 1988	388,205	540,411	399,174	41,515
Population in Mill.	283,682	283,682	283,682	283,682
Toe/capita per year	1,368	1,905	1,407	0,146
Extrapolated world consumption	6995,528	9738,293	7193,179	748,117
Remaining static resource life	30,8	19,5	1348,6	410,4
Brasil				
Consumption in 1988	47,187	2,888	10,053	0,141
Population in Mill.	144,428	144,428	144,428	144,428
Toe/capita per year	0,327	0,020	0,070	0,001
Extrapolated world consumption	1670,174	102,227	355,813	4,980
Remaining static resource life	128,8	1857,1	27263,0	61656,5
China				
Consumption in 1988	85,359	13,279	464,122	0,00
Population in Mill.	1084,310	1084,310	1084,310	1084,310
Toe/capita per year	0,079	0,012	0,428	0,00
Extrapolated world consumption	402,425	62,604	2188,111	0,00
Remaining static resource life	534,6	3032,5	4433,3	

Table 4 continued:

	Liquid fuels Mtoe	Natural gas Mtoe	Solid fuels Mtoe	Uranium Mtoe
India				
Consumption in 1988	41,302	6,215	112,207	0,832
Population in Mill.	796,596	796,596	796,596	796,596
Toe/capita per year	0,052	0,008	0,141	0,001
Extrapolated world consumption	265,048	39,881	720,063	5,337
Remaining static resource life	811,7	4760,3	13471,8	57536,8
Nigeria				
Consumption in 1988	9,996	4,028	0,069	0,00
Population in Mill.	104,957	104,957	104,957	104,957
Toe/capita per year	0,095	0,038	0,001	0,00
Extrapolated world consumption	486,862	196,177	3,341	0,00
Remaining static resource life	441,9	967,7	2903300,2	
Tansania				
Consumption in 1988	0,609	0,00	0,030	0,00
Population in Mill.	23,997	23,997	23,997	23,997
Toe/capita per year	0,025	0,00	0,001	0,00
Extrapolated world consumption	129,669	0,00	6,284	0,00
Remaining static resource life	1659,1		1543575,3	

Note: Uranium calculated at 47960 toe/t U (557,8 GWh el.output)

Sources:Statistische Jahrbücher 1990 and 1991 , IEA 1991 and 1989, Table 3

4.5 Interest Rate for Long Term Financial Investments

The interest rate for discounting the future value of the reinvestment surcharge at the point of reinvestment is derived from the real interest rate earned on public bonds in the Federal Republic of Germany on average in the last thirty years. This real interest rate was about 4.1% per annum.

Figure 1:   National per capita energy consumption of fossil fuels in 1988 (source VIK 1991)

Consumption of fossil fuels
per capita in 1988

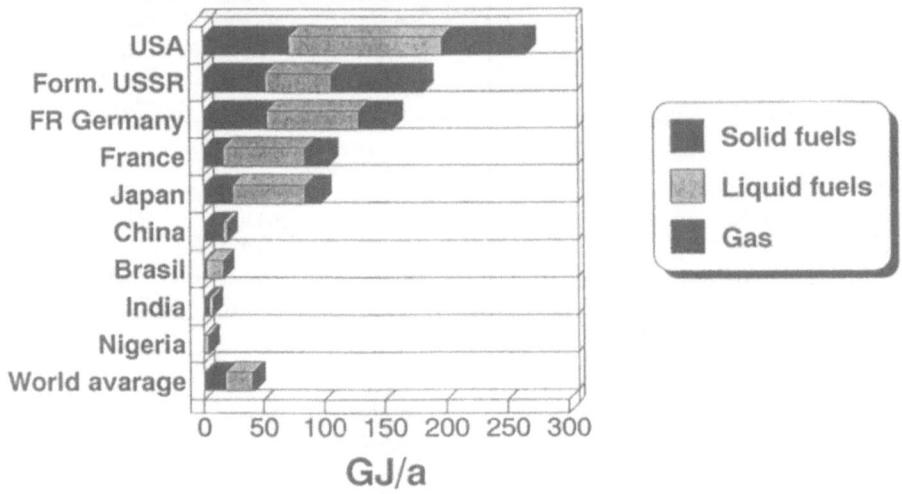

4.6   External Costs Included in the Calculations

As long as external costs of the use of non renewable energy sources are accounted for separately as we have discussed during the conference, these external costs have to be considered in calculating the reinvestment surcharges. Based on the estimated external costs for environmental and health damages estimated by the author before (see Hohmeyer 1989 or 1992) these cost elements are accounted for in the following calculations. Table 5 shows the low and high estimates used for the external costs of the different non renewable energy sources.

4.7   Global Reinvestment Surcharges

Based on the data and assumptions outlined so far global reinvestment surcharges (w_t) are calculated according to function (4). Using the global per capita consumptions of 1988 for the different fuels Table 6 shows all parameters used and the calculated value of the reinvestment surcharges for primary energy and for electricity. Furthermore the table gives the FRG market prices including the reinvestment surcharge and the external costs taken into account. Finally it shows the shares of the external costs and of the reinvestment surcharges of the total costs of energy. Global reinvestment surcharges only have an impact on the total costs of oil and gas and the derived electricity. The impact on uranium based electricity is minimal, while coal is still quite abundant and the surcharge is practically zero. If the problem of CO_2 induced global warming would not exist the coal resources could well substitute oil and gas for hundreds of years in the future. Due to the extremely high CO_2 emissions of coal such substitution will not be feasible. The necessary reinvestment surcharges change dramatically if national per capita consumption is considered.

Table 5:   External costs taken into account in the calculation of reinvestment surcharges for non renewable energy sources

Type of non renewable energy	Total damage costs 1982 Mill.$	Primary energy consumption 1982 PJ/a	Specific damage costs 1982 $/toe	Generated electricity from fossil fuels 1982 TWh	Specific damage costs 1982 c/kWh
Low estimate					
Natural gas	13.1	389.5	1.41	45.8	0.03
Mineral oil	29.8	134.5	9.29	17.3	0.17
Hard coal	262.2	1146.0	9.58	122.2	0.21
Lignite	182.6	924.3	8.27	94.0	0.19
Averages and totals	487.8	2594.3	7.87	279.3	0.17
High estimate					
Natural gas	69.8	389.5	7.50	45.8	0.15
Mineral oil	158.8	134.5	49.45	17.3	0.92
Hard coal	1395.9	1146.0	51.00	122.2	1.14
Lignite	972.1	924.3	44.04	94.0	1.03
Averages and totals	2596.7	2594.3	41.91	279.3	0.93

Note: All costs originally given in DM, conversion factor used 1.5 DM/$

Sources: Hohmeyer 1989, Statistisches Jahrbuch 1986, S.209 and own calculations

418

Table 6: Global reinvestment surcharges based on global per capita energy consumption in 1988

Function: $w(t) = (s(T) - (m(t) + c(t) + q(t))) * (1+r) ** (-(T-t))$

Parameter (q max)	Substitution of primary energy					Electricity generated on the basis of					
	Natural gas $/toe	Heavy fuel oil $/toe	Hard coal imported $/toe	Hard coal domestic $/toe	Lignite housholds $/toe	Natural gas c/kWh	Heavy fuel oil c/kWh	Hard coal imported c/kWh	Hard coal domestic c/kWh	Lignite c/kWh	Uraniu c/kWh
Market price $m(t) + c(t)$	183.98	124.49	70.23	273.62	210.41	22.22	15.49	6.73	10.97	7.07	7.07
External costs $q(t)$	8.95	58.98	60.82	60.82	52.52	0.18	1.09	1.36	1.36	1.23	8.74
Intrest rate r	4.1	4.1	4.1	4.1	4.1	4.1	4.1	4.1	4.1	4.1	4.1
Years til exhaustion $T-t$ (world)	130.40	73.40	2286.50	2286.50	1438.50	130.40	73.40	2286.50	2286.50	1438.50	177.70
Backstop costs $s(T)$	2576.14	2576.14	2576.14	2576.14	2576.14	42.73	42.73	42.73	42.73	42.73	42.73
Reinvestment surcharge $w(t)$	12.64	125.31	3E-37	3E-37	2E-22	0.11	1.37	4E-39	4E-39	3E-24	0.02
Market prices plus reinvestm. surcharges $m(t)+c(t)+w(t)$	196.62	249.80	70.23	273.62	210.41	22.33	16.86	6.73	10.97	7.07	7.09
Total costs of the base year $z(t)=m(t)+c(t)+q(t)+w(t)$	205.57	308.78	131.06	334.44	262.93	22.51	17.95	8.10	12.34	8.31	15.83
Shares in % of $z(t)$											
- market price	89.50%	40.32%	53.59%	81.81%	80.02%	98.71%	86.28%	83.17%	88.95%	85.15%	44.68%
- external costs	4.35%	19.10%	46.41%	18.19%	19.98%	0.81%	6.08%	16.83%	11.05%	14.85%	55.19%
- reinvest. surcharge	6.15%	40.58%	0.00%	0.00%	0.00%	0.48%	7.63%	0.00%	0.00%	0.00%	0.13%

Notes:
- all prices 1988
- base year all consumption 1988
- base year external costs 1982
- estimated uranium resources 6.7 Mt U
- all cost and price figures originally given in DM, conversion factor used 1.5 DM/$

Sources: Tables 1,2,3 and 5

4.8 National Reinvestment Surcharges

As Figure 1 has shown vast discrepancies in per capita energy consumption exist internationally. The rather low global reinvestment surcharges calculated above will lead to a situation where the industrialized countries and some East European states will keep up their totally oversized energy consumption during the next ten to twenty years at the expense of all other countries and their future access to cheap energy sources. This consumption will go on far beyond the point at which the high energy consumption countries have used up their fair share of the cheap non renewable energy resources of the world. Thus, the citizens of the presently poor nations will never have access to their fair share of these non renewable and comparatively cheap energy resources.

If we calculate national reinvestment surcharges based on national per capita energy consumption the picture changes dramatically. Table 7 shows the time spans until reinvestment will be necessary for a number of different nations. Furthermore, it shows the backstop costs based on photovoltaics in the year of the necessary reinvestment. If every person in the world would use as much energy as the industrialized nations listed in the table the world oil resources would be exhausted within the next thirty years. The situations for natural gas and uranium are almost as bad. During the time span remaining for the industrialized countries the backstop costs would not be reduced as much as we have assumed for the global reinvestment regime.

Table 8 gives the resulting national reinvestment surcharges for ten different countries. The calculations were based on the internal and external energy costs in the FRG in 1988. This facilitates a direct comparison of the absolute magnitude of the national surcharges calculated. Nevertheless, in reality these surcharges need to be based on the national cost situation. We see that for natural gas and mineral oil as well as for oil, gas, and uranium based electricity these surcharges reach a tremendous magnitude for the industrialized countries and the former USSR.

The reinvestment value of the non renewable energy resources consumed by the Federal Republic of Germany in 1988 amounted to more than 420 billion US $. The discounted value of this necessary reinvestment into backstop technologies for keeping up the energy resource stock still amounted up to 200 billion US $. This discounted value of the German reduction of our global non renewable energy resource capital correspond to about 15 % of the total GDP of the Federal Republic of Germany. This shows at which pace we are wasting away our inherited stock of energy resource capital.

Table 7: Time span left until the necessary reinvestment into backstop technologies based on national per capita consumption figures of 1988 and the estimated backstop costs at the time of reinvestment

Years until exhaustion (T-t) Backstop costs s(t) in $/toe or c/kWh	Substitution of primary energy					Electricity produced on the basis of					
	Natural gas $/toe	Heavy fuel oil $/toe	Hard coal imported $/toe	Hard coal domestic $/toe	Lignite households $/toe	Natural gas c/kWh	Heavy fuel oil c/kWh	Hard coal imported c/kWh	Hard coal domestic c/kWh	Lignite c/kWh	Uranium c/kWh
FRG											
(T-t) years	52	25	1579	1579	1579	52	25	1579	1579	1579	100
s(T)	3052	4933	2291	2291	2291	57	92	43	43	43	59
Frankreich											
(T-t) years	79	30	7084	7084	7084	79	30	7084	7084	7084	48
s(T)	2422	4069	2291	2291	2291	45	76	43	43	43	59
USA											
(T-t) years	21	14	956	956	956	21	14	956	956	956	103
s(T)	5752	7919	2291	2291	2291	107	147	43	43	43	43
Japan											
(T-t) years	109	27	2947	2947	2947	109	27	2947	2947	2947	162
s(T)	2291	4589	2291	2291	2291	43	85	43	43	43	43
Former USSR											
(T-t) years	20	31	1349	1349	1349	20	31	1349	1349	1349	410
s(T)	6060	3901	2291	2291	2291	113	73	43	43	43	43
Brasil											
(T-t) years	1857	129	27263	27263	27263	1857	129	27263	27263	27263	61657
s(T)	2291	2291	2291	2291	2291	43	43	43	43	43	43
China											
(T-t) years	3033	535	4433	4433	4433	3033	535	4433	4433	4433	-
s(T)	2291	2291	2291	2291	2291	43	43	43	43	43	43
Indien											
(T-t) years	4760	812	13472	13472	13472	4760	812	13472	13472	13472	57537
s(T)	2291	2291	2291	2291	2291	43	43	43	43	43	43
Nigeria											
(T-t) years	968	442	2903300	2903300	2903300	968	442	2903300	2903300	2903300	-
s(T)	2291	2291	2291	2291	2291	43	43	43	43	43	43
Tansania											
(T-t) years	-	1659	1543575	1543575	1543575	-	1659	1543575	1543575	1543575	-
s(T)	2291	2291	2291	2291	2291	43	43	43	43	43	43

Note: All costs originally given in DM, conversion factor used 1.5 DM/$

Sources: Table 1 and 2

Table 8:   National reinvestment surcharges

w(t)	Substitution of primary energy					Electricity produced on the basis of					
	Natural gas $/toe	Heavy fuel oil $/toe	Hard coal imported $/toe	Hard coal domestic $/toe	Lignite households $/toe	Natural gas c/kWh	Heavy fuel oil c/kWh	Hard coal imported c/kWh	Hard coal domestic c/kWh	Lignite c/kWh	Uranium c/kWh
m(t)+c(t)+q(t)	193	183	131	334	263	22	17	8	12	8	16
FRG	361	1739	6,14E-25	5,56E-25	5,77E-25	4,4	27,6	9,85E-27	8,64E-27	9,79E-27	0,48
Frankreich	95	1150	5,19E-121	4,70E-121	4,87E-52	1,0	17,5	8,32E-123	7,31E-123	8,27E-123	6,46
USA	2391	4461	4,57E-14	4,14E-14	4,29E-14	36,4	75,5	7,33E-16	6,43E-16	7,28E-16	0,42
Japan	26	1483	8,10E-49	7,34E-49	7,61E-49	0,3	23,2	1,30E-50	1,14E-50	1,29E-47	0,04
UDSSR	2680	1078	6,31E-21	5,71E-21	5,92E-21	41,4	16,2	1,01E-22	8,89E-23	1,00E-22	1,85E-06
Brasilien	8,20E-30	12	0	0	0	0	0,1	0	0	0	0
China	2,52E-50	9,87E-07	9,33E-75	8,45E-75	8,76E-75	4,17E-52	1,62E-08	1,49E-76	1,31E-76	1,50E-76	0
Indien	1,78E-80	1,44E-11	0	0	0	2,94E-82	2,37E-13	0	0	0	0
Nigeria	2,72E-14	4,09E-05	0	0	0	4,49E-16	6,73E-07	0	0	0	0
Tansania	0	2,35E-26	0	0	0	0	3,86E-28	0	0	0	0

Notes:-in each calculation the high estimate of the external costs q(t) have been used. Thus, w(t) tends to be too low.
- all cost figures originally given in DM, conversion factor used 1.5 DM/$

Sources: Tables 1,2,3,4,5 and 7

4.9 Control Over and Use of the Collected Reinvestment Funds

The fundamental requirement for the use of the collected funds can be derived directly from theory (special case of 'Hartwick's-Rule'). To achieve sustainability and intertemporal equity the collected funds (wt) have to be invested into an expansion of the capital base of the renewable energy resource stock.

The simplest way to assure collect the funds is that every energy consuming country levies the national energy surcharge either at the point of final use or at the point of import. The surcharge is paid into a national fund lending money for secure long term investments. These loans are limited to the point in time when reinvestment into the backstop technologies is due. If the international money markets function, no major disturbance of the financial markets is to be expected. Ideally the resource rents, which may be collected by the resource owner today, are transferred to the country of final energy use. Thus, the functional resource stock is slowly transferred to the county of final use.

If the assumptions of neoclassical resource theory about the functioning of the resource markets and the reinvestment of the resource rents hold, the transfer effect is the only change that will occur. If neoclassical theory is wrong, the introduction of the reinvestment fonds will lead to an overdue reduction in the international interest rates as money is shifted from oversized consumption to investment with larger investment funds available.

The most important condition for a functioning of the suggested model is that the collected funds will not be used to expand public budgets and consumption oriented spending. Such use of the funds would jeopardize the aim of the surcharge of reaching sustainability as well as intertemporal and international equity in the use of energy. The control over the fund should therefore be trusted with an independent body, which should only be responsible for the adequate use of the fund. At the same time very stringent public control of the conduct this independent body needs to be installed.

5 Conclusion

It has been shown that it is possible to achieve a sustainable energy future under the simple reinvestment surcharge regime suggested. This regime can secure intergenerational and international equity at the same time.

The political will to install such a regime will make the difference between a wasteful non sustainable energy future with extreme levels of international injustice and a sustainable energy use including intergenerational and international equity.

6 References

Baumol, Wiliam J.: On the Possibility of Continuing Expansion of Finite Resources, in: Kyklos, Vol.39, 2/1986, S. 167-179

Brand, Michael und Eberhard Jochem: Betriebs- und volkswirtschaftliche Beurteilung unterschiedlicher Möglichkeiten zur Lastvergleichmäßigung sowie der derzeitigen Strompreisgestaltung im Hinblick auf Investitionen in Kraftwerke und rationellere Energieverwendung. Karlsruhe 1986

Hartwick, J.M.: Intergenerational Equity and the Investing of Rents from Exhaustible Resources, in: American Economic Review, Vol. 67, Dezember 1977, S. 972-974

Hartwick, J.M.: Intergenerational Equity and Substitution Among Exhaustible Resources, in: Review of Economic Studies, Vol. 45, 1978, S. 347-354

Heal, Geoffrey: Intertemporal Allocation and Intergenerational Equity, in: Horst Siebert (Hrsg.): Erschöpfbare Ressourcen. Verhandlungen auf der Arbeitstagung der Gesellschaft für Wirtschafts- und Sozialwissenschaften - Verein für Sozialpolitik - in Mannheim vom 24. - 26. September 1979. Berlin 1980, S. 37-73

Hohmeyer, Olav: Soziale Kosten des Energieverbauchs. 1.Auflage, Berlin, Heidelberg, New York 1988

Hohmeyer, Olav: Soziale Kosten des Energieverbrauchs. 2.Auflage, Berlin, Heidelberg, New York 1989

Hohmeyer, Olav: Adäquate Berücksichtigung der Erschöpfbarkeit nicht erneuerbarer Ressourcen. Report for Prognos AG, Basel, Karlsruhe 1992

Hotelling, Harold: The Economics of Exhaustible Resources, in: The Journal of Political Economy, Vol. 39, 2/1931, S. 137-175

L'Institut d'Economie et de Politique de l'Energie: Energie Internationale 1989 - 1990. Grenoble 1989

International Energy Agency (IEA): IEA Statistics, World Energy Statistics and Balances 1971 - 1987. Paris 1989

International Energy Agency (IEA): IEA Statistics, Energy Balances of OECD Countries 1980 - 1989. Paris 1991

International Energy Agency (IEA): Energy Prices and Taxes. Third Quarter 1991. Paris 1992

Jochem, Eberhard und Helmut Schaefer: Emissionsminderung durch rationelle Energieverwendung, in: Energiewirtschaftliche Tagesfragen, Heft 4/1991, S. 207-215

Kay, John A. und James A. Mirrlees: The Desirability of Natural Resource Depletion, in: D. W. Pearce (Hrsg.): The Economics of Natural Resource Depletion. New York 1975

OECD/NEA: Uranium-Resources, Production and Demand. Paris 1973, 1975, 1977, 1979, 1982 und 1986

Palz, Wolfgang und J. Schmid: Electricity production costs from photovoltaic systems at several selected sites within the European Community, in: International Journal of Solar Energy, Vol. 8, 1990, S. 227-231

Pearce, David W. und R. Kerry Turner: Economics of Natural Resources and the Environment. New York, London 1990

Prognos AG und ISI (Masuhr, Bradke et al.): Konsistenzprüfung einer denkbaren zukünftigen Wasserstoffwirtschaft. Basel 1991

Rawls, J.: A Theory of Justice. Oxford 1971

Solow, Robert M.: Intergenerational Equity and Exhaustible Resources, in: Review of Economic Studies, 1974, S. 29-45

Solow, Robert M.: On the Intergenerational Allocation of Natural Resources, in: Scandinavian Journal of Economics, Vol. 88, 1/1986, S. 141-149

Statistisches Bundesamt (Hrsg.): Statistisches Jahrbuch für die Bundesrepublik Deutschland. Stuttgart und Mainz 1986, 1990, 1991

Ströbele, Wolfgang: Rohstoffökonomik - Theorie natürlicher Ressourcen mit Anwendungsbeispielen Öl, Kupfer, Uran und Fischerei. München 1987

Welsch, Heinz: Erschöpfbare Energieressourcen und die sozialen Kosten der Elektrizitätserzeugung, in: Zeitschrift für Energiewirtschaft 3/1989, S. 208-213

Participants of the Workshop

USA

Mr Stephen Bernow
Stockholm Environmental Institute, Boston Center, 89 Broad St., Boston, MA 02110

Mr Stephen Brick
MSB Energy Associates, Inc., 7507 Hubbard Ave., Suite 200, Middleton, WI 53562

Mr Ashley C. Brown
Commissioner, Public Utilities Commission of Ohio, 180 E. Broad St., Columbus, OH 43215

Mr Paul L. Chernick
President, Resource Insight, Inc., 18 Tremont St., Suite 1000, Boston, MA 02108

Mr Roger Dower
Program Director, World Resources Institute, 1709 New York Ave., NW, Suite 700, Washington, DC 20006

Mr Daniel J. Dudek
Senior Economist, Environmental Defense Fund, 257 Park Ave., South, New York, NY 10010

Mr Ashok Gupta
Senior Energy Analyst, Natural Resources Defense Council, 40 W. 20th St., New York, NY 10011

Mr David Harrison
Vice President, National Economic, Research Associates, One Main St., 5th Fl., Cambridge, MA 02142

Professor David Hodas
Widener University School of Law, 4601 Concord Pike, PO Box 7474, Wilmington, DE 19803

Mr Florentin Krause
Energy and Environment Division, Lawrence Berkeley Laboratory, Berkeley, CA 94720

Mr Alan Krupnik
Senior Fellow, Resources for the Future, 1616 P St., NW, Washington, DC 20046

Ms Carolyn Lang
Senior Associate, RCG / Hagler Bailly, Inc., PO Drawer O, Boulder, CO 80306-1906

Mr Russell Lee
Oak Ridge National Laboratory, PO Box 20008, Bldg. 4500N, Oak Ridge, TN 37831-6205

Mr Anil Markandya
Harvard Institute for International Development, One Eliot St., Cambridge, MA 02138

Mr Richard L. Ottinger
Co-Director, Pace University Center for Environmental Legal Studies, 78 N. Broadway, White Plains, NY 10603

Mr Sury Putta
New York State Department of Public Service, 3 Empire State Plaza, Albany, NY 12223

Mr Ajay K. Sanghi
Chief, IAU, New York State Energy Office, 2 Rockefeller Plaza, Albany, NY 12223

Mr Robert B. Shelton
Director, Energy Division, Oak Ridge National Laboratory, PO Box 2008, Oak Ridge, TN 37831-6187

Mr Hilary H. Smith
U. S. Department of Energy, 1000 Independence Ave., SW, Washington, DC 20585

Mr John H. Smolinsky
Office of Energy Effic. and Env., New York State Department of Public Service, 3 Empire State Plaza, Albany, NY 12223

Mr Phil Sparks
Communications Consortium Media Center, 1333 H St., NW, Suite 700, Washington, DC 20005

Ms Anita Sprenger

Director / Electric Division, Public Service Commission, Wisconsin, PO Box 7854, Madison, WI 53707-7854

Mr Jeffrey D. Tranen
Vice President, New England Electric, , 25 Research Dr., Westboro MA 01582

Mr Stephen Wiel
Washington D. C. Office, Energy & Environment Div., Lawrence Berkeley Laboratory, 1523 New Hampshire Ave NW, Washington, CA 20036

Mr Derek Winstanley
Director, National Acid Precipitation, Assessment Program, 722 Jackson Pl., NW, Washington, DC 20503

Germany

Mr Rainer Friedrich
Institute for Energy Economics and the Rational Use of Energy, Universität Stuttgart, Hessbrühlstr. 49a, 7000 Stuttgart 80

Mr Uwe R. Fritsche
Coordinator, Energy Division, Öko-Institut e. V., Institut für angewandte Ökologie, Bunsenstr. 16, 6100 Darmstadt

Eberhard Moths
Regierungsdirektor, Bundesministerium für Wirtschaft, Postfach 14 02 60, 5300 Bonn

Dipl.-Volksw. Klaus Rennings
Institut für Verkehrswissenschaft, Universität Münster, Am Stadtgraben 9, 4400 Münster

United Kingdom

Professor Robert Hill
Newcastle Photovoltaics, Applications Centre, University of Northumbria, Ellison Place, Newcastle Upon Tyne NE1 8ST

Mr Michael R. Holland
Energy Technology Support Unit, B156 Harwell Laboratory, OX11 ORA

Switzerland

Klaus P. Masuhr
Prognos AG, Missionsstr. 62, 4012 Basel

Belgium

Mr Pierre Valette
Commission of the European Communities, DG XII, 200, rue de la Loi, 1049 Brussels

Springer-Verlag
and the Environment

We at Springer-Verlag firmly believe that an international science publisher has a special obligation to the environment, and our corporate policies consistently reflect this conviction.

We also expect our business partners – paper mills, printers, packaging manufacturers, etc. – to commit themselves to using environmentally friendly materials and production processes.

The paper in this book is made from low- or no-chlorine pulp and is acid free, in conformance with international standards for paper permanency.